Environmental Microbiology

Environmental Microbiology

Editor

Ram Prasad

Assistant Professor

Amity Institute of Microbial Technology

Amity University

Noida (Uttar Pradesh)

I.K. International Pvt. Ltd.

NEW DELHI

Published by

I.K. International Pvt. Ltd.
S-25, Green Park Extension
Uphaar Cinema Market
New Delhi–110 016 (India)
E-mail: info@ikinternational.com
Website: www.ikbooks.com

ISBN: 978-93-84588-52-6

Published by Krishan Makhijani for I.K. International Pvt. Ltd., S-25, Green Park Extension, Uphaar Cinema Market, New Delhi–110 016 and Printed by Rekha Printers Pvt. Ltd., Okhla Industrial Area, Phase II, New Delhi–110 020.

Preface

Microbes are everywhere, all over the world and in every imaginable environment. For example in soil, just one gram contains billions of microorganisms and all their associated activities. "Environmental microbiology" is the study of microbes within all earthly habitats, and their beneficial and detrimental effects on human health and welfare. This field has expanded to the study of environmental factors, including the interaction of microbes with organic and inorganic pollutants, behaviour of pathogens, discovery and applications of new microbes and their products to benefit human health and welfare. This book incorporates the most recent advances in environmental microbiology. The use of microorganisms to improve environmental quality is the basis of wastewater management, vermicompost and pest control. These improvements include preventing the discharge of pollutants to the environment, cleaning up contaminated environments and generating valuable resources for human society. Environmental microbiology is competing with great success against traditional technologies in recent days and for providing solutions to acute problems through the so-called bioremediation. This book provides a natural way of addressing mainly environmental problems ranging from the identification of biohazards to bioremediation of industrial, agricultural and domestic waste like municipal effluents and residues.

This book is designed for undergraduate, postgraduate and research students working or interested in environmental microbiology and to serve as a reference for scientists, professionals and engineers in this field. The overall objectives of the text are to define the important microbes involved in environmental microbiology, the nature of the different environments in which microbes are situated, and the methodologies used to monitor the microbes and their activities and, finally, to evaluate the success achieved in the application of microbes in solving environmental problems through manipulating their activities. This book represents a joint effort led by the editor and authors who have diverse yet complementary backgrounds in environmental microbiology. The authors are actively involved in environmental microbiology or related research. They have worked together extensively on a variety of practical problems using advanced, interdisciplinary approaches. I thank all the contributors who spared their precious time, thoughts and expertise in the writing of the chapters and hope that their contribution will stimulate further high-quality teaching and research. It has been a pleasure to compile this book, primarily due to the stimulating cooperation of the contributors.

I welcome comments, suggestions and feedback from readers of this book. I take this opportunity to thank all colleagues, who have inspired me to look further and deeper and for their valuable guidance. I specially thank Professor Dr. Ishan Barman, Biophotonics Laboratory, Whiting School of Engineering, Department of Mechanical Engineering, Johns Hopkins University, USA for providing necessary facilities for editing and finalizing the book during my visit to his institute. I am also thankful to UICC and American Cancer Society for financial assistance to visit the Barman Laboratory. I gratefully acknowledge the cooperation, encouragement and support of my wife Dr. Avita throughout the preparation of this book.

Ram Prasad

Contributors

A.A. Haleem Khan, Dept of Botany, University College, Telangana University, Nizamabad-503322 (Telangana) India

Ajit Varma, Amity Institute of Microbial Technology, Amity University Uttar Pradesh, Noida-201303, India

B.Vidya Vardhini, Dept of Pharmaceutical Chemistry, University College, Telangana University, Nizamabad-503322, (Telangana) India

Chandrama Prakash Upadhyaya, Department of Biotechnology, Dr. Hari Singh Gour Central University, Sagar (M.P.) India

Charu Gupta, Amity Institute for Herbal Research and Studies, Amity University Uttar Pradesh, Noida-201303, India

D.D. Joshi, Institute of Bioresources & Sustainable Development (IBSD), (Department of Biotechnology, Govt. of India), Takyelpat, Imphal-795001 (Manipur), India

Devarajan Thangadurai, Department of Botany, Karnatak University, Dharwad-580003, (Karnataka) India

Dhan Prakash, Amity Institute for Herbal Research and Studies, Amity University Uttar Pradesh, Noida-201303, India

Kanchan Singh, Botany Department, Feroze Gandhi College, Rae Bareli-229 001 (Uttar Pradesh) India

M. Yaqub Bhat, University Department of Botany, University of Kashmir, Srinagar, (J & K) India

Mamta Baunthiyal, Department of Biotechnology, G.B. Pant Engineering College, Pauri-246194 (Uttarakhand) India

Manoj Kaushal, Department of Basic Science, Dr. Y. S. Parmar University of Horticulture and Forestry, Nauni-173230, Solan, (Himachal Pradesh) India

Manoj Kumar, Amity Institute of Microbial Technology, Amity University Uttar Pradesh Noida-201303, India

Mohd Ashaq, Department of Botany, Govt. College for Women, Gandhi Nagar-180001, Jammu, (J & K) India

Munawar Fazal, Department of Botany, S. Sinha College, Aurangabad, (Bihar) India

Narendra Kumar, School of Studies in Microbiology, Jiwaji University, Gwalior-474011, (M.P.) India

Naseem, Dept of Pharmaceutical Chemistry, University College, Telangana University, Nizamabad-503322, (Telangana) India

Neeraj, Botany Department, Feroze Gandhi College, Rae Bareli-229 001, (Uttar Pradesh) India

Nivedita Sharma, School of Studies in Microbiology, Jiwaji University, Gwalior-474011, (M.P.) India

Peter Twumasi, Department of Biochemistry and Biotechnology, Kwame Nkrumah University of Science and Technology (KNUST), Kumasi, Ghana

Priyanku Teotia, Department of Biotechnology, Division of Biosciences, College of Applied Education & Health Sciences, Meerut-250003, (Uttar Pradesh) India

Rajesh Kaushal, Department of Basic Science, Dr. Y. S. Parmar University of Horticulture and Forestry, Nauni-173230, Solan, (Himachal Pradesh) India

Rajesh Kumar Pandey, Department of Botany, Bundelkhand University, Jhansi-284128, (Uttar Pradesh) India

Rajni Singh, Amity Institute of Microbial Biotechnology, Amity University Uttar Pradesh, Noida-201303, India

Rajshree Saxena, Amity Institute of Microbial Biotechnology, Amity University Uttar Pradesh Noida-201303, India

Ram Prasad, Amity Institute of Microbial Technology, Amity University Uttar Pradesh, Noida-201303, India

Ramesh Kumar, Virology Section, Department of Microbiology, All India Institute of Medical Sciences, New Delhi-110029, India

Santosh Kumar Tiwari, Microbiology and Plant Pathology Research Laboratory, University Department of Botany, T. M. Bhagalpur University, Bhagalpur-812007, (Bihar) India

Shweta Ranghar, Department of Biotechnology G.B. Pant Engineering College, Pauri-246194 (Uttarakhand) India

Sneh Gupta, Department of Zoology, RGPG College, Meerut-250001 (UP), India

Sonali Gupta, Amity Institute of Microbial Biotechnology, Amity University Uttar Pradesh, Noida-201303, India

Sunita J. Varjani, Science and Technology Department, Gujarat National Law University, Gandhinagar-382007 (Gujarat) India

Surendra Singh, Department of Botany, Banaras Hindu University, Varanasi-221005, (Uttar Pradesh) India

U.S. Bagde, Department of Botany, Dr. Hari Singh Gour Central University, Sagar (M.P.) India

Vinay Sharma, Department of Bioscience & Biotechnology, Banasthali University Banasthali Vidyapith-304 022 (Rajasthan), India

Vivek Kumar, Amity Institute of Microbial Technology, Amity University Uttar Pradesh, Noida-201303, India

Wangkheirakpam Sujata, Institute of Bioresources & Sustainable Development (IBSD), (Department of Biotechnology, Govt. of India), Takyelpat, Imphal-795001 (Manipur), India

Contents

Preface *v*
Contributors *vii*

1. **Evolutionary History of Microbiology** **1**
   Ram Prasad, Chandrama Prakash Upadhyaya, Manoj Kumar, U.S. Bagde, Peter Twumasi, Devarajan Thangadurai and Ajit Varma
2. **Prospects of Vermicompost for Organic Cultivation in India** **11**
   Santosh Kumar Tiwari
3. **Role of Blue-Green Algae in Environment Management** **22**
   Charu Gupta, Dhan Prakash and Sneh Gupta
4. **Medicinal Plants for Microbial Control** **45**
   D.D. Joshi, Ram Prasad and Wangkheirakpam Sujata
5. **Microorganisms and Their Role in Sustainable Environment** **60**
   A.A. Haleem Khan, Naseem and B.Vidya Vardhini
6. **Fluoride Tolerant Microorganisms and Their Possible Role in Bioremediation** **89**
   Mamta Baunthiyal, Shweta Ranghar and Vinay Sharma
7. **PGPR Bio-inoculants for Sustainable Crop Production** **98**
   Manoj Kaushal and Rajesh Kaushal
8. **Mass Production of Arbuscular Mycorrhiza** **108**
   Munawar Fazal, M. Yaqub Bhat and Mohd Ashaq
9. **Physicochemical and Bacteriological Studies of Wetland Soils of Keoladeo National Park, Bharatpur, Rajasthan (India)** **136**
   Narendra Kumar, Nivedita Sharma and Surendra Singh
10. **Production of Bacterial Proteases Using Agro-industrial Wastes by Statistical Experimental Designs** **152**
    Rajni Singh, Sonali Gupta and Rajshree Saxena
11. **Adverse Effect of Indiscriminate use of Chemical Pesticides and Its Alternative Ecofriendly Means for Management of Diseases and Pests Infecting Agricultural Crops** **167**
    Rajesh Kumar Pandey and Ram Prasad

12. **Arbuscular Mycorrhizal Fungi and Its Interactions with Environment** 211
Neeraj and Kanchan Singh

13. **Arbuscular Mycorrhiza as Bioprotectant: A Review of the Mechanisms Involved** 234
Munawar Fazal, M. Yaqub Bhat and Mohd Ashaq

14. **Resistance: Origin, Evolution and Challenges in Future** 259
Ramesh Kumar

15. **Algae and Wastewater Remediation: Current and Future Scenario** 269
Priyanku Teotia and Vivek Kumar

16. **Screening of Thiosulphate Utilizing Bacteria from Mining Site of Ambaji, Gujarat, India** 289
*Sunita J. Varjani**

17. **Biodegradable Polymers from Microbes** 300
Charu Gupta, Dhan Prakash and Sneh Gupta

Index 325

1
Evolutionary History of Microbiology

Ram Prasad*[1,2], Chandrama Prakash Upadhyaya[3], Manoj Kumar[1], U.S. Bagde[4], Peter Twumasi[5], Devarajan Thangadurai[6] and Ajit Varma[1]

[1]*Amity Institute of Microbial Technology, Amity University Uttar Pradesh Noida-201303, India*

[2]*Department of Mechanical Engineering, Johns Hopkins University Baltimore-21218, Maryland, USA*

[3]*Department of Biotechnology, Dr. Hari Singh Gour Central University Sagar (M.P.) India*

[4]*Department of Botany, Dr. Hari Singh Gour Central University Sagar (M.P.) India*

[5]*Department of Biochemistry and Biotechnology, Kwame Nkrumah University of Science and Technology (KNUST), Kumasi, Ghana*

[6]*Department of Botany, Karnatak University, Dharwad-580003, Karnataka, India*

Email: rprasad@amity.edu; rpjnu2001@gmail.com

In the field of observation, chance favors only prepared minds

– Louis Pasteur

1 INTRODUCTION

Microorganisms are present nearly everywhere in nature. Some microorganisms can live in hot springs and others in frozen sea ice. Microorganisms have a critical role in biogeochemical cycles, particularly the carbon, nitrogen and sulphur cycles. Biological activity of micro-organism is thought to be responsible for production of sufficient amounts of oxygen in the earth's atmosphere to support life well over two billion years ago. Microorganisms are the most important components of plant and soil ecosystems. There is one to ten million microorganisms present in each gram of soil. Nearly same amounts are found in plants and animals. The scavenging role of microorganisms which involves the breakdown of dead plant and animal tissues result in release of nutrients including carbon and nitrogen which support plant growth. Besides many benefits, some microorganisms have cause harm to humans with manifestation of disrupted societies over the millennia. These pathogenic microorganisms not only cause diseases in humans but also in other animals and plant hosts. Some of these diseases are by direct effect of the microorganisms living in or on their host or through intermediate toxins such as in the case of food spoilage.

2 MICROBIOLOGY AND THE ORIGIN OF LIFE

Life originated in the sea following millions of years of a chemical evolutionary process. According to this evolutional theory, inorganic compounds of the atmosphere, under the influence of ultraviolet light, electrical discharges and/or high temperatures, interacted to form organic compounds which precipitated in the sea, where they accumulated. These organic compounds, subjected to additional physical effects of the environment, combined to form amino acids. The amino acids interacted to form peptides, polypeptides and other more complex organic substances which served as the precursor of the first form of life.

Table 1 The time scale of chemical evolution, biological evolution and the emergence of microbial life

Millions of years ago	*Geological era*	*Approximate time of origin*	
	Cenozoic	*Homo sapiens*	
	Mesozoic	Mammals and birds	
1000-	Palaeozoic Proteozoic	Terrestrial plants Fishes, invertebrates Multicellular organisms Eukaryotic cells	Biological evolution
2000-	Archaean	Aerobic bacteria Prokaryotic cells (Aerobic and anaerobic bacteria)	
3000-		First fossil evidence of life	
4000-		• First living "cells" • Protocell: membranes enclosing prototypes of nucleic acids • Coacervate formation: Concentration and aggregation of molecules • Organic soup: Synthesis of amino acids, sugars and peptides	Chemical evolution

Despite their small size, microbes play an enormous role in the cycle of matter and in the metabolism of this planet. With the discovery of the huge microbial biota in subsoil fissures and better estimates of microbial life in the oceans, the microbial biota on the earth is thought to exceed in weight all other living things combined. Bacteria alone account for 50% of the biomass of carbon and over 90% of the biomass of nitrogen and phosphorus on our planet. At least as much photosynthesis is carried out by marine microbes as by terrestrial plants. It has been estimated that if the action of microbes on the nitrogen cycle were to cease, the amount of nitrogen available to plants would become too low to sustain life within about one week. Microbes are also critical for assimilating dissolved organic material into the particulate organic matter used by eukaryotes. Microbes are responsible for enormous geochemical activities, such as formation of carbon and silica deposits and the shaping of limestone caverns. Microorganisms can be found in unexpected niches, such as thermal ocean vents and fissures in deep rocks.

The history of microbiology or the development of science of microbiology is full of serendipities, diligently made inventions, circumstantial discoveries and efforts.

Table 2 Some important events in the development of microbiology

Date	*Microbiological History*
1546	Fracastoro suggests that invisible organisms cause diseases
1590–1608	Jansen develops first useful compound microscope
1676	Leeuwenhoek discovers "animalcules"
1688	Redi publishes work on spontaneous generation of maggots
1765–1776	Spallanzani attacks spontaneous generation
1786	Müller produces first classification of bacteria
1798	Edward Jenner introduces cowpox vaccination for smallpox
1838–1839	Schwann and Schleiden, the Cell Theory
1835–1844	Agostino Bassi discovered that silkworm disease is caused by a fungus and proposes that many diseases are microbial in origin
1847–1850	Ignaz Semmelweis shows that childbed fever is transmitted by physicians and introduces the use of antiseptics to prevent the disease
1849	Snow studies the epidemiology of a cholera epidemic in London
1857	Louis Pasteur determines that yeast causes fermentation of wine and develops the process of pasteurization that saves the French wine industry. This marks the beginning of the Golden Age of Microbiology—a period of explosive growth of knowledge of microbes (1857–1914)
1858	Virchow states that all cells come from cells
1861	Louis Pasteur shows that microorganisms do not arise by spontaneous generation Louis Pasteur, studying fermentation by yeast, coins the terms aerobic and anaerobic
1867	Joseph Lister used phenol as a disinfectant to treat wounds and disinfection of instruments used for antiseptic surgery
1869	Miescher discovered nucleic acids
1872	Ferdinand Cohn describes the roles of bacteria in the cycling of elements in nature
1876–1877	Koch demonstrates that anthrax is caused by *Bacillus anthracis*
1879	Neisser identifies the causative agent of gonorrhea (*Neisseria gonorrhoeae*). This may be the first case where a microbe is implicated as the cause of a chronic disease

Contd...

Contd...

1880	Pasteur develops a vaccine for chicken cholera. This is the first attenuated vaccine C. Laveran discovered *Plasmodium,* the cause of malaria
1881	Koch developed solid media (Gelatin) and methods to obtain pure cultures of bacteria Pasteur develops anthrax vaccine
1882	Koch discovered tubercle bacillus, *Mycobacterium tuberculosis* Angelina Fannie Hesse and Walther Hesse while working in Koch's laboratory introduced the use of agar in solidifying media
1883	Koch identifies *Vibrio cholerae* as the causative agent of cholera
1884	Koch's postulates first published Metchnikoff describes phagocytosis Autoclave developed Hans Christian Gram stain developed method for differential staining for the identification of bacteria — The Gram's staining technique
1885	Pasteur develops a vaccine for rabies Escherich discovered *Escherichia coli,* a cause of diarrhea
1886	Fraenkel discovered *Streptococcus pneumoniae,* a cause of pneumonia
1887	Petri dish (plate) developed by Richard Petri
1887–1890	Winogradsky studies sulfur and nitrifying bacteria
1889	Beijerinck isolates root nodule bacteria
1890	Von Behring prepared antitoxins for diphtheria and tetanus
1891	Paul Ehrlich proposes that antibodies are responsible for immunity
1892	Ivanowsky provides evidence for virus causation of tobacco mosaic disease Welch and Nuttall identify *Clostridium perfringens* as the causative agent of gangrene
1894	Kitasato and Yersin discovered *Yersinia pestis,* the cause of plague
1895	Bordet discovered complement
1896	Van Ermengem discovered *Clostridium botulinum,* the cause of botulism
1897	Buchner prepares extract of yeast that ferments Ross shows that malaria parasite is carried by the mosquito
1899	Beijerinck proves that a virus particle causes the tobacco mosaic disease The Society of American Bacteriologists (later renamed the American Society of Microbiologists) is organized. This is the oldest scientific society in America
1900	Walter Reed proves that yellow fever is transmitted by the mosquito
1902	Landsteiner discovered blood groups
1903	Wright and others discovered antibodies in the blood of immunized animals
1904	Koning suggests that fungi play important roles in the decomposition of organic matter and the formation of humus (topsoil)

Contd...

Contd...

1905	Schaudinn and Hoffmann show *Treponema pallidum* causes syphilis Ishiwata discovered that silkworm disease is caused by the bacterium *Bacillus thuringiensis* (now an important insecticide)
1906	Wassermann develops complement fixation test for syphilis Sohngen shows that bacteria can use methane as an energy and carbon source
1907	Smith and Townsend discovered *Agrobacterium* is responsible for the plant tumor disease called crown gall
1909	Ricketts shows that Rocky Mountain spotted fever is transmitted by ticks and caused by a microbe (*Rickettsia rickettsii*), an organisms intermediate in size between viruses and bacteria
1910	Paul Ehrlich develops chemotherapeutic agent for syphilis Nicolle shows that typhus fever is transmitted more than 20 million people from person to person by body lice. This information is used to reduce incidence of typhus in both world wars
1911	Rous discovered a virus that causes cancer in chickens
1915–1917	D'Herelle and Twort discovered bacterial viruses (bacteriophases)
1915	McCrady develops the fecal coliform test as a method for bacteriological analysis of water quality
1918	Worldwide influenza epidemic killed at least 40 million people
1919	Blood agar is used for the first time. Hemolytic ability is used to separate three types of *Streptococcus* species
1921	Fleming discovered lysozyme
1923	First edition of *Bergey's Manual* on the characterization and classification of bacteria
1928	Frederick Griffith discovered bacterial transformation
1929	Alexander Fleming discovered penicillin (first antibiotic)
1931	Van Niel shows that photosynthetic bacteria use reduced compounds, for example H_2S, as a source of electrons. He posits that plants use H_2O as an electron source. He also begins the first General Microbiology course
1933	Knoll and Ruska develops first transmission electron microscope
1935	Stanley crystallizes the tobacco mosaic virus Domagk discovered sulfa drugs (antimetabolic antibiotic) Prontosil, as a treatment against *Streptococcus*
1937	Chatton divides living organisms into prokaryotes and eukaryotes
1938	Theiler produces a successful vaccine against yellow fever
1940	Selman Waksman discovered actinomycin—the first antibiotic from the actinomycetes. Later he discovered tetracycline and streptomycin
1941	Beadle and Tatum proposed the one gene-one enzyme hypothesis
1944	Avery shows that DNA carries information during transformation Selman Waksman discovered streptomycin, a very effective drug against tuberculosis

Contd...

Contd...

1946	Lederberg and Tatum describe bacterial conjugation
1949	Enders, Weller and Robbins grow poliovirus in human tissue cultures
1950	Lwoff induces lysogenic bacteriophages
1952	Hershey and Chase show that bacteriophages inject DNA into host cells Zinder and Lederberg discovered generalized transduction Lederberg and Lederberg develop the method of replica plating
1953	Phase-contrast microscope developed Medawar discovered immune tolerance Watson and Crick propose the double helix structure for DNA
1955	Jacob and Wollman discovered the F- factor is a plasmid Jerne and Burnet propose the clonal selection theory
1958	Meselson and Stahl show the semiconservative nature of DNA replication
1959	Yalow develops the radioimmunoassay technique
1961	Jacob and Monod propose the operon model of gene regulation Nirenberg and Matthaei synthesize a poly-U RNA. They find this RNA directs the synthesis of a protein containing only phenylalanine. They conclude that UUU encodes phenylalanine. This is the start of the efforts to decipher the genetic code
1961–1966	Nirenberg, Khorana, and others elucidate the genetic code
1962	Porter proposes the basic structure for immunoglobulin G First quinolone antimicrobial (nalidixic acid) synthesized
1965	Linus Pauling suggests that DNA can be used to understand evolutionary relationships and to identify microorganisms
1967	Smallpox eradication programme by World Health Organization started Theodor O. Diener discovered viroids
1970	Discovery of restriction endonucleases by Arber and Smith Discovery of reverse transcriptase in retroviruses by Temin and Baltimore
1973	Ames develops a bacterial assay for the detection of mutagens Cohen, Boyer, Chang and Helling use plasmid vectors to clone genes in bacteria
1975	Kohler and Milstein develop technique for the production of monoclonal antibodies Lyme disease discovered
1977	Recognition of archaea as a distinct microbial group by Woese and Fox Gilbert and Sanger develop techniques for DNA sequencing Bishop and Varmus discover retroviral oncogenes (cancer-causing genes)
1979	Insulin synthesized using recombinant DNA techniques Smallpox declared officially eliminated
1980	Development of the scanning tunneling microscope
1981	Margulis formalizes the endosymbiotic theory of organelle evolution
1982	Recombinant hepatitis B vaccine developed Prusiner discovers a new type of infectious agent—an infectious protein called a prion—that is responsible for the sheep disease, scrapie

Contd...

Contd...

1982–1983	Discovery of catalytic RNA by Cech and Altman
1983–1984	The human immunodeficiency virus isolated and identified by Montagnier and Gallo The polymerase chain reaction developed by Kary Mullis
1984	Marshall demonstrates that a bacterium, *Helicobacter pylori,* causes peptic ulcers
1986	First vaccine (hepatitis B vaccine) produced by genetic engineering approved for human use
1993	Schopf demonstrates that cyanobacteria—like organisms had evolved 3.46 billion years ago W. French Anderson leads first successful gene therapy
1994	Cano cultures bacteria from endospores in 40 million year old material
1995	Chickenpox vaccine approved for U.S. use Venter and colleagues elucidate the first complete genome sequenced *Haemophilus influenzae*
1996	*Methanococcus jannaschii* genome sequenced Yeast genome sequenced NASA scientists find bacteria-like fossils in rocks from Mars
1997	Discovery of *Thiomargarita namibiensis,* the largest known bacterium *Escherichia coli* genome sequenced
2000	Discovery that *Vibrio cholerae* has two separate chromosomes The human genome project began in 1990 is completed; entire human genome is deciphered Artnzen and colleagues develop the first edible vaccine – a genetically-modified potato containing part of the Norwalk virus
2003	A new, mysterious respiratory disease (SARS) that kills healthy young adults becomes epidemic
2006	Bird Flu– A potential pandemic

3 FUTURE PERSPECTIVES OF MICROBIOLOGY

Microbes are the most significant life forms sharing this planet with humans because of their pervasive presence and utilization of diverse available food source, including humans whose defenses may be breached. Depending on the food source, microbes may either have beneficial roles in maintaining life or undesirable roles in causing human, animal and plant disease. Science writer Bernard Dixon is very optimistic about microbiology's future for two reasons. First, microbiology has a clearer mission than do many other scientific disciplines. Second, it is confident of its value because of its practical significance. Dixon notes that microbiology is required both to face the threat of new and re-emerging human infectious diseases and to develop industrial technologies that are more efficient and environmentally eco-friendly.

Among the leading areas of microbiological research presumed to have significant relevance on human life include the following:

- New infection and re-emergence of diseases: New infectious diseases are continually arising and old diseases are once again becoming widespread and destructive. AIDS, cancer, hepatitis and tuberculosis are excellent examples of new and re-emerging infectious diseases. Microbiologists will have to respond to these threats, many of them presently unknown.
- New modes of infection: New ways should be established for microbial spread and infectious diseases. Increases in antibiotic resistance will be a continuing problem, particularly the spread of multiple drug resistance that can render a pathogen impervious to current medical treatment. Microbiologists therefore need to create new and effective drugs, and find ways to prevent or slow the spread of drug resistance. Recombinant vaccines with identified high efficacy must be developed to protect against diseases such as AIDS. It will also be necessary to bring to fore molecular biology techniques and recombinant DNA technology in solving infection menace.
- Research is needed on the association between infectious agents and chronic diseases such as autoimmune (rheumatoid arthritis) and cardiovascular diseases.
- We are only now beginning to understand how pathogens interact with host cells and the ways in which diseases arise. There is a lot to learn about how the host resists pathogen invasions. For example, the presence of the normal microbial biota in the intestine of mice promotes proper development of the vascular tissue as well as host immunity to invading pathogens.
- Microbes play an increasing role in industrial production of pharmaceuticals and other compounds. For example, microorganisms can (a) serve as sources of high-quality food, dairy products and other practical products such as antibiotics, vitamins, enzymes for industrial applications; (b) fermentation processes have the potential to decrease our dependence on fossil fuels, and bacterial systems are being developed as nano-biosensors, nano-batteries, and nano-computers; (c) degrade pollutants and toxic wastes; and (d) be used as vectors to treat diseases and enhance agricultural productivity. There also is a continuing need to prevent microbial spoilage of food and work on methods to increase soil fertility and crop yields.
- Microbial diversity is another area requiring considerable research. Indeed, it is estimated that less than 1% of the earth's

microbial population has been cultured. We must develop new isolation techniques and an adequate classification method for microorganisms, one which includes those microorganisms that cannot be cultivated in the laboratory. Much work needs to be done on microorganisms living in extreme environments. This has greatly expanded our awareness of the diversity of types and total number of microbes, even embracing a new domain of life, the *Archaea,* a division that we did not know existed during the two previous golden ages. The discovery of new microorganisms may well lead to further advances in industrial processes and enhanced environmental control.

- The field of microbial ecology is closely related to agricultural microbiology. Microbial ecologists study the relationship between microorganisms and their habitats. Microbial ecologists are employing microorganisms in bioremediation to reduce pollution effects. In general, microbe-microbe interactions have not yet been extensively explored.
- The genomes of many microorganisms already have been sequenced (*Haemophilus influenzae, Methanococcus jannaschii, Escherichia coli*), and many more will be determined in the coming years. These sequences are ideal for learning how the genomes are related to cell structure which can provide information about the minimum assortment of genes necessary for sustenance of life. Analysis of the genome and its activity will require continuing advances in the field of bioinformatics and computational biology to investigate important biological problems.
- Further research on unusual microorganisms and microbial ecology is expected to lead to a better understanding of the interactions between microorganisms and the inanimate world. Microbial and host interactions are very useful for designing effective programmes for the control of environmental pollution, disease control in plants, livestock, and humans.
- Because of their relative simplicity, microorganisms are excellent subjects for the study of a variety of fundamental questions in biology. For example, studies involving elucidation of development of complex cellular structures and cell-to-cell communication and responses to environment are best understood in microbial systems as compared to higher organisms.
- Finally, microbiologists will be challenged to carefully assess the implications of new discoveries and technological developments. A better judgment and communication of a balanced view concerning positive and negative short- and long-term impacts of these events on society. The future of microbial biotechnology is bright.

4 CONCLUSION

Despite the great discoveries and impacts of microbial technologies in the past, the area remains very bright. The microbiologist René Dubos summarizes the excitement and promise of microbiology when he stated that "How extraordinary that, all over the world, microbiologists are now involved in activities as different as the study of gene structure, the control of disease, and the industrial processes based on the phenomenal ability of microorganisms to decompose and synthesize complex organic molecules". Microbiology is among the most rewarding professions because it gives its practitioners the opportunity to be interacting with diverse fields of natural sciences that continuously focus on solving challenges in man's environment for betterment of human life.

References

Gould SJ (1996). Planet of the bacteria. *Washington Post Horizon* 119: 344.

Lin Y and Tanaka S (2006). Ethanol fermentation from biomass resources: current state and prospects. *Appl Microbiol Biotechnol* 69: 627–642.

Mir J (2004). Industrial microbiology. A new challenge. *Int Microbiol* 7: 81-82.

Pelczar MJ, Chan ECS and Krieg NR (2005). Microbiology. Tata McGraw-Hill Publishing Company Limited, New Delhi.

Prescott LM, Harley JP and Klein DA (2007) Microbiology, 5th edition, McGraw Hill, Publishing Company Limited.

Schaechter M (2003). Integrative microbiology– the third golden age. *J Biosci* 28: 149–154.

Scholz F and Schroder U (2003). Bacterial batteries. *Nat Biotechnol* 21: 1151-1152.

Teske AP (2005). The deep subsurface biosphere is alive and well. Trends Microbiol 13: 402–404.

van den Burg B (2003). Extremophiles as a source for novel enzymes. *Curr Opin Microbiol* 6: 213–218.

Wood AP, Aurikko JP and Kelly DP (2004). A challenge for 21st century molecular biology and biochemistry: what are the causes of obligate autotrophy and methanotrophy? *FEMS Microbiol Rev* 28: 335–352.

Yokobayashi Y, Weiss R and Arnold FH (2002). Directed evolution of a genetic circuit. *Proc Natl Acad Sci USA* 99: 16587–16591.

2

Prospects of Vermicompost for Organic Cultivation in India

Santosh Kumar Tiwari

Microbiology and Plant Pathology Research Laboratory, University Department of Botany, T. M. Bhagalpur University, Bhagalpur-812007

Email: santosh_fertilizer@yahoo.co.in

1 INTRODUCTION

Huge quantity of agricultural wastes is produced around the world as residues of different crops. They are mainly destroyed by burning or disposed of in the environment uncontrolled. This biomass could be usefully recycled in agriculture as safe soil conditioners following the simple recipe: "What comes from soil must be returned to the soil" (Sequi 1990) that tunes up with the exhortation: "Now time has come not to expect from the earth to produce more, but to expect man to do more with what the earth produces" (Pauli 1996).

On the other hand, with uncontrolled rise in populations (over 1.0 billion) a continuous fall in landholdings size is well understandable due to family level division of properties. Today, 91.3 per cent of total holdings are classified as marginal, small and semi-medium (1-2 ha) and 75 per cent of the 100 million farming families operate own holdings of less than a hectare and the situation is likely to worsen in coming generations at this rate. Smaller landholdings combined with dry land, rain-fed areas, culturable waste land, large family size, shrinking natural resources, increasing input costs, low net returns from conventional crops and unreliable weather conditions make agriculture an uneconomical proposition leading to grower's inclination towards urbanisation.

To meet the requirement of conventional fertilizers, India imports large quantity of N, P and K fertilizers which is a heavy burden on foreign exchange. High cost of conventional fertilizers is also a deterrent in increasing use of chemical fertilizers. The nutrient consumption in India in 2001-2002 was 52.69 kg N, 15.54 kg P_2O_5 and 6.20 kg K_2O per hectare. The projected demand for chemical fertilizers in 2025 is expected to be 28.8 m tonnes. About 25% value of this demand can be met by utilizing a huge quantity of various organic resources such as farmyard manure (200 m tonnes), crop residues (30 m tonnes), urban and rural wastes (910 m tonnes) and green manures (925 m tonnes). All such organic resources are composted with the help of earthworms as vermicompost that way may achieve the demand of nutrient supply, which is cost-wise also beneficial for crop production. Vermicompost is the basic ingredient for successful organic farming in India. More than 85% of organic crop cultivation depends on it (Prabhu 2007). Vermicompost technology or vermitechnology has solved such problems wherever it has been implemented (Tiwari 2007; Roy and Tiwari 2008).

2 CONCEPT OF VERMICOMPOST

Vermicompost technology or vermitechnology is a modified natural way of handling biodegradable wastes. It is very simple, almost an aerobic process and is equally effective for all kinds of organic wastes. In this process earthworms of selected species are fed upon pre-decomposed organic wastes, soil, etc., and which excretes small pellets known as vermicompost, rich in nitrogen, phosphorus and potassium which are essential nutrients for plants. The production of vermicompost involves the breeding of earthworms in a prepared mixture of cow dung, soil and crop residues/kitchen garbage/other biodegradable wastes till the whole mass is converted into casts. Vermicomposting is a natural aerobic process of converting all biodegradable wastes into plant nutrient rich organic manure (Edwards and Burrows 1988; Bhawalkar 1991) through feeding the earthworms. Principal requirements are to make vermiculture pit, collection and processing of organic waste and crop fields, procuring earthworms, maintaining the shed, watering, sieving the vermicasts and their application in crop fields.

2.1 Beneficial Role of Earthworm and Vermicompost

Earthworms have been known as farmer's friend since long (Darwin 1881). As worms ingest and digest decomposing plant material, their digestive tracts process the organic matter and important nutrients are retuned to the soil through castings. Worm waste or vermicasts have micronized organic particles containing several enzymes, microorganisms, minerals and plant nutrients. Vermicast production is the direct outcome of earthworms' activities which plays a significant role in the nutrient cycle of soil, they also help to increase the percentage of micronutrient (Umamaheshwari 2005) and improve the soil structure such as soil porosity, soil aggregation, water and nutrient conservation in the soil (Ellerbrock et al., 1999).

It is supposed that earthworms augment the rate of microbial decomposition by fragmenting the large organic matter during its passage through gut, thus increasing surface area for establishment of microorganisms (Peterson and Luxton 1982; Anderson and Inesion 1984; Anderson et al., 1985). Ross and Gainus (1982) recorded increased respiration and enzyme activity in cast. Kale et al. (1991) also recorded increased cellulolytic microorganisms in earthworm worked soil. Further, they emphasized that a heterogeneous substrate like a mixture of agricultural wastes with more lignin content would encourage the build-up of not only cellulolytic but also lignolytic microflora in vermicast. Microorganisms are actively involved in solubilization of soil phosphorus. Favourable condition in earthworm cast for the plant growth may partly be attributed to the microorganisms present in the cast. Earthworms due to their varied activities like charring, breaking, gliding and assimilating create congenial atmosphere for growth and multiplication of microorganisms involved in the nutrient cycling. Further, due to its digestive activity and supporting the growth of antibiotic microbes, it may help in the elimination of plant pathogens.

A variety of fungi, bacteria and actinomycetes are associated both externally as well as in the gastrointestinal tracts of the earthworms. Thus, the role of earthworms and microorganism is of special interest from agricultural point of view. Many earlier workers including Edward and Fletcher (1988) and Ghosh et al. (1989) have reported active association of a variety of fungi in vermicast. Earthworm activities promote aeration and water retention in soil. Many of these beneficial activities of earthworms may be directly or indirectly attributed to microorganisms associated with it. It has been reported earlier that the earthworms feed on several fungi and other microorganisms for their feed. The microbes which survive with the digestive activities of worms are excreted in cast, when they readily get germinated and grow fast due to increased nutrient level (Rabatin and Stinner 1988). Krishnamurthy and Vajranabhaiah (1986) recorded a correlation between plant growth promotion by vermicast which may be attributed to the growth promoting substances such as auxins, cytokinins produced by microorganisms present in the cast. Satchell et al. (1984), Reddy and Reddy (1991) reported phosphate activities of microbes in vermicast. In totality, vermicompost is a beneficial compost to create an agricultural land with several natural amendments.

2.2 Vermicomposting Practices in India

The University of Agriculture Science, Bangalore, firstly initiated vermiculture in India with *Perionyx excavatus* and later in 1982 *Eudrilus eugeniae* was brought to the laboratory to test the possibilities of their establishment in this continent which has similar climatic conditions as that of its place of origin, Nigeria. It was observed that *Eudrilus* has successfully established its in the climatic conditions of India and surpassed feeding and reproductive rate of *Perionyx excavatus*. *Eisenia foetida* also entered the laboratory from Germany and found as preferred species for composting activity in the urban areas. Nowadays several standard projects for preparing vermicompost are available in CAPART (Council of Advancement of Peoples Action and Rural Technology), NABARD (National Bank for Agricuture and Rural Development), National Horticulture Mission (NHM), Khadi and Village Industries Commission (KVIC), Departments of Ministry of Science and Technology, Government of India, New Delhi and elsewhere. JNAVV Agriculture College, Indore and Kasturba Gram, Indore transfer the vermitechnology to farming community especially in the state of Madhya Pradesh since 1990.

In Uttarakhand, it is proposed to enhance manufacturing of NADEP compost, BD/EM compost, vermicompost, liquid manure, etc. For this KVIC, NHM, KVK, etc., providing technical aids as well as margin money to the farmers. Uttar Pradesh has also taken lead and somersault to opt organic farming.

Government of Madhya Pradesh is also promoting farmers on a large scale for green manuring use of biofertilizers, NADEP composting, vermicomposting, phosphocomposting, etc. In this respect dry dairy concept is also becoming popular. In Maharashtra regarding setting up of vermicompost units the State Govt. assisted financially ₹ 5000 per unit. In Rajasthan, Morarka Foundation, Jaipur and Society for Organic Agriculture Movement (SOAM) are effectively doing technology extension programme for popularizing vermitechnology.

In Kerala, Kerala Agricultural University has done good research and development work on vermitechnology. Sikkim, Pondichery and many North Eastern states have declared their state as 100 per cent organic farming states using vermicompost as the main source of organic compost. Several NGOs like M.S. Swaminathan Foundation (Tamil Nadu), Association for Promotion of Organic Farming (APOF, Banglore), Manipur Small Farmers Agri Business Consortium (Imphal), etc., are doing good promotional job for vermicomposting. Greencross Society of Mumbai, India, in collaboration with the Bombay Municipal Corporation daily vermicomposts 4 tonnes of slaughter house waste and 20 tonnes of vegetable waste in their own unit.

The Indian Aluminium Co. Ltd., Belgaum, India since July 1995 is vermicomposting the garbage and sewage generated by 500 houses. However, Biogenic System, Mumbai and Bhawalkar Earth Worms Research Institute, Pune have done good commercial as well as research and development work on vermitechnology.

In West Bengal Farmers Training Institute, Kalyani, Matar-Beni Trust, Hatsimul, Burdwan, Nimpith, Narendrapur, South 24 Paragana and Vishwa Bharti Shanti Nekatan are doing very good job of extending vermitechnology.

In Bihar, University Department of Botany, T.M. Bhagalpur University has established three vermitechnology demonstration centres and has created 180 composting units since 2001 under Bhagalpur Divisions and has conducted more than 100 demonstrations in eastern Bihar for popularizing vermitechnology among rural masses with the economic help of Department of Biotechnology, Govt. of India. Nowaday some colleges, KVK, NHM, KVIC, etc., are extending their hand for vermicomposting in Bihar. Krishak Vermicompost Udyog, Banka; Vermicompost Vikash Kendra, Begusarai; Vinayak Vermicompost, Samastipur; etc., are well established vermi-cottage industries in Bihar. Their cyclic production capacity is more than 100 tonnes.

2.3 Need of Agriculture

Chemical fertilisers are responsible for deterioration of soil friability (crumbliness) and destruction of beneficial soil life such as earthworms, bacteria and micro-arthropods. Moreover, plants grown on chemical fertilizers have low and contaminated nutrient value in comparison to grown naturally or organic way. Organic cultivation requires less irrigation water in comparison to chemically grown crops. The International Conference on Water and the Environment, Dublin the Millennium Summit, 2000 and the United Nations Conference on Environment and Development and the Earth Summit at Reo De Janero, held in 1992, have drawn the world's attention to this crisis. India accounts for 16 per cent of the world's human population and nearly 30 per cent of the cattle with only 2.4 per cent of the land area and 4 per cent of water resources. Even if full irrigation potential is exploited, about 50 per cent of the country's cultivated area will remain un-irrigated, particularly with current level of irrigation efficiency.

Automatically it creates a situation for farmers to go for organic farming in order to meet quality food demand by using organic manures or vermicompost which also enrich the soil with nutrients and improve waterholding capacity. Therefore, viewing the conditions of farming, the first thing is to utilize organic composts for nutrient supply and maintaining the moisture of soils and another thing is to grow such crops that may provide a sustainable means of produce as return to farmers.

2.4 Vermicomposting Methods in India

Production of vermicompost is not much more technical, any farmer can produce this after a short training or with the help of literature. In India different vermicomposting methods have been applied from time to time, which are detailed below:

2.4.1 Platform method

In this method a well plastered platform of size 100′ × 40′ × 1.5′ is made. On the upper surface of the platform a boundary wall of 10″ width and 3′ height is made and at one of the wider sides 20′ surface is left for the entrance of tractor. This platform is covered by roof or some kind of shade. The passage of wall and roof is covered by wire mesh so as to prevent birds, snakes, lizard, large insects, etc., from entering into the production unit and some darkness is also maintained. At the joint of the platform and boundary wall after each 10 feet a 2″ diameter hole is left through which excess of water may pass out. In order to collect the vermiwash, a 5″ wide and 5″ high cemented ditch is made using one outer wall of the platform and is linked to a cemented tank where it is gradually collected. If the production unit is attacked by ants the production unit may be protected by constructing another drainage which is 24 hours filled with water. In this method of vermicomposting processed mixture of cow dung and waste is deposited in 4 rows for recycling by earthworms and after processing closer to the old row another fresh row of composting mixture is deposited through which 80% earthworms automatically migrate into new heap and in this way manpower is saved.

2.4.2 Open bed method

Open bed method of vermicomposting follows the same process as done in case of platform method of vermicomposting, but it requires tree shade at the place of cemented platform. The earth surface is labelled and or a polyethylene sheet is covered in the area of vermicomposting and then biodegradable processed feeding mixture of earthworm is loaded to precede vermicomposting (Fig. 1). This process requires a lot of precautions and constant watch on composting process. Open bed vermicomposting is well practised in Bihar by Krishak Vermicompost Udyog, Banka; Vermicompost Vikash Kendra, Begusarai; Vinayak Vermicompost, Samastipur, etc.

Fig. 1 Open bed method vermicomposting
(*Source:* Krishak Vermicompost Udyog, Banka, Bihar)

2.4.3 *Pit/Tank method or composite pit method*

Pit or tank method of vermicomposting is suitable for small and marginal farmers. A 10′ × 3′ × 1.5′ sized cemented brick tank of 5″ width wall is most suitable for faster bioconversion of the waste; its wall does not need plastering. The number of tank depends on the availability of waste per day. When more than one pit is constructed in a unit it is known as composite pit vermicomposting (Fig. 2). While constructing tank the floor is strengthened with cemented mortar. The shade, drainage and protection ditch is also made just like platform method.

(a) (b)

Fig. 2 Vermicomposting methods in India: (a) an ideal composite pit vermicomposting unit; (b) internal cemented structure

2.4.4 *Bamboo pit method*

This method has been demonstrated in some village of Samastipur district of Bihar by Rajendra Agricultural University. In this method required size of bamboo logs (for 10′ × 3′ × 3′ size pit) are split longitudinally in 1″-2″ width and then with the help of nails these are arranged in the shape of a pit. After making pit shape this is hanged on 4-6 bamboo pillars so as the lower side does not touch the soil surface and maintains a distance of 6″ to 12″ from the soil. The inner and outer wall of the pit is layered by soil or by the mixture of sand and cement (Fig. 3). Generally, farmers make

Fig. 3 Bamboo pit method of vermicomposting
(*Source:* R.A.U. Pusa, Bihar)

roof by using its pillar bamboo which is insufficient for protecting the pit from the force of angular rainfall and sunlight. This is the main drawback of this composting method.

2.4.5 Bag method

Recently vermicompost production in specially designed HDPE bag (Fig. 4) has been recommended by several companies like Organic Farming Solution, Bengaluru, Karnataka; Surya Structurals, Solan, Himanchal Pradesh, etc. In the market it is available by its trade name, i.e., TETRA VERMI BED and Ready Made Vermibeds respectively (Fig. 4). Producers and promoters of these vermi-bags say that it is economical and mobile, easy to install and having provision to collect vermiwash liquid. This rectangular bag is 10-12 ft long, 3-4 ft width and 2 ft height that takes load of 1200–1400 kg biodegradable waste and produces 800–1000 kg vermicompost in a cycle.

Fig. 4 HDPE bag and bamboo formulated tetra vermi bed
(*Source:* Organic Farming Solution, Bengaluru, Karnataka)

In spite of these methods some more methods are practised for the production of vermicompost with some modifications. The merits and demerits of above-mentioned methods are furnished in Table 1.

Table 1 Merits and demerits of various vermicomposting methods

Vermicomposting Methods	*Merits*	*Demerits*
Platform Method	Large scale production, easy and quality vermiwash collection, less monitoring, less manpower requirement	One-time capital investment
Open Bed Method	Large scale production, require less manpower and capital investment	Continuous monitoring, not suitable in rains

Contd...

Contd...

Pit Method	Small scale production, easy and quality vermiwash collection, less monitoring	None
Bamboo Pit Method	Suitable for bamboo growing areas as low cost pit vermicomposting	Require skilled farmers in making the bamboo pit. Small-scale production, vermiwash collection tough, excess monitoring and only well trained and experienced farmers can do it.
Bag Method	Quick setup and production	Damage of HDPE bags by rats and dogs. Heating of vermibed and costly for farmers

2.5 Application in Crop Field

Normally application of any fertilizing agent depends on the soil quality, climatic condition, manure and crop type, therefore, the dose of same manure or biofertilizer may vary. Likewise vermicompost as manure has been applied in different doses for different crops. In India Vijaya and Aliya (2003) have applied vermicompost in 1:1:1 ratio (soil + vermicompost + farm yard manure) for seed germination of *Alibizia lebbeck, Azadirachta indica, Pterocarpus santalinus* and *Tarmarindus indica*. In the condition of field Parthasarathi and Ranganathan (2002) has observed significant seed germination percentage after applying press mud vermicompost @ 2 tonnes/acre. Sinha et al. (2007) have demonstrated to apply 5 kg vermicompost/plantlets of *Aloe barbadensis* as six months dose. However, Tiwari (2010) has applied vermicompost in 3:1, 2:1 and 1:1 ratios (soil + vermicompost) for medicinal plants like *Gloriosa superba, Costus speciosus* and *Vitex negundo* and stated that 3:1 ratio of soil and vermicompost have good growth response and increased phytochemical ingredients. In floriculture, Roy (2008) has demonstrated the use of equal quantity of vermicompost, soil and sand as most suitable potting media. Further, he has advised to apply 600 kg vermicompost/acre for paddy and wheat and 1000 kg/acre for maize crop in Bihar.

3 CONCLUSION

In India availability of sufficient biodegradable waste, better possibility of its bioconversion as vermicompost and the best use to adopt organic cultivation may open up a new avenue of income to farmers. Pit method, platform method and open bed method vermicomposting not only provide quality vermicompost but also will play a vital role in replacing chemical fertilizer and minimize organic pollution. Agricultural practices with vermicompost will provide natural nutrients to soil, increase rain-fed agriculture with quality produce and may save irrigation water.

References

Anderson JM and Inesion (1984). Invertebrate microbial interactions (In: Anderson JM, Rayner ADM and Walton DWM (eds), Cambridge University Press, Cambridge, 58–59.

Anderson JM, Huish SA, Inesion P, Leonard MA and Splat PR (1985). In: Ecological interaction in soil (Filter AH, Atkinson D, Read DF and Usher MB (eds), Blackwell Scientific Publications Oxford, 377–392.

Bhawalkar US (1991). Vermiculture biotechnology for LEISA. *Seminar on Low external input sustainable agriculture.* Amsterdam. Netherlands

Darwin CR (1881). The formation of vegetable mould through the action of worms with observations on their habits. Murray London. 298–299.

Ellerbrock RH, Hoen A and Rogasik J (1999). Functional analysis of soil organic matter with respect to soil management. *European J Soil Science* 50 (1): 65–71.

Edwards CA and Fletcher KE (1988), Interactions between earthworms and microorganisms in organic-matter breakdown. *Agric Eco Syst Environ* 24: 235–247.

Edwards CA and Burrows I (1988) The potential of earthworms compost as plant growth media. In: Edwards CA and Neuhauser E (eds) Earthworms in waste and environmental management. SPB Academic Press. The Hague. The Netherlands, pp. 21–32.

Ghosh N, Basu S and Bahera N (1989). Microfungi in the gut and cast of *Perionyx millardi* a tropical earthworm. *Journal of Soil Biology* 9: 46–50.

Kale D Radha, Kubra Banu, Sreenivasa MN, Vinayak K and Bagyaraj DJ (1991). In. Advances in Management and Conservation of Soil Fauna (Veeresh GK, Rajagopal D and Vikramathi CA (eds) Oxford & INH Publishing Co., New Delhi, pp. 599-603.

Pauli G (1996). Breakthroughs: What business can offer society. Epsilon press, Surrey, UK, 241.

Prabu MJ (2007). Reducing input costs. In: *The Hindu Survey of Indian Agriculture*, 121–128.

Peterson H and Luxton M (1982), Quantitative ecology of microfungi and animals in soil and litter. *Oikos* 39: 287–388.

Parthasarathi K and Ranganathan LS (2002). Supplementation of pressmud vermicasts with NPK enhances growth and yield in leguminous crops (*Vigna munga* and *Arachis hypogaea*). *J Curr Sci* 2 (1): 35–41.

Rabatin SC and Stinner BR (1988). Indirect effects of interactions between VAM fungi and soil-inhabiting invertebrates on plant processes. *Agric Eco Environ* 241: 135–146.

Roy AK and Tiwari SK (2008). Economic empowerment of farm women through Vermitechnology— A case study of tribal areas. In: Sinha AK (ed) New dimensions of women empowerment. Deep and Deep Publishers, New Delhi, India 268–275.

Roy AK (2008). A training and extension programme for vermiculture and vermicomposting for socioeconomic upliftment of the SC/ST population in tribal area. In: *Final Report of Dept. of Biotech., New Delhi, India Sponsored Project* (July 2004 - December 2007).

Satchell JE, Martin K and Krishamoorty RV (1984). Stimulation of microbial phosphatase production by earthworm activity. *Soil Biol Biochem* 16: 195

Sequi P (1990), The role of agriculture in nutrient cycling. Alma Mater Studiorum 3(2): 155–162.

Sinha J, Biswas CK, Ghosh A and Haque N (2007). Vermicompost: Its proper and successful application in the cultivation of *Aloe Barbadensis*. *Indian J Environmental and Ecoplanning* 14: 177–182.

Tiwari SK (2007). Extension of low cost and eco-friendly rural biotechnologies for the sustainable development of rain-fed agriculture. In: *International Symposium*, Organized by S.K.M. University, Dumka, Jharkhand, India.

Tiwari SK (2010). Studies on production of quality vermicompost with reference to growth performance of some common medicinal plants. In: Ph.D. Thesis of Univ. Dept. of Botany, T. M. Bhagalpur University, Bhagalpur.

Umamaheshwari S (2005). Vermicomposting of mango litter using earthworm *Perionyx excavatus*. *J Ecotoxic Monit* 15 (1): 69–72.

Vijaya T and Aliya S (2003). Evaluation of nutritional status of vermicompost and growth response of some forest trees species to its application in a nursery. *J Curr Sci* 3 (2): 449–452.

3

Role of Blue-Green Algae in Environment Management

Charu Gupta[1]*, Dhan Prakash[1] and Sneh Gupta[2]

[1]*Amity Institute for Herbal Research and Studies, Amity University, Uttar Pradesh Noida-201303, India*

[2]*Department of Zoology, RGPG College, Meerut-250001 (UP), India*

**Email: charumicro@gmail.com*

1 INTRODUCTION

The term algae refers to microscopically small, unicellular organisms, some of which form colonies and thus reach sizes visible to the naked eye as minute green particles. These organisms are usually finely dispersed throughout water and may cause considerable turbidity if they attain high densities. Cyanobacteria are organisms with some characteristics of bacteria and some of algae. They are similar to algae in size and in external appearance and requirements for light, nutrients and carbon dioxide, but unlike other bacteria; they contain blue-green and green pigments and can perform photosynthesis. Therefore, they are also termed blue-green algae (although they usually appear more green than blue). Blue-green algae occur naturally in habitats such as marine waters, rivers, lakes, damp soil, tree trunks, hot springs and snow. They can vary considerably in shape, colour and size. Blue-green algae are also known as cyanophytes, cyanobacteria and most recently cyanoprokaryotes.

Human activities (e.g., agricultural runoff, inadequate sewage treatment, runoff from roads) have led to excessive fertilization (eutrophication) of many water bodies. These have led to the excessive proliferation of algae and cyanobacteria in fresh water and thus have a considerable impact upon recreational water quality. In temperate climates, cyanobacterial dominance is most pronounced during the summer months, which coincides with the period when the demand for recreational water is highest.

Although many species of freshwater algae proliferate quite intensively in eutrophic water, they do not accumulate to form dense surface scums (often termed blooms) of extremely high cell density, as do some cyanobacteria. The toxins that freshwater algae may contain are therefore not accumulated to concentrations likely to become hazardous to human health or livestock. There are more than fifty major types of freshwater blue-green algae, but only about one-third (1/3) of them are capable of producing toxins.

2 GENERAL FEATURES OF BLUE-GREEN ALGAE

Blue-green algae, scientifically known as cyanobacteria, are microscopic single-celled organisms that grow naturally in fresh and salt waters. They can be seen with the aid of a microscope as single cells, accumulations of cells (colonies) or filaments of cells (trichomes). They are not algae (eukaryotes), but are a type of bacteria (prokaryotes), possessing the ability to synthesize chlorophyll *a*. Therefore, they act like plants by using sunlight to manufacture carbohydrates from carbon dioxide and water, process known as photosynthesis. Blue-green algae have vesicles or gas pockets inside vacuoles within their cells that they inflate with gas, thus able to regulate their buoyancy in response to environmental conditions. This buoyancy-regulating mechanism gives the blue-green algae a competitive advantage in obtaining light and nutrients and they move to where nutrient and light levels are at their highest.

2.1 Nutrient Requirements of Phytoplankton and N: P Ratio

In addition to light and carbon, growth of phytoplankton (all photosynthetic aquatic microorganisms including algae and blue-green algae) consumes 'nutrients'. Every replication of an algal cell roughly demands the uptake and assimilation of a quota of inorganic nutrients similar to that in the mother cell. In addition to carbon, the living protoplast comprises 19 other elements. The elements/nutrients most often implicated in the constraint of algal growth are: nitrogen, phosphorus, iron, and one or two of other trace elements, together with silicon – the well-known constraint on diatom skeletal growth.

Since the early twentieth century (1934) it has been recognized (primarily through the late Harvard University scientist Alfred Redfield's work on nitrogen: phosphorus ratios) that the elemental composition of phytoplankton was similar to that of the ocean: 16N : 1P. Scientists have accepted this as a constant called the Redfield ratio. However, the canonical Redfield N : P ratio of 16 for phytoplankton is not a universal optimal value but instead represents an average for a diverse phytoplankton assemblage growing under a variety of different conditions and employing a range of growth strategies. The N : P ratio is not fixed in the environment and this is mainly due to the inflow of nutrients from anthropogenic sources such as fertilizers and runoff containing nutrient rich waste (e.g., effluent).

Different cellular components of phytoplankton cells have their own unique stoichiometric properties. Most notably, resource (light or nutrients) acquisition machinery, such as proteins and chlorophyll, is high in nitrogen but low in phosphorus, whereas growth machinery, such as ribosomal RNA, is high in both nitrogen and phosphorus. Because these components make up a large proportion of cellular material, changes in their relative proportions have a marked effect on bulk cellular C : N : P stoichiometry.

During exponential growth, bloom-forming phytoplankton optimally increases their allocation of nutrients towards production of growth machinery, reducing their N : P ratio to ~8, far below the Redfield value of 16 (Fig. 1). However, when nutrients are scarce, slow-growing phytoplankton that can synthesize additional nutrients acquisition machinery are favoured. This allocation of nutrients results in optimal N : P ratios ranging from 36 – 45 (Fig. 1).

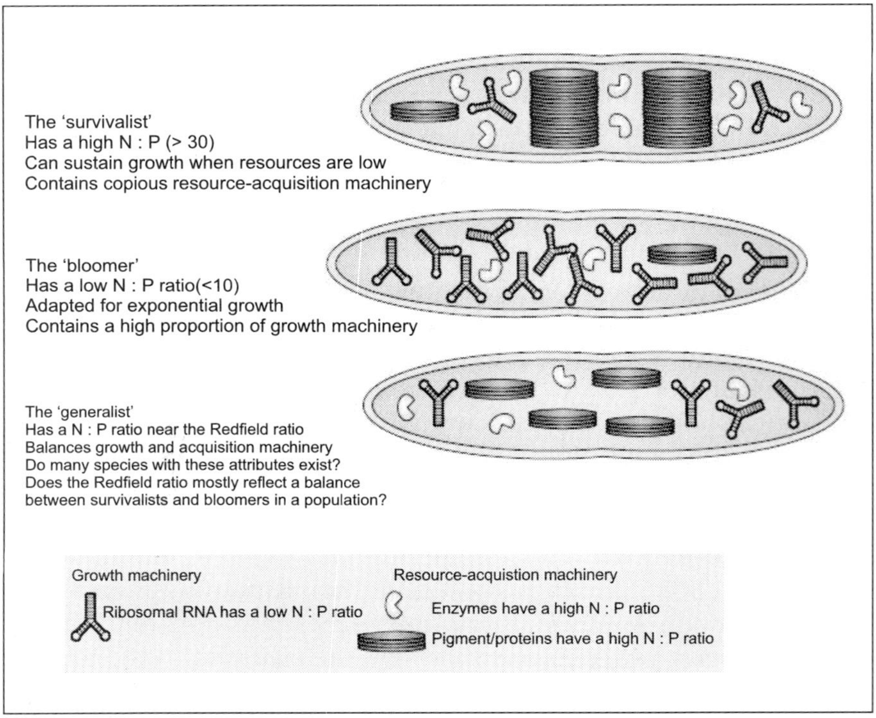

Fig. 1 Three different phytoplankton growth strategies and their resulting cellular N : P ratios (Arrigo 2005)

The optimal N:P ratio will vary from 8.2 to 45.0, depending on the ecological conditions. Nitrogen-fixing species (e.g., nitrogen-fixing blue-green algae) often have a higher N : P stoichiometry than non-fixing species. For example, nitrogen-fixing, Trichodesmium blooms have N : P ratios ranging from 42 to 125. The differences in N : P ratios between phyla and superfamilies are also significantly different. For example, green algae required N : P ~30 whereas diatom required ~10 and Dinophyceae required ~12 (Quigg et al., 2003) and red algae required N:P ~10 (Arrigo 2005).

2.2 Nutrient Dynamics in Aquaculture Ponds

In intensive aquaculture ponds, natural carrying capacities are greatly exceeded, and heavily laden artificial ecology is established among the various organisms and the environment they live. Due to the high densities of fish or prawn stock for profitable commercial aquaculture, ponds receive large nutrient inputs from uneaten fish feeds, fish excretion, and sediment mineralization/re-suspension. Nutrient budget in aquaculture ponds revealed that fish could only assimilate 20–27% nitrogen and 8–24% phosphorus of the total inputs. It was reported that fish feed accounted for 90–98% nitrogen and 97–98% phosphorus of the total inputs in aquaculture ponds. The major nutrients sank into the sediment, which accounted for 54–77% nitrogen and 72–89% phosphorus of the total inputs.

2.3 Growth and Proliferation of Blue-Green Algae in Aquaculture Ponds

A large portion of nutrients from fish feed is chemically or biologically transformed and then released into the water and taken up by the algae (including blue-green algae), resulting in the excessive production of algae.

Because of "static" pond systems and high amounts of nutrients added daily, algal blooms are encouraged to grow and proliferate. Blue-green algae are advantageous over other algae because of their ability to control buoyancy to access areas of increased nutrients and light. Therefore, their growth rate is much higher than other non-harmful algae in aquaculture ponds. Although phytoplankton/algae are important to the aquatic food chains because they are primary producers, blue-green algae are generally not eaten by other aquatic organisms, because they produce secondary metabolites (e.g., toxins).

Therefore, blue-green algae are not an important part of the aquatic food chain and food web. Uneaten excessive blue-green algae can produce blooms in aquaculture ponds.

Blue-green algal blooms in aquaculture ponds are also promoted by zooplankton's predatory activity. The relationship between algae and their zooplanktonic predators typically involves consumption of nutrients by algae, grazing of the algae by zooplankton which, in turn, enhances predator biomass, controls algal growth and regenerates nutrients. However, the daily feed supply for aquaculture raises nutrient levels, but does not simply increase normal predator–prey activity; rather, harmful algal bloom events develop often with serious ecological and aesthetic implications. In the absence of the predator (zooplankton or fish), the non-harmful algal species outgrow the harmful algal species and inhibit its growth. However, in the presence of the predator (e.g., zooplankton, fish or prawns in aquaculture ponds) it completely overturns by grazing out the non-harmful species, for which it demonstrates a preference. The dynamics of regeneration of the limiting nutrients versus nutrient consumption by algae is critical for bloom formation, as is the role of predators that exhibit prey selectivity (eat harmless algae only). When the supply of limiting nutrients exceeds the demand, the harmful algal biomass increases but nutrient status and palatability remain depressed.

2.4 Occurrence and Abundance of Cyanobacteria

Cyanobacteria are oxygen evolving and nitrogen fixing prokaryotes using sunlight as the sole energy source for the fixation of nitrogen (Stewart 1980). These occur abundantly in waterlogged rice fields of Japan, Thailand, China, Phillipines, Bangladesh and India (Roger et al., 1980; Venkataraman 1981). In India, on an average, BGA accounted for 33% of the 2213 samples and according to some reports up to 50% of total algae were BGA in some of the southern and eastern states (Venkataraman 1975; Kaushik 2005). These were most abundant in paddy soil than in other cultivated soils and the predominant nitrogen fixing genera are *Anabaena, Nostoc, Aulosira, Calothrix, Tolypothrix, Aphanothece* and *Gloeotrichia*. The nitrogen fixing potential of cyanobacteria can be determined by evaluating its biomass, N-content and nitrogen fixing activity. Phosphorus nutrition is known to induce the native blue-green algae. Top dressing with single super-phosphate @ 5 kg/ha, 10–15 days after transplanting induces the native nitrogen fixing algal flora. Additions of bone meal or superphosphate to rice soils enrich the soil with both phosphorus and nitrogen. Environmental conditions that prevail during dry season are more favourable for multiplication and establishment of BGA. Algal populations colonize well in rice straw.

Bio-volume is the measure of space occupied by the algae. It is used as a quantitative measure of the volume of cell material of algae in the sample. The size of cells can vary within and between species and toxin concentration relates more closely to the amount of dry matter in a sample rather than to the number of cells.

The use of bio-volume for determining compliance has been included to account for mixed species or small algae.

Cell counts should be used as the primary source for determining risks from recreational activities. The bio-volume should be used for determining the risks associated with mixed species where a known toxin producer is dominant or small algae such as *Aphanocapsa* spp., *Aphanothece* spp., *Cyanodictyon* spp., *Chroococcus* spp., *Radiocystis* spp. are dominant.

2.5 Nitrogen Contribution by Cyanobacteria

The annual turnover of nitrogen in biosphere varies from estimated 100 to 200 million metric tons of which 2/3rd comes from biological sources.

Nitrogen fixation is found only in prokaryotic organisms, i.e., bacteria and blue-green algae. The contribution of cyanobacteria to the total nitrogen fixed in paddy fields varies widely and is mainly dependent on physico-chemical properties of soil and many other climatic and biotic factors. Nitrogen fixed by cyanobacteria becomes available to rice plants only after its release extracellularly into the surroundings, either as extracellular products or by mineralization of their intracellular contents through microbial decomposition after death. It has been found that nitrogen fixation by blue-green algae and its release in the soil water system may become more useful for crop production during vegetative growth stage of rice plant than at the later stages (Ghosh and Saha 1993; Roger et al., 1993). Nitrogen can also be released in the soil from rapid decomposition of fresh or dry algal mass incorporated into the soil (Saha et al., 1982; Mian and Stewart 1985). In inoculated soil, recovery of cyanobacterial fixed nitrogen by rice varied from 13–50% depending upon the nature of inoculum, method of application and the absence of soil fauna (Tirol et al., 1982). Inoculation of soil with Aulosira (60 kg fresh wt/ha) significantly changes the nitrogen content of soil. Ghosh and Saha (1997) reported that the inoculation of soil with soil based mixed culture of *Aulosira fertilissima, Nostoc commune, N. muscorum* and *Anabaena* sp. significantly increased the release of inorganic nitrogen in the soil and the nitrogen content of soil was higher in exposed light incubated soil than unexposed soil due to N gain by blue green algae (Singh et al., 1987).

2.6 Blue-Green Algae as Biofertilizers

The fertility status of Indian soils with respect to the available nitrogen is generally low, low to medium in phosphorus and medium to high in potash. Since soil is a non-renewable resource, it should be managed properly; otherwise essential plant nutrients and organic matter will deplete. The modern-day intensive crop cultivation requires the use of mineral fertilizers which have posed a serious threat to soil health, environment and sustainability of food grain production. At this critical juncture, there is a need to use biofertilizers, which are low cost inputs, renewable and pollution free. Biofertilizers have crucial role in augmenting nutrient supply to crops by increasing the nutrient availability through exploitation of natural processes like biological nitrogen fixation, solubilization of insoluble phosphorus, decomposition of organic wastes, etc.

Cyanobacteria or blue-green algae find a highly favourable abode in the waterlogged conditions of rice fields. They provide cheap nitrogen to plants besides increasing crop yield by making soil vital, fertile and productive. BGA biofertilizer in rice popularly known as 'Algalization' helps in creating an environmentally safe agro-ecosystem that ensures economic viability in paddy cultivation while saving energy-intensive inputs. The agricultural importance of these organisms lies in their capacity to metabolize the molecular nitrogen, liberation of part of fixed nitrogen and growth promoting substances such as extra metabolites, solubilizing the insoluble phosphates, addition of organic matter and improving the physical and chemical nature of soil. Cyanobacteria are most simple photosynthetic, oxygen evolving, non-vascular plants having characteristics of both prokaryotes and higher plants. These are known to be the primary colonizers in most inhospitable environments. Trophic independence with regard to carbon and nitrogen combined with a great adaptability to variations in edaphic factors permit these to be ubiquitous and at the same time give them a unique potential to contribute to productivity in a variety of agricultural and ecological situations (Fogg et al., 1973). Most paddy soils have a natural population of cyanobacteria that provides free of cost, a potential source of combined nitrogen.

2.7 Cyanobacteria and Soil Fertility

Nitrogen fixing blue green algae/Cyanobacteria have been shown to be most important in maintaining and improving the productivity of rice fields (Roger et al., 1993). Cyanobacteria are colonizing microorganisms which remarkably adapt to a wide range of environment conditions. These are the most abundant microbial constituents of microbiotic crusts that exert a mechanical effect on soil particles as they form a gluing mesh and bind soil particles on the surface of their polysaccharide sheath material (Belnap 1993; Malam Issa et al., 1999, 2001). Cyanobacteria also excrete extracellular polymeric secretions (EPS) mainly composed of polysaccharides which ensure the role of binding agent of soil particles (Lynch and Bragg 1985). These improve the stability of soil surface and protect it from erosion. Cyanobacterial sheath and EPS also play a significant role in water storage due to the hygroscopic properties of polysaccharides and contribute to increased water retention capacity of the soil. Subsequent increase in water stable aggregates as a result of algal growth is important because soil aggregation and their arrangement influence infiltration rate, aeration, and soil temperature, thereby improving the physical environment of the crop. Thus, Cyanobacteria have been used as inoculants to improve soil structure, increase soil fertility or recover damaged soil crusts. BGA can affect rice plants by producing plant growth regulators (PGRs). The PGR like effects of BGA culture are similar to those of vitamin B_{12}, gibberellins and amino acids. Production of vitamins and growth substances by Cyanobacteria has been shown to have a positive effect on crop growth and yield. They also synthesize auxin-like substances which stimulate the root growth of seedlings. The biologically fixed nitrogen is gradually released through exudation and autolysis ensuring linear availability throughout the crop growth cycle. In addition, they add substantial amount of organic matter to the soil

(Goyal 2002). Polysaccharide sheath binds the soil particles together and increases the soil aggregates size which, in turn, reduces soil compaction. The cyanobacterial sheath also plays an important role in protecting the organism from toxic effects of metal ions (Rai 2001).

Phosphorus is another major nutrient after nitrogen in terms of quantitative requirements for crop plants which most of the Indian soils are deficient. In organic matter rich soils, the availability of P is greatly enhanced through microbial activity. Cyanobacteria also play an important role in solubilizing phosphorus making it available to plants. Nitrogen fixing Cyanobacteria, *Tolypothrix tenuis, Hapalosiphon fontinalis* solubilize Mussorie Rock Phosphate (MRP), a source of P_2O_5. Similarly, *Westiellopsis prolifica* and *Anabaena variabilis* were able to solubilize insoluble phosphate sources like MRP and Tricalcium phosphate (TCP) which can be exploited for the efficient utilization of low cost, low grade rock phosphate fertilizers (Yandigeri and Pabbi 2005). A major role of BGA comprises organic matter accumulation in soil which becomes available to the subsequent crops. Organic matter acts as a storehouse of nutrients like nitrogen, phosphorus and micronutrients, increases the exchange capacity of soils, releases CO_2, increases water holding capacity of soil, buffers the soil against rapid changes in pH, etc. Cyanobacteria also contribute in improving soil health by producing humus on soil surface after death and dissolve certain soil minerals, maintaining a reservoir of elements in a semi-available form for higher plants.

Reclamation of saline, saline-sodic and sodic soils with BGA has been advocated (Singh 1961; Thomas 1977; Kaushik 1995). The use of Cyanobacteria has several advantages over chemical amendments (gypsum) by enriching the soil with organic carbon, nitrogen and increased constituents on coming in contact with salt affected soils like salt stress protein and amino acids (Kaushik 1985). Cyanobacteria also colonize the calcareous nodules that are richly present in saline soils. During the dissolution process, the calcium ions are released which replace sodium ions on soil complex and thus reducing Na:Ca ratio in soil. Application of selected strains of Cyanobacteria (*Nostoc punctiforme, Tolypothrix ceylonica, Westeillopsis prolifica* and *Calothrix* sp.) followed by incubation and irrigation has brought about reclamation of these soils (Kaushik 1995).

2.8 Occurrence of Toxic Cyanobacteria

Toxic cyanobacteria are found worldwide in inland and coastal water environments. At least 46 species have been shown to cause toxic effects in vertebrates (Sivonen and Jones 1999). The most common toxic cyanobacteria in fresh water are *Microcystis* spp., *Cylindrospermopsis raciborskii, Planktothrix* (*syn. Oscillatoria*) *rubescens, Synechococcus* spp., *Planktothrix* (*syn. Oscillatoria*) *agardhii, Gloeotrichia* spp., *Anabaena* spp.,

Lyngbya spp., *Aphanizomenon* spp., *Nostoc* spp., some *Oscillatoria* spp., *Schizothrix* spp. and *Synechocystis* spp. Toxicity cannot be excluded for further species and genera. As research broadens and covers more regions over the globe, additional toxic species are likely to be found. Therefore, it is prudent to presume a toxic potential in any cyanobacterial population. The most widespread cyanobacterial toxins are microcystins and neurotoxins.

Field populations of the most common bloom-forming genus, *Microcystis*, are almost always toxic (Carmichael 1995). Generally, toxicity is not a trait specific for certain species; rather, most species comprise toxic and nontoxic strains. For microcystins, toxicity of a strain depends on whether or not it contains the gene for microcystin production (Rouhiainen et al., 1995; Dittmann et al., 1996) and that field populations are a mixture of both genotypes with and without this gene (Kurmayer et al., 2002). Experience with cyanobacterial cultures also shows that microcystin production is a fairly constant trait of a given strain or genotype, only somewhat modified by environmental conditions (Chorus 2001). While conditions leading to cyanobacterial proliferation are well understood (the physiological or biochemical function of toxins for the cyanobacteria is the subject of many hypotheses, (Chorus and Bartram 1999), the factors leading to the dominance of toxic strains over non-toxic ones are not.

Worldwide, about 60% of cyanobacterial samples investigated contain toxins. The toxicity of a single bloom may change in both time and space. Demonstrations of toxicity of the cyanobacterial population in a given lake do not necessarily imply an environmental or human hazard as long as the cells remain thinly dispersed. Mass developments and especially surface scums pose the risks.

2.8.1 Toxins of BGA

Blue-green algae can produce three different types of toxins with different modes of action.

Hepatotoxins: They are the most common of the blue-green toxins. They attack the liver and other internal organs. They may also cause visual disturbances, gastroenteritis, nausea, vomiting and muscle weakness. They are slower acting than neurotoxins. They can be produced by such species as *Anabaena, Cylindrospermopsis, microcysfis,* and *Nodularia.*

Neurotoxins: These act as neuromuscular blocking agents. They produce death by paralysis of peripheral skeletal muscles, then respiratory muscles leading to respiratory arrest. Neurotoxins are produced by species of *Anabaena, Aphanizomenon, Nostoc* and *Oscillafora.*

Endotoxins: The outer walls of all blue-green algae contain lipopolysaccharides. These are mainly contact irritants and can cause severe dermatitis and conjunctivitis in people coming into contact with the algae through swimming or showering. They may also cause stomach cramps, nausea, fever and headache and if swallowed may also cause irritation to airways and breathing difficulties. Lipopolysaccharides from blue-green algae can produce irritant and allergic responses in humans and animals that come into contact with the compounds.

There have been reported cases of adverse health consequences for swimmers exposed to blue-green algal blooms. The symptoms include itchy contact dermatitis, hay fever like symptoms, conjunctivitis and asthma. Gastroenteritis may result from accidentally ingesting the water.

Table 1 Summarizes the symptoms produced by different toxins produced by species of cyanobacteria

Type of Toxin	*Species Producing these Toxins*	*Toxins Produced*	*Symptoms*
Neurotoxins (affecting the nervous system)	*Anabaena, Oscillatoria*	Anatoxin-a Anatoxin-a(s) Saxitoxin	Muscle cramps, twitching, paralysis, cardiac or respiratory failure, death in animals
Hepatotoxins (affecting the liver)	*Microcystis, Cylindrospermopsis*	Microcystin Cylindrospermopsin	Nausea, vomiting, acute liver failure

Table 2 List below is to provide an indication of potential toxicity of certain genera

High risk	*Potential risk*	*Lower risk*
Microcystis sp.	*Aphanizomenon* sp.	*Cyanodictyon* sp.
Anabaena sp.	*Planktothrix* sp.	*Aphanocapsa* sp.
	Pseudoanabaena sp.	*Chrococcus* sp.
	Phormidium sp.	

2.9 Blue-green Algal Blooms

Algal bloom is a common term used to describe an increase in the number of algal cells to a point where they can seriously reduce the water quality. Blooms can discolour water, form surface scums, produce unpleasant tastes and odors, and create problems for aquatic life. They may vary in colour from green to blue, red, brown, dark green or black. Some will be easily seen as a scum while others may be evenly spread throughout the water or concentrated at depth. Blue-green algae blooms are likely to occur more often in warmer months. Many blooms are composed of non-harmful species; however it is best if a sample can be analyzed in a laboratory to determine if blue-green algae are present, and whether they are producing toxins.

Species of blue-green algae may dominate and increase excessively in water when the:

- nutrient levels, particularly phosphorus and nitrogen are sufficient to support the population growth;
- water is still and turbulence is low (lack of mixing);
- weather patterns are stable for a long time; and
- weather is warm (although blooms can occur in cooler weather too).

The longer the period of calm weather conditions, the greater the "bloom-forming potential". Such potential exists mostly in slow flowing rivers or in lakes, dams, weirs and reservoirs.

Blooms can persist for several weeks or sometimes months. Cooler, windy weather or increased flow may reduce or stop them fairly quickly. As bloom ages or begins to die, concentration of toxins may increase. Some toxins may persist for more than three months before they are degraded by sunlight and microbial activity. However, the time taken to degrade may increase if a chemical agent such as copper sulphate is added to reduce the number of microorganisms.

All blue-green algal blooms are not toxic. Those that are may change their toxicity slowly over a period of weeks to months.

Exposure: There are three potential routes of exposure to the toxins:

1. Direct contact of exposed parts of the body, including sensitive areas such as the ears, eyes, mouth and throat.
2. Accidental swallowing.
3. Inhalation of water.

Intake of toxins through swallowing, contact with nasal mucosa or by inhalation is an important route of exposure to toxins during water contact sports. There have been reported cases of people becoming sick after exposure via inhalation and resorption through nasal and pharyngeal mucous membranes in water sports involving submersion of the head, pumping from diving boards, sail boarding, canoe capsizing, swimming and inhalation of aerosols (water skiing).

Risks of ingestion are particularly high for children playing in shallow near shore water where scums tend to accumulate.

2.10 Blue-green Algae can affect Humans

Water contaminated by blue-green algae constitutes a health risk to water users in the following ways.

2.10.1 Drinking water

Due to the toxins they may produce, blue-green algae can affect water consumers in a number of ways. They have been associated with nausea, headache, vomiting, abdominal pain, diarrhoea, gastroenteritis, muscle weakness, pneumonia and paralysis. The occurrence and severity of symptoms is strongly related to the type of toxin, the toxin's concentration in water and to the amount of water consumed.

Boiling the water will not destroy toxins. Boiling will kill algae but in doing so will release toxins into the water.

In fact some toxins become more dangerous as a result of boiling the water. Also, laboratory studies with experimental animals indicate that blue-green algae may promote skin, liver and gut tumors. Further research is needed to establish whether they have similar affects on humans.

2.10.2 *Skin contact with blue-green algae*

Contact with blue-green algae (toxic and non-toxic forms) through water-based activities can cause problems such as skin rashes, swollen lips, eye irritation and redness, ear ache and itchiness, sore throat, hay-fever symptoms, asthma and possibly promotion of skin tumors. The risk of problems is likely to grow as contact time increases.

The risk from various activities ranges from zero where there is no contact, to high depending on the likelihood of an activity bringing an individual into contact with the algae. Wearing wet suits may result in greater risk of skin irritation because algal material trapped inside the wet suit may be in close contact with the skin for long periods.

2.10.3 *Eating fish from algae-affected waters*

The identification of "Paralytic Shellfish Poisoning" (PSP) toxins in one of the species of blue-green algae that has highlighted concerns about possible neurotoxin bioaccumulation in edible mussels and other shellfish. In marine waters consumption of shellfish contaminated by PSP has caused many human deaths. Cylindrospermopsin, another blue-green algal toxin common in Queensland freshwaters, has been identified in the edible flesh of crayfish. Therefore, mussels, yabbies and other shellfish caught in algae infested waters should not be consumed. The liver and gastro-intestinal tract of fish caught in algae infested waters are likely to be poisonous and should not be consumed.

2.10.4 *Irrigating with algae-affected water*

Research is currently being undertaken in relation to the uptake and accumulation of toxins in various plants. Toxins such as microcystin can be detrimental to both plants and animals. Based on the current knowledge, an affected water source should not be used for watering edible plants if there is an alternate supply.

If there is no alternative supply, direct spray on the edible part of the plants should be avoided. This is particularly important with plants such as cabbages, lettuces, tomatoes, and strawberries. Prior to use, the vegetables and fruit should always be thoroughly washed and rinsed with non-toxic water. Research has shown that dried blue-green algal cells can remain toxic for several months.

Irrigated pastures may cause problems with stock because of the potential for long-term poisoning over many months or years. The effects of sunlight may reduce the level of some toxins; however other toxins can be persistent and remain active for many months. Contact with the outer layer of blue-green algal cells may cause allergic reactions. The presence of algae and blue-green algae in water can cause clogging of filters, meters, valves and sprinklers.

2.10.5 Household usage — washing clothes and dishes

Contaminated water should not be used for washing clothes or dishes where a safe source of water is available. If the sources are not available, the following precautions are recommended:

- use rubber gloves to avoid contact with the water
- rinse dishes with non-toxic water or remove surplus water with a tea towel
- Give clothes a final rinse with non-toxic water and dry them in open air exposed to the sun.

Blue-green algal blooms often persist for several weeks, sometimes months, depending mainly on the weather or flow conditions. Cooler, windy weather or increased flow may reduce or prevent blooms from occurring.

As the bloom dies, the cells tend to become 'leaky'. If the bloom contains species that produce toxins, these will be released into the surrounding water. Once released, some toxins may persist for more than three months before sunlight and the natural population of bacteria in the water degrades them.

Other areas of concern

Aesthetic effects: Blooms may discolour the water and form unsightly and sometimes smelly scums on the surface and along the shore. A number of blue-green algae and other organisms can also produce substances that may impart a pungent earthy taste and odour to the water. While these may not be a health risk they may make the water unpalatable.

Animal health: Sometimes blooms wash onto the shore of a water body where a crust of dead algae can form and dry. In the dry crust, toxins may be active for months. Consumption of such material by animals may cause illness or even death.

If an algal bloom is suspected, farmers should prevent animal access to the affected water and provide alternate supplies where possible. It is most important that animals are not fenced into small areas where the only source of water has a covering of algal scum.

Ecological effects: Blue-green algae have been implicated in the poisoning of wild animals including rodents, amphibians, fish, pelicans, waterfowl, bats, and zooplankton. However, nitrogen-fixing blooms provide a valuable source of nitrogen in tropical and semi-arid zone water bodies. In some areas, a large diversity of microscopic animals (including rare forms) only exists during the blooms.

Environmental Impacts: All water bodies contain algae; however, the concentration and type of algae varies greatly. Because most algae blooms do not produce toxins, significant environmental impacts are infrequent. When algal blooms die off, they sink to the bottom and absorb available oxygen in the water. In some high concentrations, the depletion of oxygen may be great enough to result in fish kills. In extremely rare instances, toxin-producing blue-green algae have resulted in the sickness or death of some other animals, including livestock. High-density algae blooms can create taste and odor issues for public drinking water supplies. However, these are considered secondary drinking water standards. In some cases, these algae blooms can impair drinking water supplies that use a surface water source.

Water supply operation: Blue-green algae affects water and will require additional water treatment. Modern water treatment processes are capable of dealing effectively with blue green algae. Chlorination and activated carbon are two methods of treatment.

2.11 Possible Health Effects Related to Blue-green Algae Blooms

1. Exposure to blue-green algae can cause rashes, skin and eye irritation, allergic reactions, gastrointestinal upset, and other effects. At high levels, exposure can result in serious illness or death.
2. Depending on the particular cyanobacteria and the amount to which one is exposed; blue-green algae have the potential to cause a variety of adverse health effects, including liver toxicity (e.g., *Microcystis aeruginosa*) and neurotoxicity (e.g., *Anabaena circinalis*). Microcystin toxins may also promote tumor growth.

2.12 Blue-green Algae affect Water Quality and Promote Disease

Maintenance of appropriate water quality is essential in aquaculture farm management to ensure optimal growth and survival of aquaculture species. Poor water quality is linked with disease problems contributing to the collapse of several prawn farming industries in some South East Asian countries. Unmanaged blue-green algal growths in aquaculture ponds cause poor water quality following algal degradation. When algae reach their maximum growth phase, they flourish for a period and then die. This is known as an algal "crash". After a crash or periodical collapse of algal populations, the decomposition of these dead algae utilizes a large amount of oxygen and can cause oxygen deficits and increased concentration of toxic ammonia.

Insufficient oxygen and high ammonia concentrations may, in turn, kill aquaculture species, promote disease, and/or temporarily reduce the feeding and growth rates of fish/prawns.

2.13 Blue-green Algae Produce Toxins and Cause Stress to Aquaculture Species

Blue-green algae produce toxins which may cause toxic and physical stress to fish/prawns. Exposure to harmful algal neurotoxins can alter swimming and social behavior in fish and alter neuronal activity. Microcystins (a toxin produced by blue-green algae) can accumulate in the tissues of fish (particularly in the viscera-liver, kidney, etc.) and in shellfish. Blue-green algae produce taste and odour compounds and cause off-flavour problems.

The objectionable odours or tastes detected in aquaculture products are referred as 'off-flavour'. Off-flavour is one of the most economically significant problems encountered in catfish and prawn aquaculture. Most common "earthy" and "musty" off-flavors are caused primarily to the absorption of odorous compounds such as geosmin (GSM) and 2-methylisoborneol (2-MIB) from the water through skin and gills. These compounds accumulate in the fatty tissue of the aquaculture species. These odorous compounds are metabolites produced by certain species of blue-green algae. Off-flavour problems cause inconsistent product quality that may adversely affect consumer demand and ultimately, hinder industry development and decrease profits for producers and processors by increasing production costs. Off-flavour compounds are soluble in lipid and tend to concentrate in shrimp heads. Geosmin is also found to be the main compound associated with the seasonal occurrence of earthy-taints in UK farmed trout.

3 WHAT CAN BE DONE?

It is extremely unlikely that blue-green algal blooms can ever be prevented from occurring in aquatic systems. There are a widely occurring group of organisms that are opportunistic and will develop high populations rapidly under favourable conditions. To exclude only the blue-green algae without eliminating a wide range of other organisms is almost impossible. The objectives of control and treatment are therefore to:

- Minimize nutrient concentrations in streams and storages
- Encourage water conservation measures
- Develop or improve techniques to treat or detoxify contaminated water.

3.1 Immediate Action

If you suspect that blue-green algae have infested a water source you use for drinking, stock watering or domestic purposes, you should refrain from using water from that source until a suitably trained person has identified the algae.

It should be noted that boiling the water will not solve the problem.

3.2 Short-Term Action

If the presence of blue-green algae is suspected, alternative water sources should be used. These might include bottled, carted or tank water, or good quality groundwater. Treatment of the water with herbicides, copper or other algaecides should not be carried out; as such treatments break open algal cells and release toxins into the water. Recent studies have demonstrated that some blue-green algal toxins when released to water form more toxic products that may persist for many months.

If there is no alternative source available, filtration of the water through activated carbon may be used for domestic applications. The activated carbon filtration system will remove some or all of the toxins, but the results in particular applications cannot be guaranteed. They depend upon the adsorption qualities of the carbon and efficient maintenance of the system. Carbon filtration systems can be obtained from manufacturers of water treatment equipment.

3.3 Long-Term Action

There are a number of strategies currently being used to control blue-green algae. These include:

- Integrated Catchment Management aimed at reducing and managing nutrient flows into waterways from natural and artificial sources.
- Promoting natural algal grazers through selective fish stocking, aquatic plant restoration, and reduction in the contamination of waterways with pesticides.
- Supporting the maintenance and restoration of riparian vegetation (i.e., the plants near or at the edges of water bodies).
- Community involvement in tackling catchment management issues.
- Research programmes focused on toxin identification, rapid testing procedures for field use, toxin stability and degradation under different environmental conditions.

Prevention of blooms is the primary aim but as they form part of the natural ecosystem, this may not always be possible. It is important, therefore, to be aware of the potential dangers associated with blooms. By adhering to the procedures as outlined above, dangers and risks can be minimized.

4 CONTROL OF ALGAL BLOOMS

4.1 Using Algicides to Control Blue-green Algae in Aquaculture Ponds

Currently, there is no non-toxic, sustainable method to control blue-green algae in aquaculture ponds. However, using chemical algicides (copper sulphate and certain chelated or complexed copper products) may prevent blue-green algae. But copper algicides are not suited to aquaculture. They are not selectively toxic to odour-producing blue-green algae. Copper also interacts strongly with other water quality variables, and one important consequence of those interactions is that copper products become more toxic to fish and algae as hardness and alkalinity of water decrease. Copper sulphate treatments cause poor water quality. Increased aeration is required in copper treated ponds than in untreated ponds. Copper can accumulate in the sediments of treated ponds, with a relatively large fraction of the copper in the sediment initially present in a potentially bio-available form. In addition, due to their potential adverse environmental impacts, there may be local regulations in place, restricting the use of algicides.

Moreover, algicides or chlorination may cause the algal cells to burst and this can cause toxins to be released into the water. In addition, when algae decompose, the phosphorus stored in the algal cells is released into the water. The released phosphorus is available to fuel new algae growth.

4.2 Limiting Nutrients to Control Blue-green Algae in Aquaculture Ponds

The definition of a limiting nutrient is one that is necessary for algal growth, but available in a concentration insufficient to support continued growth. Once supply of this nutrient is exhausted, algal growth ceases independently of other remaining nutrient quantities. Any nutrients, for example, nitrogen, phosphorus or certain metals can become limiting nutrients for phytoplankton growth. Limitation of nitrogen is an expensive process, which require high energy and chemical costs and specialized equipment. Certain microorganisms, including algae, are able to fix atmospheric nitrogen opportunistically. Removal of metals may disrupt the local ecology, especially that of aquatic plants.

Therefore, phosphorus limitation is the cheapest and most practical means of preventing the growth of phytoplankton, particularly toxic blue-green algae.

Phosphorus is an essential requirement of living, functional algae. Phosphorus is a component of nucleic acids governing protein synthesis and of the adenosine phosphate transformations that power intracellular transport.

4.3 How Phosphorus Limitations Control Blue-green Algae in Aquaculture Ponds?

In thc events of phosphorus availability, phytoplankton intake excess phosphorus then their requirements which is called 'luxury uptake'. As a result, the cell may contain 8 – 16 times more phosphorus than the minimum required quota. As a consequence, it is theoretically able to sustain three or possibly four cell doublings without taking up any more phosphorus form the environment. It has been suggested that phosphorus storage in blue-green algae may be larger than in other algae providing them with a competitive advantage.

The phytoplankton biomass and the cyanobacterial component respond to phosphorus remedial action in four stages:

4.3.1 No biomass reduction if phosphorus in excess to requirements

This stage occurs in nutrient rich water bodies where nutrients are never growth limiting and there is an unused fraction of the total phosphorus. In these cases there is no immediate effect on phytoplankton biomass or species composition.

4.3.2 Declining amount of unused phosphorus, small reduction in biomass

This stage of recovery depends upon the behaviour of the phytoplankton in those water bodies where phytoplankton community is dominated by motile algae such as dinoflagellates and buoyant cyanobacteria. As a consequence of the reduced nutrient load, these phytoplanktons move to greater depths as they seek additional nutrients.

4.3.3 Phytoplankton biomass falls, minimal unused phosphorus

The phosphorus concentration continues to decline as a consequence of both the internal and external reduction of the phosphorus-loading. The overall result is a significant decrease in the phytoplanktons biomass as P-limitation begins to take effect.

4.3.4 Further decline in biomass and changes in composition of the phytoplankton

The fourth stage of recovery occurs once the water body reaches its new equilibrium state with a change in species composition. In this stage, the N : P ratio increases and the algal speciation shifts from toxic harmful blue-green algae to harmless beneficiary green-algae because green-algae required N : P ~30 whereas red algae required ~10, diatom required ~10, Dinophyceae required ~12 and blue-green algae required less than 29.

5 PHOSLOCK

Phoslock is a natural product which controls and manages blue-green algae in aquaculture ponds in environmental friendly and sustainable way. Phoslock is produced from modified bentonite clay that was invented and developed by the CSIRO (Commonwealth Scientific and Industrial Research Organisation) Land and Water Division, Australia to significantly reduce the amount of bioavailable phosphorus (Filterable Reactive Phosphorus (FRP) or Soluble Reactive Phosphorus (SRP)) present in the water column and in the sediment pore water of a water body.

5.1 How does Phoslock Work?

Phoslock locks up the soluble inorganic phosphate (PO_4) in the matrix of Phoslock clay. The phosphate is rendered bio-unavailable and cannot be used to fuel algal growth. When applied as slurry, Phoslock moves down through the water column and up to 95% of the Filterable reactive phosphorus (FRP) is rapidly removed and adsorbed onto the surface, forming an insoluble complex within the clay structure. As the Phoslock settles on the sediment-water interface it forms a 1–3 mm layer that is capable of adsorbing the FRP (on available binding sites) that is released from the sediment layer. Once the FRP is bound to Phoslock, it is no longer bioavailable for use by algae for assimilation and growth.

5.1.1 Benefits of Phoslock to control blue-green algae in aquaculture ponds

The reduction of phosphate (after locking by Phoslock) in the aquaculture ponds will have a direct impact on the proliferation of blue-green algae. Reductions in phosphate will increase the N : P ratio and algal speciation will shift from toxic blue-green algae to beneficiary algae. Remaining beneficiary algae will be eaten by zooplankton, which are a natural food source for aquaculture species.

Growth and proliferation of blue-green algae are not limited by the concentration of nitrogen in a water body. Some bloom forming blue-green algae are capable of fixing and storing nitrogen from the atmosphere. Their proliferation is related to the competitive advantage they have over other phytoplankton groups where excess phosphate (FRP) is available in the water. Therefore, phosphorus limitation by using 'Phoslock' is the most practical means of preventing the growth of toxic blue-green algae in intensive aquaculture ponds where nutrients regularly enter the water.

5.1.2 Potential application of Phoslock in the catfish aquaculture ponds to control off-flavour problems

The off-flavour problem is a considerable economic burden to the U.S. Catfish aquaculture industry (Channel Catfish) and other aquaculture industries throughout the world. The catfish aquaculture industry is the fastest growing sector of the United States and some Asian countries such as Vietnam, China, Thailand, Malaysia, Indonesia, India and Bangladesh. When catfish have an off-flavour, they cannot be marketed and this results in increased production costs. Phoslock is capable of controlling the growth and proliferation of blue-green algae in catfish aquaculture ponds in a sustainable way. Controlling blue-green algal growth will reduce or eliminate the off-flavour problems from aquaculture ponds.

5.1.3 Use of cyanobacteria for the production of ethanol and natural products research

Recent findings have shown that the photosynthetic assimilation products of cyanobacteria and other photosynthetic microorganisms can be processed into ethanol. The yield of ethanol from cyanobacteria compared to traditional plant feed stocks is higher because faster growth under optimal conditions can be achieved. Cyanobacteria have been isolated from various habitats both from freshwater and marine systems. Since cyanobacteria show an impressive variation in physiological properties a collection of cyanobacteria from various habitats also provides an excellent platform for the experimental selection of suitable strains.

Cyanobacteria are extremely rich in the production of secondary metabolites with interesting properties, for example, the UV-sunscreen compound scytonemin. Scytonemin is produced by cyanobacteria and absorbs sunlight in the wavelength from 325–1425 nm. It is responsible for the black colour of the so-called "Tintenstriche" on the rocks in the Alps. Cyanobacteria also produce various extracellular polysaccharides that are used for sequestration of nutrients, protection against grazers, and resistance to desiccation. Other cyanobacteria contain various sugar binding proteins, for example, cyanovirin-N that is under development as an antiviral agent due to its activity against HIV.

Ethanol is currently one of the most promising alternative energy sources to gasoline and has up to date been produced via fermentation mainly from corn, sugar cane and other agricultural plants (Marris 2006).

5.2 Citizen's Role

Citizen's involvement plays a vital component in the effort to manage algae blooms. There are a number of actions every citizen can take to reduce the influx of nutrients into the watershed and improve overall water quality:

- Do not over-fertilize. Most established lawns need a few nutrients to be healthy.
- Check soil nutrient levels prior to applying fertilizer to ensure correct application. Soil test kits can be purchased from some local hardware stores and through online distributors.
- If applying fertilizer, use phosphorus-free lawn fertilizers. Lawn-fertilizer packaging is labelled with three numbers for nutrient content. Look for a zero (0) as the middle number (phosphorus content) to indicate phosphorus-free fertilizer.

- Do not fertilize up to the edge of a waterway. Check with your local government for any specific setback requirements.
- Do not dispose of grass clippings or leaves in or near a waterway.
- To prevent nitrogen input from human waste, have your septic system inspected and tank pumped out at least every two years
- If conducting land disturbing activity, prevent soil and organic matter from washing into waterway as soil can carry nutrients into the waterway.

6 CONCLUSION

Thus, from the above paragraphs it is quite pertinent that blue-green algae though useful to mankind through its applications in the production of ethanol and being rich in other secondary metabolites, it is also menace to the society in the form of toxic algal-blooms, it affects the water quality and promote diseases, and it affects animal health and produces harmful toxins including neuro- and hepato-toxins. Thus, cyanobacteria should be judiciously used for the benefit of mankind and suitable measures and precautions should be adopted where it is affects the ecology of our environment.

References

Arrigo KR (2005). Marine microorganisms and global nutrient cycles. *Nature* 437: 349–355.

Belnap J (1993). Recovery rates of crypto biotic crusts: inoculant use and assessment methods. *Great Basin Naturalist* 53: 80–95

Beman M, Arrigo KR and Matson PA (2005). Agricultural run-off fuels large phytoplankton blooms in vulnerable areas of the ocean. *Nature* 434: 211–214.

Carmichael W (1995). Toxic Microcystis and the environment. In: Watanabe M, Harada K, Carmichael W, Fujiki H (eds) Toxic Microcystis. Boca Raton, FL, CRC Press, 1-12.

Chorus I and Bartram J (1999). Toxic cyanobacteria in water. A guide to their public health consequences, monitoring and management. Published by E & FN Spon on behalf of the World Health Organization.

Chorus I and Fastner J (2001). Recreational exposure to cyanotoxins. In: Chorus I (ed) Cyanotoxins, occurrence, causes, consequences. Heidelberg, Springer, 190–199.

Dittmann E, Meissner K and Börner T (1996). Conserved sequences of peptide synthetase genes in the cyanobacterium Microcystis aeruginosa. *Phycologia 35 (Suppl. 6):* 62–67.

Fogg GE, Stewart WDP, Fay P and Walsby AE (1973). The Blue Green Algae. Academic Press London and New York, 459.

Ghosh TK and Saha KC (1993). Effects of inoculation with nitrogen fixing Cyanobacteria on the nitrogenase activity in soil and rhizosphere of wetland rice (*Oryza sativa* L.), *Biology and Fertility of Soils* 16: 16–20.

Ghosh TK and Saha KC (1997). Effects of inoculation of Cyanobacteria on nitrogen status and nutrition of rice (*Oryza sativa* L.) in an entisol amended with chemical and organic sources of nitrogen. *Biology and Fertility of Soils* 24: 123–128.

Goyal SK (2002). Algal inoculants for sustainability in rice cultivation. In: Arora JK, Marwaha SS and Grover R (eds) Biotechnology in agriculture and environment. Asitech Publishers, New Delhi. 233–243.

Grond S and Meurer G (2007). Exploiting green treasures. *Chemistry and Biology* 14: 469–471.

Kaushik BD (1985). Effect of native algal flora on nutritional and physico-chemical properties of sodic soils. *Acta Botanica Indica,* 13: 143–147.

Kaushik BD (1995). Cyanobacterial response to salinity and amelioration technology. In: Kannaiyan S (ed) Rice management biotechnology, Associated Publishing Co., New Delhi, 323–329.

Kaushik BD (2005). Blue green algae (Cyanobacteria) in agriculture. In: *Advances in Microbiology at Indian Agricultural Research Institute.* 1961-2004. 19–34.

Kurmayer R, Dittmann, E, Fastner J and Chorus I (2002). Diversity of microcystin genes within a population of the toxic cyanobacterium Microcystis spp. in Lake Wannsee (Berlin, Germany). *Microbial Ecology* 43: 107–118.

Lynch JM and Bragg, E (1985). Microorganisms and aggregate stability. In: *Advances in Soil Sciences,* Springer Verlag, New York. 134–170.

Malam Issa O, Bissonnais YL, Defarge C and Trichet J (2001). Role of a microbial cover on structural stability of a sandy soil in Sahelian part of western Niger. *Geoderma* 101: 15–30.

Malam Issa O, Trichet J, Defarge C, Coute A and Valentine C (1999). Morphology and microstructure of microbiotic soil crusts on a tiger bush sequence (Niger, Sahel). *Catena* 37: 175–196.

Marris E (2006). Drink the Best and Drive the Rest. *Nature* 444: 670–672.

Mian MH and Stewart WDP (1985). Fate of nitrogen applied as Azolla and blue green algae (Cyanobacteria) in water logged rice soils- a ^{15}N tracer study. *Plant and Soil* 83: 363–370.

Quigg A, Irwin AJ and Finkel ZV (2003). The evolutionary inheritance of elemental stoichiometry in marine phytoplankton. *Nature* 425: 291–294.

Rai LC (2001). Cyanobacterial ecology and environment management. In: PK Singh PK, Dhar DW, Pabbi S, Prasanna R, Arora A (eds) Recent advances in the exploitation of blue green algae and azolla. *NCCUBGA,* IARI. New Delhi. 31–44.

Roger PA and Kulasooriya SA (1980). Blue green algae and rice. IRRI, Las Banos, Laguna Philippines. 112.

Roger PA, Zimmerman WJ and Lumpkin TA (1993). Microbiological management of wetland rice fields. In: Metting B (ed) Soil Microbiological Ecology – application in agricultural and microbiological management, Marcel Dekker, New York. 417–455.

Rouhiainen L, Buikema W, Paulin L and Haselkorn R (1995). Cloning and characterization of peptide synthetase genes from a hepatotoxic Anabaena strain. Oral presentation at the 1st International Congress on Toxic Cyanobacteria, Bornholm.

Saha KC, Panigrahi BC and Singh PK (1982). Blue-green algae, Azolla, additions on nitrogen and phosphorus availability and redox potential of a flooded soil. *Soil Biology and Biochemistry* 14: 23–26.

Sell D, Puls J and Ulber R (2007). White biotechnology energy solutions for the future? *Chemie in unserer Zeit* 41: 108–116.

Sielaff H, Christiansen G and Schwecke T (2006). Natural products from cyanobacteria: exploiting a new source for drug discovery. *I Drugs* 9: 119–127.

Singh AL and Singh PK (1987). Nitrogen fixation and balance studies of rice soils. *Biology and Fertility of Soils* 4: 15–19.

Singh RN (1961), Role of blue green algae in nitrogen economy of Indian agriculture. ICAR, New Delhi, India 175.

Sivonen K and Jones J (1999). Cyanobacterial toxins. In: Chorus I and Bartram J (eds) Toxic Cyanobacteria in water. A guide to their public health consequences, monitoring and management. Published by E & FN Spon on behalf of the World Health Organization, 41–112.

Stewart WDP (1980), Some aspects of structure and function in nitrogen fixing Cyanobacteria. *Annual Review of Microbiology* 34: 497–536.

Thomas J (1977). Biological nitrogen fixation. *Nuclear India* 15: 2-8.

Tirol AC, Roger PA and Watanabe I (1982). Fate of nitrogen from blue green alga in a flooded rice soil. *Soil Science and Plant Nutrition* 28: 559–569.

Venkataraman GS (1975). The role of blue green algae in tropical rice cultivation In: WDP Stewart (ed) Nitrogen Fixation by Free Living Microorganisms. Cambridge Univ. Press, 207–218.

Venkataraman GS (1981). Blue green algae production for rice. *FAO Soils Bulletin* 46: pp 102.

Yandigeri MS and Pabbi S (2005). Response of diazotrophic Cyanobacteria to alternative sources of phosphorus. *Indian Journal of Microbiology* 45(2): 132–134.

Useful Links:

- http://aidsinfo.nih.gov/DrugsNew/DrugDetailNT.aspx?int_id=0395prod
- http://ngm.nationalgeographic.com/ngm/2007-10/biofuels/biofuels.html

4

Medicinal Plants for Microbial Control

D.D. Joshi*[1], Ram Prasad[2, 3] and Wangkheirakpam Sujata[1]

[1]*Institute of Bioresources & Sustainable Development (IBSD), (Department of Biotechnology, Govt. of India), Takyelpat, Imphal-795001 (Manipur), India*

[2]*Amity Institute of Microbial Technology, Amity University Uttar Pradesh Noida-201303, India*

[3]*Department of Mechanical Engineering, Johns Hopkins University Baltimore-21218, Maryland, USA*

**Email: drddjoshi.ibsd@nic.in*

1 INTRODUCTION

The relationship between humans and plants have been very close throughout the development of different civilizations. The evidence of the use of plants for various ailments by our great ancestors, indicates the depth of association and respect to the Mother Nature. At present a large number of these natural medicines are under modern scientific studies, to understand plants and their chemical constituents. The gradual sophistication of phytochemistry, pharmacology, etc., has helped to decipher these plants for better understanding from ethno-medical uses. Medicinal plants have provided the basis for the great medical systems in human history of Hippocrates, Galen and the great Ayurveda of Indian subcontinent, traditional Chinese medicine (TCM), Islamic medical system and many other cultural traditions (that were often hybrids of the various systems of medicine), all these systems have formed a large part of the respective Materia Medica. Many plants are in use as traditional medicine for the treatment of bacterial infections worldwide, some of these have also been subjected to *in vitro* screening but the efficacy of such herbal medicines has seldom been rigorously tested in controlled clinical trials. Conventional drugs usually provide effective antibiotic therapy for bacterial infections but there is an increasing problem of antibiotic resistance and a continuing need for new solutions. The world market for plant-derived chemicals for pharmaceuticals, fragrances, flavours and colour ingredients, alone exceeds several billion dollars per year. Classic examples of phytochemicals in biology and medicine include taxol, vincristine, vinblastine, colchicines and artemisinin, etc. It is estimated that the global trade in medicinal plants is US$100 million per year, which is growing at the rate 10-15% annually (Joshi 2006).

2 ETHNOBOTANICAL KNOWLEDGE

Use of *Emblica officinalis* (Amla) is well known in Indian society for dandruff problem. Dandruff is a mycotic infection on the skin by the dermatophytes may be categorized into superficial and deep fungal infections. *Malassezia furfur* formerly known as *Pityrosporum ovale*, affects the hair and causes the disease called dandruff. People who suffer from dandruff have over-active sebaceous glands, which make their scalp oily. *Malassezia furfur*, a lipophilic, dimorphic and yeast-like fungus, occurring in human skin is opportunistic pathogen. As the fungus requires fat to grow, it is most common in areas with many sebaceous glands, on the scalp, face and upper part of the body. When the fungus grows too rapidly, the natural renewal of cells is disturbed and dandruff appears with itching (a similar process may also occur with other fungi or bacteria). Papaya (*Carica papaya*) has been traditionally used for the treatment is supported by the chemistry and pharmacology that it has cleansing, softening and soothing properties that act on flaky scalps. Its carotenoids soothe itching whilst, antiseptic protein purifies and prevents the re-appearance of dandruff. Avocado, papaya and rosemary components can purify the living environment of the scalp, balance sebum secretion, prevent and heal the cause of dandruff, abate growth of fungus and quickly remove exfoliated dandruff, achieving the function of sedating scalp and diminishing inflammation. Traditional use of papaya leaves as soap substitute, is presently in shampoos and face-lifting preparations (Silja et al., 2008).

Aloe vera is frequently cited as being used in herbal medicines and has a long history of use as a therapeutic agent with many reported medicinal properties. Amongst its therapeutic properties, it has been shown to have anti-inflammatory activity, immuno-stimulatory activity and cell growth stimulatory activity. Furthermore, activity against a variety of infectious agents has been attributed to *Aloe vera*; for instance, antibacterial, antiviral and anti-fungal. *Aloe vera* leaf gel can inhibit the growth of the two Gram-positive bacteria also. Specific plant compounds such as anthraquinones, dihydroxy-anthraquinones and saponins have been reported to have direct antimicrobial activity. Acemannan, a polysaccharide component from whole plant material, has been proposed to have indirect antimicrobial activity through its ability to stimulate phagocytic leukocytes have reported on the effect of the anthraquinone aloe emodin on arylamine N-acetyl transferase activity in *Helicobacter pylori*, and hence its antimicrobial activity. *Aloe vera* gel has been reported to be used for dandruff cure in traditional herbal treatment system (Cock 2011). Saponins from soap nut have been used for washing for thousands of years by various peoples, such as the Native Americans. Today, soap-berries are being considered for commercial use in cosmetics and detergents, among many other products. Soap nuts, especially are used medically as an expectorant, emetic, contraceptive and for the treatment of excessive salivation, epilepsy, chlorosis, migraines and diabetes. Studies show that saponin from soap nuts inhibits tumour cell growth and the progression of AIDS. Soap nut is used in traditional herb system to treat dandruff by mixing with water and applying it on the scalp. An infusion made from soap nuts gives a shampoo which works well and effective to fight dandruff as well as gives hair a silky shimmer and vitality (Maiti et al., 1968; Ojha et al., 2003).

Herbals for infectious diseases in Ayurvedic system of medicines are as for malaria (*Caesalpinia crista, Alstonia scholaris, Swetia chirata*), filariasis (*Streblus asper*), tuberculosis (*Puerasia tuberosal*) and leprosy (*Achyranthes aspera, Centella asiatica*), etc. (Joshi and Sharma 2011) In traditional remedies, some plants are used alone, e.g., the sap of the Amazonian tree, *Croton lechleri* for wound healing. Others, such as *Phyllanthus* spp. are used as a fresh herb and combined with either *Eclipta alba* or *Silybum marianum* for better results in curing for hepatitis (Elvin-Lewis 2011). As per available records *Cinchona officinalis* bark was used to cure malaria by Spanish Missionaries, in 1630 by native of Peru. Since it was transferred into Europe for curing malaria and was so successful that it rapidly replaced others globally. To satisfy the need of raw material, plantations of *Cinchona officinalis* were developed throughout tropical areas. Eventually, isolation of its primary bio-reactive alkaloids quinine and quinidine led to the development of several pharmaceuticals (Elvin-Lewis 2011). A few medicinal plants with global acceptance as anti-microbial are given below.

Garlic (Allium sativum): Garlic has anti-bacterial, anti-parasitic and anti-fungal properties, which makes it valuable as a wound healing agent. A decoction containing this herb may help control the complications of Crohn's disease. The inflamed tissues of the intestinal region might attract bacteria and fungi, which could add to further inflammation and more severe pain. An antimicrobial agent such as garlic extract acts as a natural prophylaxis against these microorganisms to promote the healing of the tissues. Garlic extracts can be taken for at least 7–10 days. The duration of treatment depends on the severity of the symptoms (Kumar and Sharma 2010).

Neem (Azadiracta indica): Neem, a very common tree in India, is considered to be the "village pharmacy" and has played a key role in Ayurvedic medicine and agriculture since time immemorial. It is a large evergreen tree growing 10 to 11 metre tall. The leaves are divided into numerous leaflets, each resembling a full-grown leaf. There is no part of the tree that is not be useful. It has a lot more than just medicinal properties, covering everything from the roots holding water in the soil to the shade it gives. The seeds contain substantial amount of essential oil (up to 40%) known as neem oil. The bitter constituents separated from this oil are nimbin, nimbinin, nimbicidine, azadiractin, salanin, etc. Neem has so many characteristics that it is not only useful for healing the earth as well as using all these products to help heal us (Jadge et al., 2008).

Turmeric (Curcuma longa): Turmeric has anti-inflammatory properties and helps to reduce the abscess on the linings of small and large intestines. Curcumin, the active ingredient in turmeric, needs to be studied in controlled conditions to generate data as is traditionally used in reducing inflammation in affected bowel tissue. Storage of the powder in an aluminum foil bag at room temperature for four months only slightly changed the curcuminoid content, indicating the high stability of the product. Hence, curcuminoid powder could be used as a food ingredient for various health-drink products. Curcuminoid, a natural colouring agent, is recognized as a rich source of phenolic compounds, consisting of three different compounds; curcumin, de-methoxy-curcumin and bis-demethoxy-curcumin. It has potential as a pharmaceutical expient, since it possesses antioxidant, anti-inflammatory, antimutagenic and anti-HIV properties and can reduce blood glucose and LDL. Curcumin, the major cucuminoid compound, is practically insoluble in water at acidic or natural pH, and it is soluble in alkaline solutions but high decomposition rate in alkaline media. Because of these properties, the use of curcumin/curcuminoid is limited. However, it was reported that curcumin could have improved solubility in complex formations or as a result of interactions with various macromolecules. The mechanisms underlying the antitumor abilities of nanoparticles are still not clear enough. The main strategies are to enhance *in vitro* and *in vivo* bioavailability of curcumin using nanocurcumin through structural modifications of the molecule and/or new formulations. However, the limited literature evidence devoted to show improvements in curcumin bioavailability reveals that the curcumin bioavailability enhancement has not gained significant attention. Yet, novel delivery strategies including nanoparticles, complexes offer significant promise and are worthy of further exploration in attempts to enhance the bioavailability, medicinal value. Nanocurcumin acts as a weapon to enter tumour cells and initiate their destruction which could emerge as an effective cancer-fighting agent (Bisht et al., 2007).

Bhringraj (Eclipta alba): The plant *E. alba* has important role in the traditional Ayurvedic, Unani systems of holistic health and herbal medicine of the east. *E. alba* is reported to possess hepatoprotective, antimicrobial, anti-inflammatory, analgesic, immunomodulatory, antiviral, and promoter for blackening and growth of hair. Important source of chemicals is wedelolactone, demethylwedelolactone exhibit antihepatotoxic activities. Studies revealed the anti-hepatitis B virus properties of *E. alba*. The shoot extract showed antibacterial activity against *Staphylococcus aureus* and *E. coli*. The herbal drug Tefroli, containing extracts of the plant in combination with others, when administered to the patients of viral hepatitis, produced improvement and good results.

There has been clinical studies conducted that prove the effectiveness of *E. alba* therapy in jaundice in children, and Bhringaraja Ghanasatwavati on the patients of kosthashakhasrits kamala with special reference to hepatocellular jaundice (Jadhav et al., 2009).

Bhringraj powder 1 part, black sesamum seeds half part, amalaka (*Emblica officinale*) half part, classically known as Bhringraj churna, was prescribed as a rejuvenating and age-sustaining tonic. Bhringraj was also used as a detoxifying deobstruent and antiseptic herb in vitiated blood, anaemia, splenic and liver enlargements, catarrhal jaundice, hyperacidity, gastritis, dysentery, nightblindness, eye diseases, toothache and laxative. The juice of *E. alba* is used for washing wounds and soft chancre (Gadanigraha). The powder of the root and haridra (*Curcuma longa*) is applied locally on skin affections. The oil extract of leaves is prescribed by Charaka and Sushruta for pointing the head, for hair growth and for giving natural colour to grey hair. Neelibhringaadi Tailam (Sahasrayoga) is prescribed for promoting hair growth and for giving natural colour to grey hair. In Unani medicine the juice of leaves is prescribed in skin diseases, cough, rheumatism, bronchitis, allergic urticaria, inflatulence, colic and liver infections. The seeds are used in sexual debility as aphrodisiac. Externally, the paste of leaves is applied over swellings. The root is used as an emetic and purgative. A decoction is used to invigorate the liver, greying of hair, staunch bleeding, spermatorrhoea, menorrhagia. In the Caribbean, the juice is taken for asthma, bronchitis, dizziness, vertigo, blurred vision, skin problems. Studies conducted with alcoholic extract of *E. alba* shows no signs of toxicity in rats and mice. The minimum lethal dose was found to be greater than 2.0 gm/kg when given orally and intraperitonially in mice (Jadhav et al., 2009).

Gokharu (Tribulus terrestris): The bacterial activity of *T. terrestris* varies depending on the origin and plant's part used. The ethanolic extract of the fruit and leaves of Indian herb has activity against *E. coli* and *S. aurues*. But the ethanolic extract of *T. terrestris* from Yeman has no detectable anti-bacterial activity, against any of the reference bacteria. The methanolic extract of the same herb grown in Iran has anti-bacterial activity. The activity is reported due to spiro-saponins, present in the herb (Bedir and Khan 2000).

Daruharidra (Berberis aristata): Daruharidra is used for topical antimicrobial in Ayurveda. The use of extract for eye drops in conjunctivitis is widespread. The active principle of the plant is berberine, has been extensively studied. The activity of berberine against a variety of organisms, including bacteria, viruses, fungi, protozoa has been demonstrated. Berberine has been shown to bind to DNA and inhibit its cleavage. There is an urgent need to develop specially targeted drug delivery systems for berberine, to topical and systemic antimicrobial use. Local anaesthetic, pigment inducing, enzyme inhibitory, antihypertensive, antipruritic and antipyretic activities have been reported with berberine. Recently, the antiamnesic activity of berberine has been reported. Despite the diversity of activities, proper clinical documentation of the reverse pharmacology of the plant is to be done (Latif et al., 2010).

Kutki (Picrorrhiza kurroa): Kutki, a well-known formulation the 'Arogyawardhani', popularized by Late Vaidya Zandu Bhatt, is a traditional Ayurvedic formulation for the treatment of jaundice, using *Picrorrhiza kurroa*, as main ingredient. A double-blind trial with Arogyawardhani, Kutki and placebo was conducted for treatment of viral hepatitis. Significant hepatoprotective effects were observed both for Arogyawardhani and Kutaki. Later, picrosides, the active principles of Kutki were also tested *in vivo* and *in vitro*. Significant antioxidants as well as hydrocholeritic effects were noted for Kutki. These effects of hepatoprotection in CCl_4 and galactosamine models are considered the pharmacological correlates of clinical actions. However, there is a need to study, at the cellular and molecular levels, how the water outflows in the biliary microcanaliculae is enhanced. A cucurbitacin free extract of *P. kurroa* (Picroliv), which is now in Phase III trials. Thus *P. kurroa* is a good example where the traditional use, observational and experiential data together led to product supported by reasonable evidence base. Various structures including family of picrosides could serve as good scaffolds for fast track drug design, discovery and development of novel hepatoprotectives (Acharyya et al., 2009).

Punarnava (Boerhavia diffusa): The plant is rich in secondary metabolites such as tannins, alkaloids, flavonoids, phenols, steroids, and volatile oils which are responsible for therapeutic activities. It has been used as antimicrobial agent, since time past. *B. diffusa* has been reported to cure many ailments including skin diseases, small pox, jaundice, gonorrhoea, asthma, cough, scabbies and yaws. It has long since been deemed a valuable source of natural products for maintaining human health. The use of plant extracts and phytochemicals, both with known antimicrobial properties, can be of great significance to therapeutic treatments. The study was carried out to determine the antimicrobial activity of the aqueous and ethanolic extracts from the leaves of *B. diffusa* on *E. coli*, *P. aeruginosa* and *S. aureus* and establish a basis for the therapeutic uses of the plant. It has been observed that antimicrobial activity of the different extracts increased with increase in concentration and the negative control (water) in this work did not show any zone of inhibition, confirming that the inhibitory agent(s) is (are) actually from the plant extracts (Akinanibosun et al., 2009).

3 SCREENING FOR NEW ANTI-MICROBIAL FROM PLANTS

The widespread use of herbal for healthcare preparation, described in the ancient text like the Bible, TCM and Vedas, have been studied for actives, using modern high-tech art. Modern herbal drugs are evidence based, so the experimental design for new drug discovery, the microbial properties, against various infections are determined as (Fig. 1). For screening selection, plants are collected either randomly or by following leads supplied by local healers in geographical areas where the plants are found. Initial screening of plants for possible antimicrobial activities typically begins by using crude aqueous or alcohol extractions followed by various organic extraction methods. Plant material can be used fresh or dried. Essential oils, a complex natural mixtures of volatile secondary metabolites, isolated from plants by hydro or steam distillation and by expression (citrus peel oils), have high antimicrobial activity. The main constituents of essential oils (mono and sesquiterpenes), along with carbohydrates, alcohols, ethers, aldehydes and ketones, are responsible for the fragrant and biological properties of aromatic and medicinal plants (Table 1). Due to these properties, since ancient time spices and herbs had been added to food, not only as flavouring agents but also as preservatives. For centuries essential oils have been isolated from different parts of plants and are also used for similar purposes. During evaluation for therapeutics, when we find an active extract, the first task to be taken is the identification of the bioactive phytocompounds, and this can be upto either a full identification of a bioactive phytocompound(s) (after purification) or partial identification to the level of a family of known compounds.

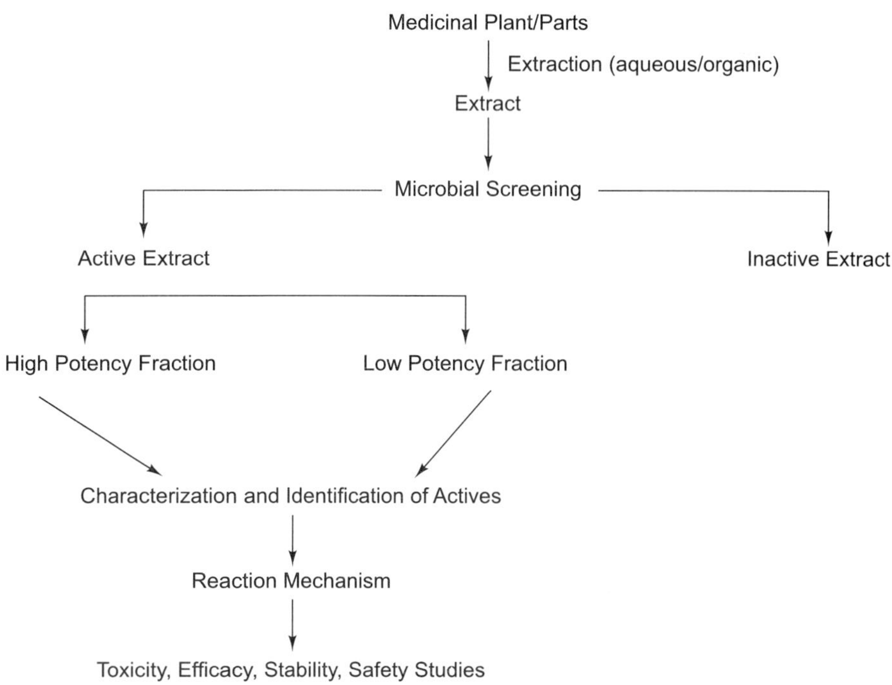

Fig. 1 Schematic presentation for screening of herbs as anti-microbial (Joshi and Sharma 2011).

Table 1 Phyto-antimicrobial compounds (Nostro 2006)

Active group	*Sub-group*	*Actives*	*Mechanism*
Alkaloids	Isoquinoline	Berberine	Intercalate into cell wall and/or DNA
	Piperidine	Piperine	Intercalate into cell wall and/or DNA
Phenolics	Flavonoids	Chrysin	Bind to adhesions
	Flavones	Apigenin	Inactivate enzymes, complex with cell wall
		Abyssinone	Inactivate enzymes, inhibit HIV reverse transcriptase
	Flavonols	Galangin	Membrane damage
	Phenolic acids	Epicatechin Cinnamic acid	Membrane disruption Membrane disruption
	Quinones	Hypericin	(i) bind to adhesins, complex with cell wall (ii) inactivate enzyme
	Simple phenols	Catechol	Substrate deprivation
	Tannins	Ellagitannin	(i) bind to proteins (ii) bind to adhesins (iii) enzyme inhibition (iv) substrate deprivation (v) complex with cell wall (vi) membrane disruption (vii) metal ion complication

4 PLANT OILS AS ANTI-MICROBIAL

Traditional systems of medicines use plant oils as anti-microbial since long. The 'Havan' in Indian system, especially Hindu ritual has the secret of fumigation of the home and surroundings, using essential oil producing items on burn. Essential oils can also be blended for a specific therapeutic application. Essential oils that are carefully blended with a specific therapeutic purpose may be referred to as an essential oil synergy. A synergistic essential oil blend is considered to be greater in total action than individual oil working independently.

Neem oil: Neem (*Azadirachta indica*) oil is highly effective as a topical drug for preventing pregnancy, used topically in the vagina. By using neem vaginal suppositories before sexual contact, it may prevent AIDS as well as prevent pregnancy. It is also very useful at the time of childbirth as it prevents possible inflammation (Sairam et al., 2008).

Apricot oil: Apricot (*Prunus armeniaca.*) seeds were historically used to treat tumours as early as AD 502. It was used against tumours and ulcers in England in the 1600s. Arab doctors were known to recommend the oil extracted from apricot stones to cure hemorrhoids, aching noses and earache. They also used it to help to regulate, moisturize and restructure the skin. Apricot kernel oil has moisturizing, nourishing and revitalizing properties and it is particularly helpful for dehydrated, delicate, mature and sensitive skin, and it helps to sooth inflammation. This finely textured oil spreads easily and is particularly rich in both oleic acid and linoleic acid. Shelf life of apricot oil is 6 months to 1 year, provided kept properly in airtight container, in dark, refrigerated after opening (Singh et al., 2010).

Calendula oil: Calendula (*Callendula officinalis.*) traditionally used for skin care in the Mediterranean regions to preserve skin freshness, protects skin from over drying and to reduce sun-induced wrinkles and sun caused aging during summer time. The anti-aging properties of calendula oil was known to ancient Egyptians, may be related to the high content and diversity of carotenes, phytosterols, polyphenols and free fatty acids. Calendula oil is good for all skin types, and is considered valuable for treating skin damage such as wounds, scars, burns, inflammation and other injuries as it is said to be an effective aid in tissue regeneration. It has anti-inflammatory, anti-spasmodic qualities and particularly blends well with hypericum. It is best when infused in light textured oil such as almond or sunflower. Calendula is a tonic, soothing, restorative oil, good for any type of skin, but especially suitable for dry and/or aging skin. On regular use it imparts a deep olive shine to the skin, and helps protect the skin from age related thinning and drying out. It is magical in its ability to prevent tearing, soothe pain, stop bleeding, promote rapid healing, and prevent scarring. There is no known toxicity in literature for calendula oil so considered safe in normal applications (Katcher 2011).

Castor oil: Castor (*Ricinus communis*) is a native of India and its seeds have fixed oil, is extracted by expression. It has been used therapeutically and medicinally in ancient India, China, Persia, Egypt, Africa, Greece and Rome. It is used mainly hair treating as conditions such as dry or brittle, damaged hair or hair loss. It is often used as an emollient and skin softener, and medicinally as a treatment of gastrointestinal problems, lacerations and other skin disorders such as psoriasis. It is an ingredient of many commercial skin care products. When it is used in the manufacture of soap it forms a clean, light-coloured soap with a stable lather, which dries and hardens well. It is considered safe in normal applications, shelf life is 1 year to 2 years, provided kept at dark places, in airtight container and refrigerated after opening (Leonardo et al., 2001).

Walnut oil: Walnut (*Juglans regia*) oil is reported to have many benefits, when used in the healing of wounds and skin problems. The oil is obtained by expression from the nut. Walnut oil is useful against fungal and parasitic infections. It may also help eliminate warts and when rubbed on the skin, walnut is reputed to be beneficial for eczema, herpes, psoriasis, and skin parasites. The oil has traditionally been used externally in the treatment of gangrene, leprosy and wounds. Walnut oil presents good moisturizing, anti-aging, regenerative and toning properties and can be used in all anti-wrinkle products and creams for dry, normal and mature skin, body and hygiene products, massage blends, and lip balms, etc. It is non-toxic, non-irritant. It is considered to be safe in all normal applications and shelf life is 6 months to a 1 year, provided kept properly in airtight container, refrigerated after opening (Saxena et al., 2009).

Wheatgerm oil: Wheat (*Triticum turgidum*) oil is made from the germ of the wheat, the heart of it. The wheat grain consists of three parts, the husk, the germ and the endosperm. It is the germ that is used in the manufacture of the oil. Although the germ only constitutes 3% of the weight of a wheat grain, it contains nearly 25% of the protein, vitamins and minerals. Wheatgerm oil is very high in vitamin E and essential fatty acids. Because of its high vitamin E content, which is a natural antioxidant, helps to prevent rancidity, is often added to other carriers to help lengthen their shelf life and to prevent them going off. When used in massage blends it may be too heavy on its own, but it has great nourishing qualities, so good choice to include when mixing a massage blend or preparing a carrier blend. It has a slight nutty aroma and is extracted by cold pressing or solvent extraction from the germ. With its high vitamin E content and essential fatty acids is used extensively in aromatherapy massage blends since it is said to promote a smoother, younger looking skin, assisting in healing scar tissue as well as stretch marks. It is too sticky to use on its own, but is wonderful when blended with a lighter carriers, such as sweet almond. When it is applied topically on to the skin it is said to promote the formation of new cells, improve circulation, and is said to help repair sun damage to the skin. It is also used to relieve the symptoms of dermatitis. This carrier is said to stimulate tissue regeneration and is often added to other blends because of its antioxidant properties (Ibrahim et al., 2011).

Sunflower oil: Sunflower (*Helianthus annuus*) contains useful amounts of essential fatty acids and a high amount of vitamin E. Since it is inexpensive it can be blended with a variety of other base oils. Generally, absorption is not as good as other carrier oils, and it does tend to leave an oily feeling on the skin. The aroma is a faint and sweet, slightly nutty scent. Sunflower oil is expressed from the seed. Sunflower oil is good for all skin types for use in beauty and skin care, however, it should be diluted with another carrier oil such as sweet almond. It may be used as a substitute for olive oil in soaps, where it provides stable, conditioning lather to bar soaps. It is very lightweight oil making it a perfect choice for bath oils, and lotions, but is not suitable for massage due to the absorption rate. The shelf life is 3-6 months (Anjali et al., 2010).

Sesame oil: Sesame (*Sesamum indicuum*) oil is a good source of vitamins E and B complex and minerals such as calcium, magnesium and phosphorus. It has protein as well as lecithin. In folk medicine it has also been used to help fade blemishes. The sesame herb is grown for its seeds and was used by the Chinese 5000 years ago. The Egyptians produced flour from its ground seeds, and Roman soldiers mixed the seeds with honey for an energy snack. It may be expressed or extracted and is also known as '*til* oil'. If the oil is extracted from raw seeds, it is a rather light colour compared to that extracted from roasted seeds. Sesame seed oil is a natural skin moisturizer so a good carrier for essential oils of microbial property (Mohamed 2011).

5 SPECIFIC PATHOGENS AND MEDICINAL PLANTS

The modern scientific knowledge has blessed the medical health by clues of specific pathogen, e.g., *Helicobacter pylori* infections and treatment thereof.

Case study: To test the efficacy of herbal products for the eradication of *H. pylori*, experiments with different medicinal plants extract are summarized as (Karen and Ernst 2003):

(i) A group of 20 *H. pylori* infected patients suffering from dyspeptic complaints for more than 2 months were recruited. A 2-week treatment with capsules containing an oil macerate of garlic (275 mg, three times a day) was compared with the combination of garlic with omeprazole (20 mg, twice a day) for the eradication of *H. spylori*. All subjects underwent endoscopy before and 1 month after the end of treatment, and the presence of *H. pylori* in biopsy specimens was confirmed by the urease test and by microscopy. Symptom scores and degree of gastritis, as judged by histological examination, were recorded pre- and post-treatment. Neither intervention resulted in the elimination of the organism, change in the severity of gastritis or a significant change in symptom scores (9.2 ± 1.55 versus 8.7 ± 1.70 in the garlic oil group, 9.0 ± 1.49 versus 8.5 ± 1.51 in the garlic oil plus omeprazole group). The team of scientists with experiment considered that low levels of the antibacterial component allicin in the garlic capsules might account for the lack of effect.

(ii) Extracts of fresh garlic and capsaicin containing peppers can inhibit *H. pylori* *in vitro* and were tested for their ability to inhibit the bacterium *in vivo* in a crossover trial involving 12 individuals infected with *H. pylori*. Test substances were included in morning, noon and evening, Mexican-style meals. Subjects participated in a minimum of 3 trial days (negative and positive controls and one experimental ingredient). At least 2 days elapsed between test substances, with bismuth always being the last intervention tested to preclude lasting anti-*H. pylori* effects. During each test meal participants received one intervention: garlic (10 freshly sliced cloves), capsaicin (six freshly sliced jalapeno peppers), two tablets of bismuth subsalicylate (Pepto-Bismol, positive control) or no additions (negative control). The urea breath test was performed before the first meal of the day, before the evening meal and the morning after each test day. The results were used to evaluate the effectiveness of the therapies. Ten subjects received garlic, six received jalapenos and 11 received bismuth. Neither garlic nor jalapenos had any effect on urease levels (median urease activity pre- and post-garlic 28.5 versus 39.8 and jalapenos 43.7 versus 46.6; $P > 0.8$) but there was a marked reduction after ingestion of bismuth (55.8 versus 14.3; $P < 0.001$). Two patients experienced nausea and diarrhoea graded as severe after eating the jalapenos and 70% of those eating garlic complained of taste disturbance and body odour. This study suffered from lack of randomization and small sample size.

(iii) The effects of extract of cinnamon (*Cinnamonum cassia*) in 23 patients who were undergoing gastroscopy and had a positive urea breath test were investigated. Thirty-two patients were randomly allocated in a 2:1 ratio to the study and control groups but only 23 could be included in the final analysis. Fifteen received 80 mg/day cinnamon and eight received placebo for 4 weeks. Cinnamon extracts *in vitro* exerted an inhibitory effect on the growth and urease activity of a number of strains of *H. pylori*. At the concentration chosen the extract was ineffective *in vivo*. A black tea leaf extract, presumed to derive its antibacterial effect from tannins and catechins, demonstrated results equivalent to antibiotic treatments for curing impetigo contagiosa. This simple and inexpensive alternative to conventional treatment may be worthy of further rigorous investigation. In conclusion, the evidence summarized above tentatively suggests possible benefits from some herbal preparations with antibacterial activity. Further, large-scale, well-designed clinical trials are required to provide more conclusive proof.

6 CONCLUSION

In the context of a modern, social and economic view of health services, there is a clear need for more research on medicinal plants for the development of new bioactive compounds/drugs. The potential use of higher plants as a source of new drugs is still poorly explored. From the whole plant kingdom, only a small percentage has been investigated phytochemically and an even smaller percentage has been properly studied in terms of their pharmacological properties, including the extensive screening. Research into the isolated plant constituents is of great importance for the development of bioactive substances from ethnic medicine. With the emergence of latest technologies and enhanced knowledge about the isolated plant constituents, characterization and analytical tools, lot of compounds are efficiently isolated from potential plants and have been of great contribution to the drug discovery from ethnic plants. Separating a medicinal herb into its constituents cannot often explain exactly the way in which it works in the natural form. The whole herb is worth more than the sum of its components. A plant contains hundreds of chemical constituents that interact in a complex way to produce therapeutic effects of the remedy. We may not understand the detailed mechanism, in which a particular medicinal plant is effective, even though its medicinal benefits are well established. With the development of standardization tools like TLC-DART, HPTLC, HPLC, LC-MS/MS, etc., several plant extracts as well as their formulations, may be standardized in a better way to enhance the bioactivity of the ethnic medicines. Validated analytical and manufacturing processes, process control, quality control, pharmacological test systems and well-designed clinical studies are essential primary symbols for successful herbal products in the competitive global market.

References

Acharyya S, Patra A and Bag PK (2009). Evaluation of the atimicrobial activity of some medicinal plants against enteric bacteria with particular reference to multi-drug resistant *Vibrio cholerae. Tropical Journal of Pharmaceutical Research* 8(3): 231–237.

Akinanibosun FI, Akinnibosun HA and Ogedegbe DE (2009). Investigations on the anti-bacterial activity of the aqueous and ethanolic extracts of the leaves of *Boerhavia diffusa L. Science World Journal* 4 (2): 15–18.

Anjali CH, Maghusmita D, Chandrasekharan N and Mukherjee A (2010). Anti bacterial activity of sunflower oil micro emulsion. *International Journal of Pharmacy and Pharmaceutical Sciences* 2: 123–128.

Bedir E and Khan IA (2000). New steroidal glycosides from the fruits of Tribulus terrestris. *Pharmazie* 57: 491–493.

Bisht S, Feldmann G, Soni S , Rav R, Karikar C, Maitra and Maitra A (2007). Polymeric nanoparticle-encapsulated curcumin (nanocurcumin): a novel strategy for human cancer therapy. *Journal of Nanobiotechnology* 5: 3–16.

Cock IE (2011). Anti-microbial activity of Aloe barbadensis miller leaf gel. The Internet Journal of Microbiology. Accessed at http://www.docstoc.com/docs/26774146/Antimicrobial-Activity-of-Aloe-barbadensis-Miller-Leaf-Gel.

Elvin-Lewis M (2011). Optimizing new therapeutic discoveries from ethnomedical and ethnobotanical data. Accessed at http://www1.uprh.edu/simpff/docs-simp-anterior/PonenciaElvinLewis.doc

Ibrahim AH. Barakat, Osama AA, Ayad S and Hassan AM (2011). Evaluation of radio protective effects of wheat germ oil in male rats. *Journal of American Science* 7(2): 664–673.

Jadge DR, Patil SV and Purohit RN (2008). Formulation of toothpaste from various forms and extracts of tender twigs of neem. *Journal of Pharmacy Research* 1: 148–152.

Jadhav VM, Thorat RM, Kadam VJ and Sathe NS (2009). Eclipta alba Linn-Kesharaja: A Review. *Journal of Pharmacy Research* 2(8): 1236–1241.

Joshi DD and Sharma RK (2011). Herbal Drugs in Patent Regime. VDM Verlag Dr. Muller ISBN: 978-3-639-32182-1. p. 51.

Karen W and Ernst E (2003). Herbal medicines for treatment of bacterial infections: a review of controlled clinical trials. *Antimicrobial Chemotherapy*. 51: 241–246.

Katcher S (2011). Calendula oil - healthful and beautiful skin. Accessed at http://ezinearticles.com/?Calendula-Oil---Healthful-and-Beautiful-Skin!&id=5681709.

Kumar A and Sharma VK (2011). Antibacterial activity of allicin from *Allium sativum* against antibiotic resistant uropathogens. Internet Journal of Infectious Diseases. 8(1): Accessed at http://www.ispub.com/journal/the_internet_journal_of_infectious_diseases/volume_8_number_1_26/article/antibacterial-activity-of-allicin-from-allium-sativum-against-antibiotic-resistant-uropathogens.html,

Latif A, Siddiqui N, Razique A and Sukul RR (2010). Spectrophotometric estimation of berberine and TLC fingerprints profile of an Unani eye drop formulation. *International Journal of Pharma Research and Development-Online*. Publication Ref No.: IJPRD/2010/PUB/ARTI/VOV-2/ISSUE-7/SEP/0062(7)

Leonardo MR, Silva, LAB, Filho, Bonifácio KC and Ito IY (2001). In vitro evaluation of the antimicrobial activity of a castor oil-based Irrigant. *Journal of Endodontics* 27: 717–719.

Maiti PC, Roy S and Roy A (1968). Chemical investigation of Indian soapnut, Sapindus laurifolius Vahl. *Cellular and Molecular Life Sciences* 24 (11): 1091.

Mohamed Saleem TS (2011). Anti-microbial activity of sesame oil. *Int J Res Phytochem Pharmacol* 1(1): 21–23.

Nostro A (2006). Activity of plant extracts and plant-derived compounds against drug-resistant microorganisms, Sample chapter of Book titled, 'Modern Phytomedicine; Turning Medicinal Plants into Drugs', WILEY-VCH Verlag GmbH & Co. KGaA, Weinheim. 199–231.

Ojha P, Maikhuri JP and Gupta G (2003). Effect of spermicides on *Lactobacillus acidophilus* in vitro-nonoxynol-9 vs. Sapindus saponins. *Contraception* 68(2): 135–138.

Saxena R, Joshi DD and Singh R (2009). Chemical composition and antimicrobial activity of walnut oil. *International Journal of Essential Oil Therapeutics* 3: 115–118.

Sairam M, Ilavazhagan G, Sharma SK, Dhanraj SA, Suresh B, Parida MM, Jana AM, Devendra K and Selvamurthy W (2000). Anti-microbial activity of a new vaginal contraceptive NIM-76 from neem oil (*Azadirachta indica*). *Journal of Ethnopharmacology* 71: 377–382.

Silja VP, Varma KS and Mohanan KV (2008). Ethnobotanical knowledge of the Mullu kuruma tribe of Wayanada district, Kerala. *Indian Journal of Traditional Knowledge* 74 (4): 604–612.

Singh R, Gupta S, Joshi DD and Nainwal N (2010). Wild apricot (Prunus armeniaca) karnel oil: A strategic alternative to value added fatty acids. *International Journal of Essential oil Therapeutics* 4: 1–4.

5

Microorganisms and Their Role in Sustainable Environment

A.A. Haleem Khan, Naseem and B.Vidya Vardhini

Dept of Botany, University College, Telangana University, Nizamabad-503322 Telangana, India

Dept of Pharmaceutical Chemistry, University College, Telangana University Nizamabad-503322, Telangana, India

**Email: aahaleemkhan@gmail.com, naseemprof@gmail.com drvidyavardhini@rediffmail.com*

1 INTRODUCTION

Improved scientific technologies and ever-growing lifestyle of mankind brought changes in routine day-to-day life to earn more than the existing output. In all walks of life man is currently interested to gain profit from the available resources around him. The lust to get more and more in short time span brought menace of machines and overuse of chemicals. Certainly extensive use of chemicals creates pollution of various components of biosphere. The unwanted changes in biosphere have caused/threatened or critically endangered living forms. Though number of conservation programmes by government and non-governmental organizations (NGOs) are putting efforts to rejuvenate the nature with autochthonous and allochthonous forms.

In this consequence microorganisms were suggested as best source to conserve and sustain the environment. Microorganisms in the nature are simple living forms with cosmopolitan distribution and densely populated amongst the biodiversity of earth. These organisms inhabit the diverse habitats by carrying several biochemical activities that could bestow the sustenance of contaminated ecosystems. In the present chapter, efforts are put to highlight various research studies carried across the globe on the role of microorganisms to preserve living standards of mankind.

The word 'sustainable development' was first introduced by *World Commission on Environment and Development* (WCED) previously known as **The Brundtland Commission**, in *The United Nations General Assembly* 1987, recognized that environmental problems were global in nature and determined as common interest of all the nations. Sustainable development was initiative to fulfil the needs of mankind that preserve the environment, consequently the necessities could be met for generations together for environment, local people, future abbreviated as ELF. The concerns of sustainable development were categorized in multiple dimensions such as environmental, social, economic and political. The critical dimension among the above categorized was Environmental sustainability. Our focus was to imply the role of microorganisms in conservation of environment.

The increase in cropland area was done by cutting down or burning of natural forests and subsequent tillage that influenced the composition and activity of the soil biota such as microbes and invertebrates. The conventional agriculture encompasses the practices, methods and systems with more and more use of chemical and mechanical energy. The soil biota changes affected the physical environment of the soil. The soil biota was significant to play role in biogeochemical cycles like N_2 cycle carried by specific organisms. The keystone organisms of decomposition and nutrient mineralization were termites, earthworms, N_2 fixers, mycorrhizae and nematodes. The mechanical agricultural practices and the agrochemicals used/substituted the structure and biological activity of soil food web. Sustainable agriculture, which was early practiced elucidated to conserve the cropping systems from application of overdoses of harmful toxic chemicals used as pesticides, herbicides, insecticides and fertilizers. Allochthonous and autochthonous microorganisms attracted the researchers to employ them in multiple ways such as to improve the soil fertility, control of pests, pathogens due to their antagonistic activities.

2 AGRICULTURE AND MICROBES

The total area of cultivated land worldwide is expanding with significant negative impact on global environment. *Green revolution* was introduced with motive to improve the crop yields by use of high yielding seed varieties, improved irrigation practices, use of mechanization and application of agrochemicals resulted in degrading the environment quality with fair increase in yields. Intense and expanded agriculture witnessed the increase in erosion, decrease in soil fertility, reduction in biodiversity, pollution of groundwater and eutrophication of surface water bodies like lakes and rivers. In India, the intensive rice-wheat cultivation in the Punjab region began serious decline in soil quality and enhanced problems of plant health. The cropping systems with high yields by monoculture influenced the abundance of the associated biota such as the pests, soil invertebrates and microorganisms that, in turn, affect plant and soil processes.

The herbivorous insects (pests), their predators and parasites, and microbial insects from both agricultural and natural ecosystems by attacking crops decreased plant productivity. The crop diversification inhibits pathogen growth such as fungi, viruses and parasitic insect vectors can be lower in polycultures, i.e., intercropping. Pest management was targeted with the use of pesticides that resulted in pesticide resistance a ubiquitous problem, spread of pesticide residues through developed human health and environmental threat.

Integrated pest management (IPM) was introduced to overcome the pest and pesticide related problems that suggest the use of host plant resistance, biological and cultural control along with pesticides. The IPM practices have been promoted for long time run with notable success. The aim to resurrect the cropping systems with alternative substrates results in practice well known as '*sustainable agriculture*'.

Benbrook defined '*sustainable agriculture*' as "The production of food and fibre by using a system that inherent productive capacity of natural and biological resources in step with demand. At the same time, it must allow farmers to earn adequate, profits provide consumers with wholesome, safe food and minimize adverse impacts on the environment."

2.1 Plant Growth Promoting Rhizobacteria (PGPR) and Crop Productivity

The discovery of microorganisms was first proved by Anton von Leeuwenhoek (1683), as 'animalcules', i.e., little animals later confirmed as microscopic prokaryotic living forms as bacteria. The utilization of bacteria to stimulate plant growth is known since ancient times. Theophrastus (372-287 BC) suggested that by mixing of different soil samples could be remedy to recover soil defects. The legume crops increase the fertility of cultivated land and improves yield of successive crops was recorded by Virgil. In grasses and legumes roots were found to get colonized by rhizosphere with atmospheric nitrogen (N_2) converted into plant utilizable form was suggested by Hellriegel and Wilfarth (1888). The area surrounding the healthy plant roots where maximum microbial activity is found compared to distant surrounding soil is known as rhizosphere by Hiltner (1904).

The experiments of Kloepper and Schroth (1978) on radishes introduced the term rhizosphere to the microbial community that competitively colonized plant roots and stimulate growth and suppress the incidence of plant pathogens (Table 2). Kloepper and Schroth (1981) termed the beneficial bacteria surrounding the root as 'plant growth-promoting rhizobacteria' (PGPR). PGPR are indispensable part of rhizosphere biota that grow in association with the host plants and benefit the host to improve its growth. PGPR get accomplished wide variety of ecosystems, show fast growth rate and metabolize the natural and xenobiotic compounds due to their innate genetic potential. The enrichment of microorganisms such as bacteria, algae, fungi, and protozoa increase the nutrients in soil from 10 to 100 fold than bulk soil.

To fulfil the concept of PGPR there are three criteria:

1. Aggressive colonization
2. Plant growth stimulation and
3. Bio-control

As per Whipps (2001), PGPR are divided into three categories based on their interactions as: neutral, negative and positive. Among the PGPR most of them are commensals and establish an innocuous interaction with healthy plant without visible effect. The phytopathogenic rhizobacteria exert negative effect by producing phytotoxic substances, i.e., hydrogen cyanide (HCN) or ethylene. In response to the deleterious or negative effect exerting bacteria there are positive effects showing bacteria improve the growth by solubilization of nutrients, nitrogen fixation, production of plant growth regulators, etc. The bacteria of this group inhabit the healthy plants in and around the roots (*Rhizosphere*), root surface (*Rhizoplane*) or the root itself (*Endophytes*).

PGPR were classified into two groups:

1. Intracellular plant growth promoting rhizobacteria (iPGPR)
2. Extracellular plant growth promoting rhizobacteria (ePGPR)

Table 1 PGPR: types and their genera

<table>
<tr><td rowspan="2">Plant growth promoting rhizobacteria (PGPR)</td><td>Intracellular plant growth promoting rhizobacteria (iPGPR)</td><td>(a) Bacteria that reside in the root tissue (endophytes)
(b) Bacteria that inhabit the specialized root structures or nodules.
Legumes – Rhizobia,
Woody plants – Frankia</td><td>Genera
Allorhizobium, Azorhizobium, Bradyrhizobium, Mesorhizobium, Sinorhizobium, Rhizobium from family Rhizobiaceae.
Frankia sp.</td></tr>
<tr><td>Extracellular plant growth promoting rhizobacteria (ePGPR)</td><td>These bacteria exist
(a) Bacteria near the roots (rhizosphere)
(b) Bacteria that colonize the root surface (Rhizoplane)</td><td>Acinetobacter, Achromobacter, Alcaligenes, Agrobacterium, Aereobacter, Arthrobacter, Azotobacter, Azomonas, Azospirillum, Bacillus, Beijerinckia, Burkholderia, Caulobacter, Citrobacter, Chromobacterium, Derxia, Erwinia, Enterobacter, Flavobacterium, Herbaspirillum, Hyphomicrobium, Klebsiella, Micrococcus, Paenibacillus, Pseudomonas, Serratia</td></tr>
</table>

Table 2 Activities of PGPR in Rhizospheric soil

Rhizosphere soil	Availability of nutrients– amino acids, fatty acids, nucleotides, organic acids, phenols, plant growth regulators, plant growth promoters, putrescine, sterols, sugars, and vitamins released from the roots by exudation, secretion and deposition.

PGPR are categorized into two groups based on the effect on plant growth as: direct promoters and indirect promoters (Table 1).

Direct PGPR are subdivided into:

- Nitrogen fixers
- Phosphate solubilizers
- Potassium solubilizers
- Sulphur oxidizers
- Iron oxidizers
- Iron chelators
- Producers of plant growth regulators

Indirect PGPR are subdivided as:

- Suppressors of the plant pathogens
- Resistors to plant pathogens

2.2 Phosphate Solubilizers

Phosphorus (P) is a major macronutrient, plant growth-limiting nutrient in soils besides nitrogen. It is required for synthesis of nucleic acids, membranes, proteins and energy molecules such as ATP, GTP and NADPH. P contributes in several metabolic processes including breakdown of sugar, photosynthesis, energy transfer, cell division, nutrient transport within the plant, expression and maintenance of genetic material, signal transduction, macromolecular synthesis and reactions in respiratory chain. The total P content of arable soil is 0.02–0.5% with approximate of 0.05% in inorganic and organic forms. The soil table possesses most of the nutrients in millimolar concentration, but P is found in micromolar quantities. The low level of P in soil is due to its high reactivity with Ca, Fe and Al. In the alkaline soils hydroxyapatite and fluoroapatite are calcium phosphates, the acidic soils possess variscite ($AIPO_4.H_2O$), strengite ($FePO_4.2H_2O$) as amorphous iron and aluminium phosphates. Though the soil has high levels of P content, there is deficiency of P due to low levels of soluble PO_4 availability to plants from soil. To meet the P requirement of plants, application of fertilizers with P content or leaf litter, plant residues or animal remains are in common practice.

In the soil there are groups of microorganisms efficient in solubilizing the phosphorus from natural ecosystems (Table 3). These organisms play a fundamental role in biogeochemical cycling of P in soil. Phosphate solubilizing microbes (PSM) can transform the insoluble phosphorus into soluble forms HPO_4^{2-} and $H2PO_4^-$ by chelation, exchange reactions, production of polymeric substances, organic acids (oxalic, citric, butyric, malonic, lactic, gluconic, acetic, glyconic, fumaric, adipic and 2-ketogluconic acid), inorganic acids, CO_2 and enzymes-acid phosphatase, phytase, phosphohydrolases, phosphonoacetate hydrolase, D-α-glycerophosphatase and C-P lyase.

Table 3 Phosphate solubilizing microbes

PO_4 solubilizing microbes (PSM)	Bacteria	*Acinetobacter, Achromobacter, Aereobacter, Agrobacterium, Alcaligenes, Arthrobacter, Azotobacter, Bradyrhizobium, Burkholderia, Chromobacterium, Citrobacter, Enterobacter, Erwinia, Escherichia, Flavobacterium, Klebsiella, Micrococcus, Pantoea, Pseudomonas, Salmonella, Serratia, Streptomyces, Thiobacillus*
	Fungi	*Aspergillus, Fusarium, Glomus, Penicillium, Rhizopus, Sclerotium, Trichoderma*

Utilization of PSM is an attractive approach to increase the availability of P content of soil that increases the crop yield and sustains the agriculture system.

2.3 Potassium Solubilizers

Potassium (K) is an essential element for plant growth and development. It is important in maintenance of cellular osmotic pressure and transport of solutes in plants. The activation of enzymes, the utilization of nitrogen (N) and the syntheses of sugar and protein are promoted by potassium. It enhances photosynthesis in plants; its deficiency causes slow plant growth and root development, yellowing of leaf margins, with burnt appearance. The soil is K rich reservoir, about 90–98% of K exists as silicate minerals (K-feldspar, illite and mica), only plant K uptake source. K is released as its mineral form at a slow rate. There are varieties of soil microbes that are released in soluble form of K from K-bearing minerals.

Potassium solubilizing bacteria (KSB) are Gram-positive, heterotrophic, aerobic bacteria. They obtain their energy and carbon source from pre-existing organic matter. These bacteria form humus in soil, cycle minerals in organic matter, maintain soil structure by formation and stabilization of water-stable soil aggregates and stimulate plant growth and inhibit the root pathogens. Among the KSB, *Bacillus mucilaginosus* and *Bacillus edaphicus* reported for the dissolution of silicate and alumino-2-silicate minerals with release of K^+ and SiO_2 by organic acids (tartaric acid and citric acid) and extracellular polysaccharide (EPS) formation. The solubilizing activity of KSB alleviates K^+ deficiency soils.

2.4 Sulphur Oxidizers

Soil is rich with reduced sulphur compounds that are exclusively oxidized by the prokaryotes. The oxidation of these compounds yield sulphate as byproduct. In the sulphur cycle around the globe, hydrogen sulphate (H_2S), sulphur, sulfite, thiosulfate and various polythionates undergo biological oxidation to form sulfate (SO_4) in acidic, alkaline and neutral conditions. Sulphur is absorbed from soils as divalent sulphate anion (SO_4) by roots and translocated to shoots through xylem. Deficiency of sulphur is rare, symptoms of deficiency are expressed as chlorosis, common in young leaves. The sulphur content is predominant in plant proteins with amino acids—cysteine and methionine, vitamins—thiamine and biotin, co-enzyme-A.

Table 4 Sulphur oxidizing bacteria and their types

Sulphur oxidizers	Aerobic-mesophilic	*Acidianus, Acidithiobacillus, Aquaspirillum, Aquifex, Bacillus, Beggiatoa, Methylobacterium, Paracoccus, Pseudomonas, Starkeya, Sulfolobus, Thermithiobacillus, Thiobacillus* and *Xanthobacter*
	Anaerobic-phototrophic (neutrophilic & mesophilic)	*Allochromatium, Chlorobium, Rhodobacter, Rhodopseudomonas, Rhodovulum, Rhodocyclus,* and *Thiocapsa*

2.5 Iron Oxidizers

Iron (Fe) is essential in certain enzymes and proteins of photosynthesis and respiration. It alternately undergoes oxidation and reduction between the Fe^{2+} to Fe^{3+} states to serve as electron carriers in proteins. Fe is accumulated in older leaves as insoluble oxide or inorganic or organic ferric-phosphate compounds. In leaves one abundant and stable iron-protein complex is called phytoferritin; this protein is a storehouse of iron in chloroplast. Fe deficiency is often found in plants of rose family, shrubs and trees, maize and sorghum. High pH and bicarbonates in soil were found to be associated with this deficiency.

In the 1830s iron oxidizing bacteria (FeOB) were recognized as chemolithoautotrophs. They grow on Fe as a sole source of energy. FeOB and iron reducing bacteria are responsible for precipitating iron deposits such as oxides (magnetite), carbonates (siderite) and sulphides (greigite) through controlled or induced mineralization process (Table 5).

Table 5 Iron oxidizing bacteria

	Acidothiobacillus ferroxidans (*Thiobacillus ferrooxidans*),
	Acidothiobacillus thiooxidans (*Thiobacillus thiooxidans*),
	Acidothiobacillus caldus,
	Acidianus ambivalensi,
	Acidianus infernus,
	Acidianus brierleyi,
	Leptospirillum ferrooxidans,
	Leptospirillum ferriphilum,
	Ferroplasma acidiphilum,
	Geobacter metallireducens,
	Geobacter sulphurreducens,
	Shewanella putrefaciens,
	Shewanella denitrificans,
	Gallionella ferruginea,
	Magnetospirillum magnetotacticum,
	Leptothrix ochracea,
	Sulfobacillus thermosulfidooxidans,
	Sulfobacillus montserratensis
	Sulfobacillus metallicus,
	Sulfolobus sp.
	Metallosphaera sp.
	Rhodobacter sp.
	Dechlorosoma sp.

2.6 Iron Chelators: Siderophores

The term siderophores is derived from Greek word for iron carriers. Siderophores are designed to form light and stable ferric iron complexes. These are ferric ion specific chelating agents synthesized by prokaryotic and eukaryotic microbes such as bacteria, actinomycetes, fungi and certain algae growing in low iron stress (Table 6). Siderophores are iron binding proteins with molecular weight ranging from 400-1500 Da. These compounds play an important role in scavenging iron from the environment. Secreted siderophores form extracellular Fe (III) complexes with stabilities. The iron charged siderophore is taken up by ferric-chelate-specific transporters or siderophore-bound Fe (III) undergoes reduction to Fe (II) that is catalyzed by ferric-chelate reductases. Most of the aerobic and facultative anaerobic microorganisms synthesize at least one siderophore. Siderophores are also found to be associated with virulence mechanisms in pathogenic microorganism of animals and plants. These compounds have applications in clinical, agricultural and environmental fields. Presently 500 siderophores are reported from different microorganisms.

The important groups of siderophores are:

- Hydroxamate (α-hydroxy-) siderophores,
- Catecholate (catecholates and phenolates) siderophores, and
- Carboxylate siderophores

Hydroxamate siderophores possess three secondary hydroxamate groups and each hydroxamate group provides two oxygen molecules that form a bidentate ligand with iron. The siderophores were reported to be produced by both bacteria and fungi.

Catecholate siderophores are produced by certain bacteria. The catecholate group provides two oxygen atoms for chelation with iron so that a hexadentate octahedral complex is formed as in the hydroxamate siderophores. The finest example for catecholate siderophore is enterobactin produced by *Escherichia coli.*

Carboxylate siderophores are produced by *Rhizobium* and *Staphylococcus* strains and fungi belonging to Mucorales, coordinating iron with carboxyl and hydroxyl groups.

Table 6 Siderophore producing microorganisms

Siderophore producing microorganisms	Bacteria	*Escherichia coli,* *Salmonella,* *Klebsiella pneumoniae,* *Vibrio cholerae,* *Vibrio anguillarum,* *Aeromonas,* *Aerobacter aerogens,* *Enterobacter,* *Yersinia and* *Mycobacterium* sp.
	Actinomycetes	*Actinomadura madurae, Nocardia asteroids,* and *Streptomyces griseus*
	Cyanobacteria	*Anabaena flos-aquae* and *Anabaena cylindrica*
	Fungi	*Aspergillus nidulans,* *A. versicolor,* *Penicillium chrysogenum,* *P. citrinum,* *Mucor,* *Rhizopus,* *Trametes versicolor.* *Ustilago sphaerogina,* *Saccharomyces cerevisiae,* *Rhodotorula minuta* and *Debaromyces* sp.

2.7 Microorganisms Producing Plant Growth Regulators

Phytohormones play an important role in regulating growth and development of plants. The conventional classification made five groups of phytohormones: auxins, gibberellins, cytokinins, ethylene and abscisic acid. The microorganisms PGPR (rhizosphere, epiphytes, and symbionts) associated with plants were reported to synthesize phytohormones, stimulate and facilitate plant growth (Table 7).

Table 7 Microorganisms producing plant growth regulators (auxins, gibberellins, cytokinins)

Auxins	Prokaryotes	Phototrophs	*Anabaena, Anabaenopsis, Calothrix, Chlorogloeopsis, Cylindrospermum, Gloeothece, Nostoc, Plectonema, Synechocystis*
		Heterotrophs	*Acetobacter, Achromobacter, Acinetobacter, Agrobacterium, Alcaligenes, Aminobacter, Arthrobacter, Azospirillum, Azotobacter, Bacillus, Bradyrhizobium, Corynebacterium, Enterobacter, Erwinia, Flavobacterium, Herbaspirillum,Klebsiella, Methylobacterium, Methylovorus, Microbacterium, Micrococcus, Mycobacterium, Paracoccus, Pseudomonas, Rhizobium, Rhodococcus, Sphingomonas, Streptomyces, Sulfolobus, Xanthomonas*
	Eukaryotes	Algae	*Chlorella, Dunaliella, Fucus*
		Fungi	*Absidia, Actinomucor, Alternaria, Amanita, Aspergillus, Colletotrichum, Fusarium, Hebeloma, Laccaria, Lentinus, Monilia, Paxillus, Penicillium, Phoma, Phymatotrichum, Phytophtora, Pisolithus, Plasmodiophora, Pythium, Rhizoctonia, Rhizopogon, Rhizopus, Saccharomyces, Sclerotium, Taphrina, Ustilago*
Gibberellins	Prokaryotes	Phototrophs	*Anabaenopsis, Cylindrospermum*
		Heterotrophs	*Acinetobacter, Agrobacterium, Arthrobacter, Azospirillum, Azotobacter, Bacillus, Clostridium, Flavobacterium, Micrococcus, Pseudomonas, Rhizobium, Xanthomonas*
	Eukaryotes	Algae	*Chara, Fucus, Porphyra, Tetraselmis*
		Fungi	*Aspergillus, Botryodiplodia, Cercospora, Fusarium, Penicillium, Phaeosphaeria, Rhizopus, Shizophillum, Sphaceloma, Ustilago, Verticillum,* as well as certain yeasts

Contd...

Contd...

Cytokinins	Prokaryotes	Phototrophs	*Anabaena, Calothrix, Chlorogloeopsis, Rhodospirillum*
		Heterotrophs	*Agrobacterium, Aminobacter, Arthrobacter, Azospirillum, Azotobacter, Bacillus, Blastobacter, Escherichia, Erwinia, Hyphomicrobium, Methiloarcula, Methylobacillus, Methylobacter, Methylobacterium, Methylomonas, Methylophylus, Methylosinus, Methylovorus, Paracoccus, Pseudomonas, Rhizobium, Rhodococcus, Streptomyces, Xanthobacter*
	Eukaryotes	Algae	Certain red and brown algae
		Fungi	*Dictyostellium, Fusarium, Paxillus, Phoma, Rhizopogon, Saccharomyces, Schizophyllum, Schizosaccharomyces, Suillus, Taphrina, Trichoderma, Uromyces*

2.8 Microorganisms Under Abiotic and Biotic Stresses

There are reports on bacteria that provide tolerance to host plants under different abiotic stress environments such as temperature (cold and heat), water (drought and flooding), humidity, light intensity, nutrients (lack of ions, over-fertilization and salinity), carbon dioxide (CO_2), oxygen (photosynthesis, respiration/photorespiration, oxidative stress and anaerobiosis).

Table 8 Microorganisms beneficial for plants under stress

Abiotic stresses	*Rhizobium*
	Bacillus
	Pseudomonas
	Pantoea
	Paenibacillus
	Burkholderia
	Achromobacter
	Azospirillum
	Microbacterium
	Methylobacterium
	Variovorax, and
	Enterobacter

Several PGPR that colonize root systems protect plants against foliar disease include *P. fluorescens, P. putida, B. pumilus,* and *S. marcescens*

(Tables 8, 9). There is a network of interconnected signalling pathways to regulate induced defenses of plants against pathogens. The primary components of the network are plant signal molecules—Salicylate (SA), Jasmonate (JA), Ethylene (ET) and Nitric oxide (NO). The signal transduction leading to ISR requires responsiveness to both JA and ethylene. Methyl jasmonate (MeJA) and the ethylene precursor, i.e., 1-aminocyclopropane-1-carboxylate (ACC) are effective in inducing resistance against phytopathogenic microflora. Mediation of ethylene levels by microbial ACC plays a key role in signal transduction viz., the ISR-inducing *P. fluorescens* bacterium enhances bacterial ACC-converting capacity and leads to potentiated level of ethylene emission in infected plants. The PGPR contain a vital enzyme, 1-aminocyclopropane-1-carboxylate (ACC) deaminase that regulates ethylene production by metabolizing ACC (a precursor of ethylene biosynthesis in higher plants) into ketobutyrate and ammonia. Inoculation with PGPR containing ACC deaminase activity could sustain plant growth and development under stress conditions by reducing stress-induced ethylene production.

Table 9 Microorganisms to overcome biotic stress

Induced systemic resistance (ISR)	*Pseudomonas fluorescens*
	P. putida
	Bacillus pumilus
	B. amyloliquefaciens
	B. subtilis, B. pasteurii
	B. cereus
	B. pumilus
	B. mycoides, and
	B. sphaericus
	Paenibacillus alvei
	Acinetobacter lwoffii
	Chryseobacterium balustinum
	Azospirillum brasilense
	Curtobacterium sp.,
	Arthrobacter oxidans
	Stenotrophomonas, and
	Serratia marcescens
Bacteria releasing volatiles against phytopathogenic fungi	*Alcaligenes,*
	Bacillus,
	Ensifer,

Contd...

Contd...

	Lysobacter
	Planomicrobium
	Sporosarcina, and
	Stenotrophomonas
	Burkholderia
	Pseudomonas
	Serratia
	Xanthomonas
	Pectobacterium, and
	Agrobacterium sp.
Microbes producing 1-aminocyclopropane-1-carboxylate (ACC) deaminase	*Agrobacterium genomovars* and *Azospirillum lipoferum*
	Alcaligenes and
	Bacillus
	Burkholderia
	Enterobacter
	Methylobacterium fujisawaense, Pseudomonas
	Ralstonia solanacearum
	Rhizobium
	Rhodococcus
	Sinorhizobium meliloti, and *Variovorax paradoxus*

2.9 Microorganisms as Inoculants

Microbial inoculants or soil inoculants use beneficial microbes (bacteria and fungi) to promote plant growth for agricultural purposes. These organisms show symbiotic relationship with crop plants where both partners get benefit. These inoculants improve plant nutrition and promote growth by stimulation of growth regulators. Nitrogen fixers and phosphate solubilizers are common inoculants that increase availability of macronutrients, nitrogen and phosphorus to plant. These bacteria are known as plant growth promoting rhizobacteria (PGPR).

Biofertilizers contain microorganisms in combination with organic fertilizers (manure) that provide high nutrient availability to host plants due to the interactions with microorganism. These fertilizers are of the following types:

1. Symbiotic nitrogen fixers (*Rhizobium* spp.) are used for legume crops like groundnut, soybean, increases crop yield from 10 to 35% by adding 50–200 kg N/ha. *Rhizobium* spp. is Gram-negative bacteria, enter roots of legume plants and form nodules in the roots and fix the atmospheric nitrogen. This bacterium improves the crop yield

of inoculated crop and also benefits the subsequent crop. *Rhizobium leguminosarum, R. melilotii, R. trifoli, R. phaseoli, R. lupinii,* and *R. japonicum* are common rhizobia are used as inoculants for legume plants.

2. Non-symbiotic nitrogen fixers (*Azotobacter, Azospirillum*) are used for cereal crops like maize, oats, sorghum, sugarcane, etc. increases yield from 10 to 15% by adding 20–25 kg N/ha. Application of *Azotobacter* to crop plants provides nitrogen through free atmospheric nitrogen fixation, synthesizes growth promoters and antibiotics. *Azospirillum* colonizes the roots and aerial parts of host plants by associative symbiosis and it is effective in farm yard manure in sorghum and millets.
3. Algal biofertilizers (Cyanobacteria, *Azolla*) are used for paddy crop, add 10 to 35 kg N/ha. Azolla supplies 30–100 kg N/ha. *Anabaena, Nostoc, Plectonema, Aulosira, Tolypothrix, Oscillatoria,* etc. are used in combination as dried flakes at the rate of 10 kg/ha after the tenth day of paddy seedlings transplantation. *Azolla* is aquatic fern that harbours *Anabaena* in leaf as symbiont that is common as green manure in paddy crop.
4. Phosphate solubilizers can be used for a variety of crops that improve 5 to 30% yield. *Pseudomonas, Bacillus* spp. are phosphate solubilizers that convert insoluble form of inorganic phosphates into soluble form that can be easily consumed by crop plants.

Mycorrhizae, used for crops and ornamental plants, increase intake of phosphorus, zinc, sulphur and water in the host plants and thereby enhances 10 to 50% yield. The symbiotic association of fungi in plant roots is called mycorrhiza. The fungal association increases nutrient uptake in host plant from soil. Ectomycorrhiza and endomycorrhiza are two types of mycorrhiza. Ectomycorrhiza are common on roots of forest trees like eucalyptus, oak, pine, etc. The association between fungi and root increases phosphorus, potassium and protect from pathogens. Endomycorrhiza are found in the roots of horticultural crops like coffee, betel vine, cardamom, etc., and they promote phosphorus nutrition.

2.10 Microorganisms in Insect Control

Pathogenic microorganisms with ability to invade and kill the insect pests were commonly known as entomopathogens. Metchnikoff (1879) and Krassilnikow (1888) were the first to introduce entomopathogens that could be used to control the pests. *Metarrhizium anisopliae*, a muscordine fungus, was first documented as pathogenic microorganism that was mass produced and applied as microbial pesticide for crops like sugar beets and grains. In 1940 White and Dutky introduced bacterium *Bacillus popilliae* spores to control Japanese beetle. *Bacillus thuringiensis* was introduced in 1960-63 as an effective bacterial insecticide and today nearly 40 types of products are popular as insecticides. Studies of Balch and Bird (1944) and Thompson and Steinhaus (1949) succeeded in the use of viruses against the insect pests. Baculoviruses are the common choice among the 1600 viruses to control insect species. The selected viruses were targeted against sawflies in forestry during 1950. Oil based formulations of entomopathogenic fungi were used to target the pests like aphids and white flies but these fungi were effective in highly humid locations. Protozoal species of microsporidia were found disease causing agents in insects.

Bacteria as Insecticides: *Bacillus popilliae*: It causes milky disease in Japanese beetle *Popillio japonica*. After ingestion, spores germinate and penetrate the alimentary canal and blood appears milky due to spores and then the larva dies. This bacterium is produced commercially by feeding or injecting in the host that multiplies in the body of the host. Then it is recovered by grinding the host with talc. This dust is expected to contain 100 million spores/g when it is ingested by the host; the mortality is due to septicemia.

Bacillus thuringiensis: It was detected in 1902 in dying larva of *Bombyx mori* by Ishiwata and Berliner in 1915 from Germany in larva of *Ephestia kuehniella*. In 1938 it was first used as microbial insecticide against Lepidopterous larvae with name 'sporeine' and this resulted in the development of several pesticide companies all over the world. It is pathogenic in Lepidopteran larvae and 4 insect orders. Bacterium is transmitted orally, after sporulation bacteria form a toxic crystal and ingestion of these spores causes cessation of feeding. The crystal activity depends on the pH of larval gut (pH 9–10.5) and the action of proteolytic enzymes. The crystal is a protoxin activated by enzymatic hydrolysis. The crystal protein is ecologically safe and successfully tested for controlling insects affecting cotton, cabbage, maize, sunflower and pigeon pea.

It is formulated under trade names such as – Thuricide (Sandoz AG), Bakthane, Dipel (Abbott labs), Biotrol, Karab, Agritol, Bactospeine (Rhone-Poulene) in dust, liquid concentrate, spray concentrate, granules and baits formulations.

Viruses as insecticides

Nuclear polyhedrosis viruses (NPV): Virus is rod in shape and outer envelope with protein crystals called polyhedra. NPV is transmitted by oral ingestion of polyhedra. Ingested polyhedra dissolve, releasing virus rods into the lumen of the insect host midgut, then larval skin darkens, hemolymph becomes turbid and prior to death, infected larva usually climbs to the highest point available and then dies.

The disease of *Heliothis armigera* is visible after 2-3 days of ingestion of virus. The cuticle of the infected larva in the first instar becomes milky white in appearance but the later infected instars turn pinkish.

The diseased larva discharges a yellowish grey fluid from their mouth. The death occurs in 3–7 days; the integument becomes fragile and bursts liberating the liquefied body contents.

Granulosis viruses (GV): GV particles are surrounded by an envelope similar to NPV envelope. The virion is rod shaped with DNA as genetic material with inclusion body, i.e., capsule or granule hence the virus is named Granulosis virus. This virus is active against rice leaf roller *Cnaphalocrocis medinalis*, castor semi-looper *Achaea janata*, hairy caterpillar *Pericallia ricini*, cut worm *Agrotis ipsilon*, internode borer *Chilo* spp. and sugarcane shoot borer *C. infuscatellus*. The fat body of Lepidopteran larvae is the primary site of infection. GV are transmitted orally and via the egg. GV infections involving the epidermis cause liquefaction of larvae similar to NPV infections.

Cytoplasmic polyhedrosis viruses (CPV): Particles are not enclosed in membranes as of NPV, but contain protein crystals similar to those of NPV. CPV infect the cytoplasm of the midgut epithelium of Lepidopterous larvae. Infected larvae appear to have small bodies and large heads.

Fungi as insecticides: More than 36 different genera of fungi contain species which cause insect diseases. Most fungi are transmitted from one host to another by a conidium. Conidia germinate and form a special structure which penetrates the insect cuticle. Fungus then grows in insect's body until the insect is filled with mycelia. Fungal infections depend on environmental conditions such as high humidity and temperature and high population densities. Examples: *Metarrhizium anisopliae, Beauveria bassiana, Entomophthora* spp.

Beauveria bassiana belongs to class fungi imperfecti and it is an actively biological control product (2.3×10^7 conidia/ml of product). Upon contact with target insect, the fungal spores attach themselves to the cuticle. The spores then secrete enzymes that dissolve the cuticle. This helps the fungal spore to penetrate hyphae into the body of insect. The production and proliferation of the fungal hyphae cause the release of toxicants from the fungus into the insects. Insect death occurs due to water loss and effects of nutrient loss. *Beauveria bassiana* is popular with trade name Dispel. It is used at the rate of 2–4 ml/L of water and is effective against coleopterous, hemipterous and lepidopterous insects.

3 MICROORGANISMS AS RENEWABLE ENERGY SOURCES

3.1 Biomining

The microorganisms solubilize the insoluble metals and could be exploited in the recovery of metals like copper, gold, iron, cobalt and uranium. This phenomenon is called bioleaching or biomining. The first process reported was conversion of insoluble copper in ore form such as covellite (CuS) or chalcocite (CU_2S) to soluble copper sulfate. The second was removal of arsenic, sulphur and iron from gold bearing arsenopyrite. In both the processes there was oxidation to recover the metal. In leaching process microorganisms form an exopolysaccharide (EPS) layer by adhering to the surface of mineral, within the layer oxidation reaction occurs.

There are two distinguished mechanisms described for oxidation of metals:

Thiosulfate mechanism: oxidation of acid insoluble metal sulphites such as FeS_2 and MoS_2.

Polysulphite mechanism: oxidation of acid soluble metal sulphites such as ZnS, $CuFeS_2$, PbS.

Oxidation reactions are produced by iron-oxidizing bacteria like *Acidithiobacillus, Leptospirillum, Sulfobacillus* (Table 10).

Table 10 Bacteria in leaching

Bacteria in leaching	Gram-negative	*Acidothiobacillus ferroxidans* (*Thiobacillus ferrooxidans*),
		Acidothiobacillus thiooxidans (*Thiobacillus thiooxidans*),
		Acidothiobacillus caldus,
		Leptospirillum ferrooxidans,
		Leptospirillum ferriphilum,
		Acidiphilium,
		Ferroplasma acidiphilum,
		Sulfobacillus thermosulfidooxidans,
		Sulfobacillus montserratensis
	Gram-positive	*Sulfobacillus metallicus,*
		Sulfolobus sp.

Commercial bioleaching is carried in two ways:

Heap leaching-low grade ores, crushed ore heap is percolated with acidic leaching solution with microbe, aeration is passive through pipe installed at the bottom.

Copper is the metal recovered by this process in large quantity in Sociedad Contractual Minera El Abra and Codelco Division Radimiro Tomic in Chile with production rate 225 000 and 180 000 tonnes of Cu per annum. In USA, Nevada, Carlin Trend region, Newmont Mining gold ore is pretreated by heap leaching.

Tank leaching-highly aerated, continous-flow, stirred tank.

In South Africa, Barberton, Fairview mine, stirred tanks are used for pretreatment of gold containing arsenopyrite concentrates, similarly at

Sansu, Ghana, West Africa, Ashanti goldfields large tank reactors are used. From both operations 55 and 960 tonnes of gold concentrate is obtained per day.

3.2 Microbial Enhanced Oil Recovery (MEOR)

It was in 1926, Beckman described for the first time about oil recovery by using microbes. In 1947, a new era of petroleum microbiology was investigated by ZoBell with patent on MEOR. ZoBell explained the mechanism for oil release from porous media due to formation of metabolites, gases, surface active agents, wetting agents by microbes that reduce the oil viscosity supporting its flow. iMEOR is tertiary recovery technique, first field test was conducted at Lisbon field, Union County, AR.

Acinetobacter, Arthrobacter, Bacillus, Brevibacterium, Clostridium, Enterobacter, Klebsiella, Leuconostoc, Methanobacterium, Mycobacterium, Pseudomonas, Peptococcus, Xanthomonas, and *Zymomonas* are reported to be anaerobic or facultative anaerobic bacteria that generate high quantities of gases, acids, solvents, polymers, surfactants, and cell biomass to release oil.

3.3 Biofuels

The limited stocks fossil fuel increased the concern about the fuel economy and many advance technologies have been introduced to overcome the shortage of fuels. The biofuels are the fuels that are not only renewable but are also capable of sequestering atmospheric carbon dioxide to maintain environmental and economic sustainability. Biofuels are rapidly being developed in order to meet global energy demand and advanced techniques have been introduced to produce biodiesel, bio-ethanol, bio-methane and bio-hydrogen.

Present biofuel technologies: Biomaterials and bioenergy have long been produced from plants. The production of methane and hydrogen from various microbes came up in the early 1950s that led to the usage of various bacteria, fungi and algae in the production of biofuels.

- First generation of biofuel production includes the plant-based biofuel production.
- Second generation is related with algae-based biofuel production.

The later generation includes the usage of genetically engineered microorganisms. Fuel biotechnology deals with the production of various biofuels by utilizing microorganisms and biomass.

Biodiesel production: Biodiesel is currently produced from oil synthesized with conventional fuel crops that harvest the sun's energy and store it as chemical energy. Apart from this the second generation of biodiesel production includes the production of biodiesel from algae. Biodiesel is a diesel-like substance obtained from the lipids accumulated in plants and algae or hydrocarbons synthesized by plants and algae. Oil seed based biodiesel production uses the lipid accumulated in seeds such as sunflower, safflower, peanut, olive, palm, jatropha.

Some algae accumulate lipids in their biomass when they are grown under mineral nutrient limitation. Under normal growth conditions most algal species have limited lipid content of 10–30% dry weight. But during nitrogen depletion the cells stop dividing and store products that result in accumulation of dry weight lipid contents to double or triple. Ex: *Botrycoccus braunii*-25–75%, *Nannochloropsis*-31–68%, *Schizochytrium* sp.-50–77%, *Neochloris oleaobundans*-35–54%, and *Nitschia* sp.-45–47%. Some plants and algae accumulate hydrocarbons in their cells, plants belonging to family Euphorbiaceae, milk weeds—*Ascelpias* sp., *Copaifera multijuga* accumulates hydrocarbons. In addition to this some algal members of both fresh and marine water accumulate hydrocarbons.

Bioethanol: It is the most widely used liquid biofuel and is produced by converting sugars directly from crops like sugarcane or sugar beets or starch from corn, wheat, potatoes or through cellulose from biomass into ethanol by fermentation. Traditional bioethanol production includes the fermentative degradation of sugary materials by *Saccharomyces cerevisiae* whereas other microbes includes *Candida* sp., *Zymomonas mobilis*, etc.

Current bioethanol production is based on the usage of genetically engineered microbes on cellulose biomass to obtain ethanol. This is the third generation approach of direct cellulose fermentation to produce ethanol through metabolic engineering that may meet the global demand of biofuels. Currently, genetically engineered *E. coli*, *Proteus*, *Klebsiella*, *Zymomonas*, and *Clostridia* are generally involved in bioethanol production.

Bio-hydrogen: Hydrogen is one of the biofuels produced from anaerobic microbial systems. The most common approach for converting biomass to hydrogen is bacterial fermentation process—essentially a truncated version of methanogenesis. A combination of moderately acidic pH and short solids retention time suppresses methanogens and accentuates hydrogen production.

Anaerobic bacteria, Cyanobacteria and photosynthetic algae produce hydrogen. *Gleocapsa alphicola* under sulphur starvation shows increased hydrogen production. *Arthrospira*, *Anabaena cylindrica* are also used for hydrogen production. Microalgae—*Scenedesmus obliqus*, *Chlamydomonas reinhardtii*—under controlled environment produces hydrogen.

Biogas: Biogas is the mixture of gases produced by microbial processes consisting of methane, carbon dioxide and hydrogen. Biogas especially, biomethane is produced by anaerobic digestion of biomass by methanogenic bacteria, hydrolytic fermentative bacteria and acetogenic bacteria.

3.4 Bioelectricity

From the times immemorial man has been depending on energy. With improvement of science and technology there is a great demand for energy expenditure as the need and use of fossil fuel has been gradually increasing. Increased utility of fossil fuels due to increased economic growth and social development has led to a large gap between energy demands and the availability of fossil fuels. These resulted in exhaustion of fossil fuels, as energy is the only driving force over the world and are in search for alternative fossil fuels that is eco-friendly and renewable. The current approaches available to energy generation are not suitable as they have pose adverse effects on environment. Thus, the world is in search of an approach that can produce huge amount of energy and natural carbon sources.

3.5 Microbial Fuel Cell (MFC)

MFC technology is used for producing electricity employing bacteria using biodegradable substrates. Wastewater from brewery, sugar industry, dairy, and paper industry are among the biodegradable substrates. Exoelectrogenic microorganisms such as *Aeromonas, Clostridium, Desulfobulbus, Geobacter, Geothrix, Geopsychrobacter, Pseudomonas, Shewanella, Ochrobactrum, Rhodopseudomonas* and *Rhodoferax* are genera that show the ability to produce electricity by MFC technology.

MFC is a technique that uses bacteria as catalyst for converting chemical energy (organic matter) into electricity. This technique consists of anode and cathode separated by a cation-specific membrane. In the anode compartment metal oxide reducing microorganisms are introduced that oxidize substrate and generate electrons and protons. External circuit is used to transfer the electrons but the protons move through the solution towards cathode, then electrons and protons combine along with oxygen and result in the formation of water molecule.

There are three proposed ways for transfer of electrons by microorganisms such as:

1. Use of exogenous mediators like thionine or neutral red.
2. Mediators produced by bacteria.
3. Using direct transfer of electrons from cytochromes (respiratory enzymes).

There are several advantages of MFCs compared to the current approaches used in electricity generation from organic matter, they are as follows:

(a) In this technique substrate (chemical energy) is directly converted to electricity with high efficiency.

(b) Operation is efficient at ambient temperature.

(c) Gas treatment is not required as the microorganisms release gases that enrich carbon dioxide.
(d) Passive aeration of cathode does not need energy input.
(e) This can be widely applied in locations that lack electrical infrastructure and also satisfy the energy requirements.

Simple carbohydrates such as glucose fructose are used that give high conversion and high efficiency compared to the complex sugars like starch, cellulose. Cellulose is preferred as substrate for MFCs as it is the most abundant biopolymer available in nature. When cellulose is used as substrate in energy generation, it is first solubilized by hydrolysis into simple sugars. Co-culture or mixed culture is performed to hydrolyze the cellulose so that it makes possible to take simple sugars by microbial cells. Energy generation is rather difficult from cellulose or cellulosic wastes than simple sugars because cellulose is highly resistant to hydrolysis and is insoluble. Lignin in combination with cellulose is very stable for microbial or enzymatic hydrolysis.

Conventional methods of isolating exo-electrogenic microorganisms are based on identification of soluble or insoluble metal oxides using organisms in their respiration on agar plates, not all the bacteria that produce electricity in MFCs may use metal oxides. Hence, electrochemically active strains of microorganisms are lost and it is a disadvantage. New methods are developed based on dilution to the extent that enriches exo-electrogenic microorganisms on anode. Hydrogen fuel cells compared to MFCs are efficient but MFCs are preferred as they produce electricity along with wastewater treatment that reduces the cost of the same.

4 MICROORGANISMS FOR POLLUTION CONTROL

4.1 Biodegradation of Pollutants

Environmental pollution is is caused by the wide range of compounds that are released as a consequence of industrial progress. Multiple hazardous waste sites have been generated as a result of accumulation of xenobiotics in soil and water. Constituents of crude oil like nitro aromatic compounds, polycyclic aromatics and other hydrocarbons along with halogenated organic compounds in combination with large and diverse group of chemicals are responsible for environmental depletion. Removal of these xenobiotics involves physico-chemical processes like excavation, incineration and landfills that are expensive and difficult to execute. Remedial strategies that clean up contaminated sites are not effective or adequate. Microorganisms are recyclers of nature that convert toxic compounds to harmless products mostly carbon dioxide and water. Hence, research is increasingly focused on the methods that degrade and eliminate the pollutants by biological means. Biodegradation is a natural cost-effective process that helps in removal of xenobiotic compounds by microorganisms from environment. These detoxifying ability possessing microbes play an imperative role in biogeochemical cycles of biosphere.

Xenobiotics contaminated sites need urgent remedial solution and search for diverse range of bacteria that utilize these pollutants as substrates. Enzymes, metabolic routes and genes involved in microbial degradation of nitro aromatic compounds, halogenated organic compounds, polycyclic aromatics and other hydrocarbons developed the broad flexibility of microorganisms in clean environment. Studies on interaction between xenobiotics and microorganisms in the environment have interconnected the disciplines of biochemical and genetic engineering. The use of microorganisms for biological clean up of polluted sites was highlighted in 1989 in accidental oil spill in Gulf of Alaska that contaminated 2000 km coast line with spilling of approximately 41000 m^3 oil. This method of cleaning or restoration of contaminated sites was found to be self-sustaining and inexpensive. Bacterial communities were characterized and their responses to pollutants, isolation of potential degraders and identification of genes involved in degradation were studied.

Biochemical pathways of pollutants biodegradation

There are two broad categories of biochemical pathways involved in the breakdown of xenobiotics, they are: aerobic or anaerobic. Wide range of bacterial species are found to be associated with aerobic degradation of xenobiotics very efficiently but the contaminated sites are anaerobic hence the metabolizing bacteria are commonly anoxic. Microbes metabolize these compounds with typical process known as co-metabolism, here the organism transforms the compound without using it as energy source. In this process organisms require the primary substrates that will support their growth. The metabolites (enzymes) released in response to primary substrate show the activity for another substrate in significant manner and these substrates are termed co-substrates.

Aerobic metabolic pathway: In carbon cycle aromatic hydrocarbons are immensely turned over by aerobic catabolic mechanisms. The aerobes degrade by releasing oxygenases that induce elemental oxygen to activate the inert xenobiotic compound. This pathway forms the intermediates such as catechol, hydroquinone, homogentisate, procatechuate and hydroxyquinol that are further metabolized to tri-carboxylic acid cycle intermediates and finally converted into carbon-dioxide and water. For alkane degradation by aerobic means occurs with oxidation of terminal methyl groups into carboxylic acid with an alcohol intermediate and finally it ends with β-oxidation. The enzymes like esterases followed by permeases act on the phthalate isomers degradation. 3, 4-dihydroxybenzoate is formed by the action of dioxygenases which are further processed into acetyl-coA and succinyl-coA as end products.

Anaerobic metabolic pathway: Clostridia, Desulfobacterium, Desulfovibrio, Methanococcus, Methanosarcina are found among the popular anaerobic bacteria that degrade xenobiotic compounds in anoxic conditions. In this process an alternative to oxygen is used to oxidize the aromatic hydrocarbons like phenols, phthalates, benzene-toluene-ethyl benzene-xylene (BTEX). These compounds themselves act as terminal electron acceptors, support microbial growth by energy generation from oxidation of simple substrates. Reactions such as dechlorination, hydrolysis, nitro reduction and dealkylation occur during anaerobic degradation of pesticides.

All halogenated pesticides including aliphatic, cyclic aliphatic, aniline based phenoxy alkanote and cyclodiene types undergo reductive dechlorination in which pesticides themselves act as terminal electron acceptors and these metabolic reactions are termed halorespiration. Anaerobic bacterium *Dehalococcus ethenogenes* dechlorinate tetra chloro ethylene to ethene by inducing novel pathway of reductive dehalogenases. Benzoyl coenzyme-A pathway is important mode of denitrifying anaerobes to degrade azo dyes, carbon tetrachloride and cyclo trimethylene trinitramine (RDX) by releasing benzoate co-A ligase that subsequently break the ring by hydratase enzyme. Methanogens like *Methanospirillum, Methanosaeta,* and *Syntrophobacter* degrade phthalate compounds with acetate and methane as end products.

Co-metabolic pathway: This concept was initiated by studies on *Pseudomonas methanica* that oxidize ethane to acetic acid, propane to propanoic acid and butane to butanoic acid. Co-metabolism is a parallel process that oxidizes co-substrate during growth on a compatible carbon and energy source. Oxidation of co-substrate does not provide energy to the degrading organism hence it needs additional carbon source. *Nocardia, Pseudomonas, Xanthomonas, Bacillus, Brevibacterium, Flavobacterium, Azotobacter, Vibrio, Achromobacter, Arthrobacter, Hydrogenomonas, Microbacterium, Micrococcus* and *Streptomyces* strains exhibit the phenomenon of co-metabolism.

4.2 Bioremediation

This is one of the important technologies that protects natural resources on earth by cycling natural and environmental chemical pollutants. This process cleans up the ecosystem threatening contaminated sites by breakdown of pollutants that cannot be degraded by traditional disposal methods.

Bioremediation can be defined as a process involved in degradation of various pollutants by employing biological methods. Microorganisms, plants and various biological catalysts are employed in degradation of toxic environmental pollutants. The use of microorganisms for cleanup of contaminated environment has been in practice since years together.

Landfill is a traditional technique used to destroy the wastes generated that contribute as practice for cleaning contaminated sites. These landfills are the natural or artificial pits used to fill with wastes then capped and often landscaped. Chemical decomposition, incineration and biological decomposition are other traditional methods of cleaning of the wastes.

Surface/subsurface environmental pollution and bioremediation practices: Vast numbers of pollutants and waste materials have been discharged into the environment per annum. Due to industrialization, modernization, globalization, urbanization and increase in the population, enhanced the utility of natural sources of environment and at the same time increased the disposal of used and waste contaminants into the environment. These contaminants cause damage to the environment and disturb the natural ecosystem. Approximately 6×10^6 chemical compounds have been synthesized with 1000 new chemicals being synthesized annually. Almost 60,000 to 95,000 chemicals are in commercial use. According to the third world network reports, more than one billion pounds of contaminants are released globally in air and water. These contaminants are causing ecological problems leading to imbalances in natural ecosystem. This environmental crisis made man to overcome with proper public awareness and the safest solution is the bioremediation.

5 CONCLUSION

Microorganisms inhabiting diverse habitats could serve the multifarious per se to fulfil mass production, formulation, interactions and signalling with the environment. The use of microorganisms and exploitation of beneficial plant-microbe interactions (PGPM) offer promising and eco-friendly strategies for conventional and organic agriculture around the globe by improving plant health and growth through conferring stress tolerance, disease resistance, nutrient availability and uptake. All the above renewable energy resources are preferred to preserve the ecological systems, in diverse fields microbes are beneficial and pollution can be cut down by the biochemical and enzymatic activities of microbes that clean up the hazardous chemicals. Bioremediation is a cost-effective treatment method for hydrocarbon contaminated soil (crude oil or petroleum) by microbial communities. These organisms and their catabolizing genes are popular to remediate the contaminated environments. The microorganisms of beneficial activities could serve as boon to preserve the biosphere after application and exploitation in appropriate locales of dire need.

References

Abou-el-Seoud II and Megeed AA (2012). Impact of rock materials and biofertilizations on P and K availability for maize (*Zea maize*) under calcareous soil conditions. *Saudi Journal of Biological Sciences* 19: 55–63.

Alexander M (1999). Biodegradation and Bioremediation, 2nd ed. Academic Press. San Diego, California.

Allard AS and Neilson AH (1997). Bioremediation of organic waste sites: a critical review of microbiological aspects. *International Biodeterioration and Biodegradation* 39: 253–285

Arthurs P, Stiver WH and Zytner RG (1995). Passive volatilisation of gasoline from soil. *Journal of Soil Contamination* 4: 123–135.

Bhardwaj D, Ansari MW, Sahoo RK, and Tuteja N (2014). Biofertilzers function as key player in sustainable agriculture by improving soil fertility, plant tolerance and grop productivity. *Microbial Cell Factories* 13: 66, 1–10.

Brahmaprakash GP and Sahu PK (2012). Biofertilzers for sustainability, *Journal of the Indian Institute of Science* 92: 1, 37–62.

Banat IM, Makkar RS and Cameotra SS (2000). Potential commercial applications of microbial surfactants. *Appl Microbiol Biotechnol* 53: 495–508.

Balch RE and Bird FT (1944). A disease of the European sawfly, *Gilpinia hercyniae* (HTG), and its place in natural control. *Sci Agric* 65: 65–80.

Berg G (2009). Plant-microbe interactions promoting plant growth and health: perspectives for controlled use of microorganisms in agriculture. *Appl Microbiol Biotechnol* 84: 11–18.

Brakstad OG and Bonaunet K (2006). Biodegradation of petroleum hydrocarbons in seawater at low temperatures (0–5°C) and bacterial communities associated with degradation. *Biodegradation* 17: 71–82.

Bredholt H, Josefsen K, Vatland A, Bruheim P and Eimhjellen K (1998). Emulsification of crude oil by an alkane-oxidizing *Rhodococcus* species isolated from seawater. *Can J Microbiol* 44: 330–340.

Bushnell DL and Haas HF (1941). The utilization of certain hydrocarbons by microorganisms. *Kansas Agricultural Experiment Station* 199: 653–673.

Carere CR, Sparling R, Cicek N and Levin DB (2008). Third generation biofuels via direct cellulose fermentation. *Int J Mol Sci* 9: 1342–1360.

Chang CH and Yamg SS (2009). Thermotolerant phosphate-solubilizing microbes for multi-functional biofertilizer preparation. *Bioresource Technology* 100: 1648–1658.

Choudhary DK (2012). Microbial rescue to plant under habitat-imposed abiotic and biotic stress. *Appl Microbiol Biotechnol* 96: 1137–1155.

Choudhary DK, Sharma KP and Gaur RK (2011). Biotechnological perspectives of microbes in agro-ecosystems. *Biotechnol Lett* 33: 1905–1910.

Calvo P, Nelson L and Kloepper JW (2014). Agricultural uses of plant biostimulants. Plant Soil DOI 10.1007/s11104-014-2131-8.

Diasa MOS, Ensinas AV, Nebrac SA, Filho RM, Rossell CEV and Maciel MRW (2009). Production of bioethanol and other bio-based materials from sugarcane bagasse: Integration to conventional bioethanol production process. *Chemical Engineering Research and Design* 87: 1206–1216.

Dien BS, Cotta MA and Jeffries TW (2003). Bacteria engineered for fuel ethanol production: current status. *Appl Microbiol Biotechnol* 63: 258–266.

Demirbas A (2005). Bioethanol from cellulosic materials: A renewable motor fuel from biomass. *Energy Sources* 21: 327–337.

Deubel A and Merbach W (2005). Influence of microorganisms on phosphorus bioavailability in soils. *Soil Biology*, Vol. 3 Microorganisms in soil: Role in Genesis and functions Ed by F Buscot and A Varma.

Duca D, Lorv J, Patten CL, Rose D and Glick BR (2014). Indole-3-acetic acid in plant-microbe interaction. *Antonie van Leeuwenhoek* 106: 85–125.

Emerson D, Fleming EJ and McBeth JM (2010). Iron-oxidizing bacteria: an Environmental and genomic perspective. *Annu Rev Microbiol* 64: 561–583.

Floodgate G (1984). Petroleum Microbiology. *In: Atlas RM (ed)*, Macmillan Publishing Company. Inc., New York, USA 355–398.

Friedrich CG, Rother D, Bardischewsky F, Quentmeier A and Fischer J (2001). Oxidation of reduced inorganic sulphur compounds by bacteria: emergence of a common mechanism? *Appl and Environ Microbiol* 67(7): 2873–2882.

Ghazali FM, Rahman RNZA, Salleh AB and Basri M (2004). Degradation of hydrocarbons in soil by microbial consortium. *Int. Biodeterior. Biodegradation* 54: 61–67.

Giovannoni SJ, DeLong EF, Olsen GJ and Pace NR (1988). Phylogenetic group-specific oligo-deoxynucleotide probes for identification of single microbial cells. *Bacteriology* 170: 720–726.

Gregory A (2001). Enhanced oil recovery— Ten years on. *Petrol Rev* 55 (655): 40–43.

Gopalakrishnan S, Sathya A, Vijayabharathi R, Varshney RK, Gowda CLL and Krishnamurthy L (2014). Plant growth promoting rhizobia: challenges and opportunities. 3 Biotech DOI 10.1007/s13205-014-0241-x.

Hommel RK (1990). Formation and physiological role of biosurfactants produced by hydrocarbon-utilizing microorganisms. *Biodegradation* 1: 107–119.

Hiltner L (1904). Uber neuere Erfahrungen und Probleme auf dem Gebiet derBodenbakteriologie und unter besonderer Berücksichtigung der Gründüngung und Brachte. Arbeiten der Deutschen Landwirtschaftlichen Gesellschaft, 98: 5978.

Jirasripongpun K (2002). The characterization of oil-degrading micro-organisms from lubricating oil contaminated (scale) soil. *Lett Appl Microbiol* 35: 296–300.

Joshi KB (2014). Microbes: mini iron factories. *Indian J Microbiol* 54(4): 483–485.

Kaplan CW and Kitts CL (2004). Bacterial succession in a petroleum land treatment unit. *Appl Environ Microbiol* 70: 1777–1786.

Kasai Y, Takahata Y, Hoaki T and Watanabe K (2005). Physiological and molecular characterization of a microbial community established in unsaturated, petroleum-contaminated soil. *Environ Microbiol* 7: 806–818.

Khan AAH, Sadguna A, Divya V, Begum S, Naseem and Siddiqui AG (2014). Potential of microorganisms in clean-up the environment. *International Journal of Multidisciplinary and Current Research* 2: 271–285.

Kosaric N (2001). Biosurfactants and their application for soil bioremediation. *Food Technol Biotechnol* 39: 295–304.

Lang S and Wullbrandt D (1999). Rhamnolipids-biosynthesis, microbial production and application potential. *Appl Microbiol Biotechnol* 51: 22–23.

Lin SC (1996). Biosurfactant: Recent advances. *J Chem Technol Biotechnol* 63: 109–120.

Larkin MJ, Kulakov LA and Allen CC (2005). Biodegradation and *Rhodococcus* masters of catabolic versatility. *Curr Opin Biotechnol* 16: 282–290.

Leahy JG and Colwell RJL (1990). Microbial degradation of hydrocarbons in the environment. *Microbiol Rev* 54: 305–315.

Liebeg EW and Cutright TJ (1999). The investigation of enhanced bioremediation through the addition of macro and micronutrients in a PAH contaminated soil. *Int Biodeterior Biodegradation* 44: 55–64.

Loh KC and Wang SJ (1998). Enhancement of biodegradation of phenol and a non growth substrate 4-chlorophenol by medium augmentation with conventional carbon sources. *Biodegradation* 8: 329–338.

Lynd LR, Cushman JH, Nichols RJ and Wyman CE (1991). Fuel ethanol from cellulosic biomass. *Science* 251: 1318–1323.

Lonhienne CP, Rentsch D, Robatzek S, Webb RI, Sagulenko E, Nasholm T, Schmidt S and Lonhienne TG (2010). Turning the table: plants consume microbes as asource of nutrients. PLoS ONE 5(7): e11915.

MacLeod CT and Daugulis AJ (2005). Interfacial effects in a two-phase partitioning bioreactor: degradation of polycyclic aromatic hydrocarbon (PAHs) by a hydrophobic mycobacterium. *Process Biochem* 40: 1799–1805.

MacNaughton SJ, Stephen JR, Venosa AD, Davis GA, Chang YJ and White DC (1999). Microbial population changes during bioremediation of an experimental oil spill. *Appl Environ Microbiol* 65: 3566–3574.

Mandri T and Lin J (2007). Isolation and characterisation of engine oil degrading indigenous microorganisms in KwaZulu-Natal, South Africa. *Afr J Biotechnol* 6: 023–027.

Marchant R, Sharkey FH, Banat IM, Rahman TJ and Perfumo A (2006). The degradation of n-hexadecane in soil by thermophilic *Geobacilli*. *FEMS Microbiol Ecol* 56: 44–54.

Mathuriya AS and Sharma VN (2009). Bioelectricity production from paper industry waste using a microbial fuel cell by *Clostridium* species. *J Biochem Tech* 1(2): 49–52.

Matson PA, Parton WJ, Power AG and Swift MJ (1997). Agricultural intensification and ecosystem properties. *Science* 277: 504–509.

Metchnikoff E (1879). Diseases of *Anisoplia austriaca* larvae. Odessa (in Russian).

Miethke M and Marahiel MA (2007). Siderophore-based iron acquisition and pathogen control. *Microbiol and Molecular Biol Reviews* 71(3): 412–451.

Mohan SV, Srikanth S and Sarma PN (2007). Bioelectricity production by mediatorless microbial fuel cell under acidophilic condition using wastewater as substrate: Influence of substrate loading rate. *Curr Sci* 92 (12): 1720–1726.

Marchesi J, Sato RT, Martin AJ, Hiam SJ and Wade W (1998). Design and evaluation of bacterium-specific PCR primers that amplify genes coding for bacterial 16S rRNA. *Appl Environ Microbiol* 64: 795–799.

Marcoux J, Deziel E, Villemur R, Lepine F, Bisaillon JG and Beauder R (2000). Optimization of high molecular-weight polycyclic aromatic hydrocarbons degradation in a two-liquid-phase bioreactor. *J Appl Microbiol* 88: 55–65.

Mukhopadhyay A, Redding AM, Rutherford BJ and Keasling JD (2008). Importance of systems biology in engineering microbes for biofuel production. *Current Opinion in Biotechnology* 19: 228–234.

Olofsson K, Bertilsson M and Liden G (2008). A short review on SSF — an interesting process option for ethanol production from lignocellulosic feedstocks. *Biotechnology for Biofuels* 1(7): 1–14.

Peralta-Yahya PP and Keasling JD (2010). Advanced biofuel production in microbes. *Biotechnol. J.* 5: 147–162.

Radakovits R, Jinkerson RE, Darzins A and Posewitz MC (2010). Genetic engineering of algae for enhanced biofuel production. *Eukaryotic Cell* 9(4): 486–501.

Rawlings DE (2005). Characteristics and adaptability of iron-and sulfur-oxidizing microorganisms used for the recovery of metals from minerals and their concentrates. *Microbial Cell Factories* 4: 13, 1–15.

Ralph Mitchell (1992). Environmental Microbiology. John Wiley and Sons Inc. New York.

Richard TW (2005). Environmental science: Toward a sustainable future, 9th Edition. Pearson Prentice Hall, New Jersey.

Rawlings DE and Johnson DB (2007). The microbiology of biomining: development and optimization of mineral-oxidizing microbial consortia. *Microbiology* 153: 315–324.

Singh JS, Pandey VC and Singh DP (2011). Efficient soil microorganisms: A new dimension for sustainable agriculture and environmental development. *Agriculture, Ecosystems and Environment* 140: 339–353.

Steinhaus EA (1949). Principles of insect pathology. McGraw Hill Co., New York.

Sydow A, Krieg T, Mayer F, Schrader J and Holtmann D (2014). Electroactive bacteria-molecular mechanisms and genetic tools. *Appl Microbiol Biotechnol* 98: 8481–8495.

Tsavkelova EA, Klimova SY, Cherdyntseva TA and Netrusov AI (2006). Microbial producers of plant growth stimulators and their practical use: A review. *Appl Biochem and Microbiol* 42(2): 117–126.

United Nations (1987). Report of the World Commission on Environment and Development, General Assembly Resolution 42/187, 11 December 1987. *Retrieved:* 2007-11-14.

Van Veen JA, Van Overbeek LS and Van Elsas JD (1997). Fate and activity of microorganisms introduced into soil. *Microbiol Mol boi Reviews* 61(2): 121–135.

Van Hamme JD, Singh A and Ward OP (2003). Recent Advances in Petroleum Microbiology. *Microbiology and Molecular biology Reviews.* 67(4): 503–549.

Villaverde JJ, Moran BS, Espana PS, Goti CL and Prados JLA (2014). Biopesticides in the framework of the European pesticide regulation (EC) No. 1107/2009. *Pest Manag Sci* 70: 2–5.

Wen Q, Wu Y, Zhao LX, Sun Q and Kong FN (2010). Electricity generation and brewery wastewater treatment from sequential anode-cathode microbial fuel cell. *J Zhejiang Univ-Sci B (Biomed & Biotechnol)* 11(2): 87–93.

White RT and Dutky SR (1940). Effect of introducing milky disease on populations of Japanese beetle larvae. *J. Econ. Entomol.* 33: 306–309.

Zhang C and Kong F (2014). Isolation and identification of potassium-solubilizing bacteria from tobacco rhizospheric soil and their effect on tobacco plants. *Applied Soil Ecology* 82: 18–25.

Zarjani JK, Aliasgharzad N, Oustan S, Emadi M and Ahmadi A (2013). Isolation and characterization of potassium solubilizing bacteria in some Iranian soils. *Archives of Agronomy and Soil Science* 59(12): 1713–1723.

6

Fluoride Tolerant Microorganisms and Their Possible Role in Bioremediation

Mamta Baunthiyal[1], Shweta Ranghar[1] and Vinay Sharma[2*]

[1]*Department of Biotechnology G.B. Pant Engineering College, Pauri (Uttarakhand) 246194 India*

[2]*Department of Bioscience & Biotechnology, Banasthali University P.O. Banasthali Vidyapith-304 022 (Rajasthan), India*

**Email: vinaysharma30@yahoo.co.uk*

ABSTRACT

Fluoride pollution as a consequence of increased industrialization imposes serious environmental problems. Over years various fluoride polluted sites have been generated in soil and water. The increased use of this pollutant has been reported to impose adverse effects on every species of ecosystem including humans. The conventional, physico chemical methods of fluoride remediation are not cost effective and adequate enough. Therefore, research should be focused on developing biological methods for removing and detoxifying such pollutants. A wide range of heavy metal resistant and tolerant microorganisms have been studied and developed for remediation of heavy metals. Using similar approach, microorganisms that are able to resist or tolerate high concentration of fluoride can be employed for detoxification of fluoride or converting fluoride to less harmful form.

1 INTRODUCTION

Fluorine, the 17th most abundant element in the earth's crust has now become one of the most widespread pollutants due to increased industrialization and its subsequent deposition in environment. It is one of the most electronegative and super reactive element which easily gets associated with other elements to form fluoride.

The sources of fluoride pollution in environment are both natural and anthropogenic. It naturally occurs in nature as a part of the mineral fluorspar, therefore erosion of fluorine containing minerals; rain cycle and emission from volcanoes are main sources of its natural occurrence. Anthropogenic causes include wastewater from sewage treatment plants, discharge from industries (mainly aluminium smelters, steel, glass, brick and ceramic manufacture), improper solid waste disposal, agro-chemicals such as phosphate containing fertilizers and pesticides, runoff from agricultural fields, and leakage from underground storage tanks (Janardhan et al., 2009).

The problem of increasing fluoride concentration in soil and groundwater is now becoming one of the most important toxicological and geo-environmental issues in many countries. Following the development of many industries, treatment of fluoride contaminated water and soil is an issue of major concern. Various physical and chemical techniques are employed for decontamination of water and soil. Unlike organic pollutants, the toxicity of fluoride ion is inherent in its atomic structure, and it cannot be further transmuted or mineralized to a totally innocuous form. However, its oxidation state, solubility and association with other inorganic and organic molecules can vary. Microbes as well as higher organisms may play a bioremidiative role by transforming and concentrating these anions so that they are less available and less dangerous (Chouhan et al., 2012).

2 TOXICITY OF FLUORIDE

2.1 Effects of Fluoride on Human Beings

In humans, fluoride ingestion in excess through water, food and air leads to serious health problem in form of a disease called "Fluorosis". Fluoride ingested with water goes on accumulating in bones up to age of 55 years. At high doses it can interfere with carbohydrates, lipid protein, vitamins, enzymes and mineral metabolism. Long-term consumption of water containing more than 1.5 mg of fluoride per litre leads to dental fluorosis. Skeletal fluorosis has been observed in persons when water contains more than 3–6 mg/L of fluoride. It affects both young and old one alike. Studies have shown that many diseases such as osteoporosis, arthritis, brittle bones, cancer, infertility, brain damage, Alzheimer syndrome, and thyroid disorder can attack human body through excessive intake of fluoride (Sharma and Gupta 2014).

2.2 Effects of Fluoride on Animals

In aquatic animals, fluoride tends to be accumulated in the exoskeleton of invertebrates and in the bone tissue of fishes. The toxicity to aquatic invertebrates and fishes increases with increasing fluoride concentration, exposure time and water temperature, and decreases with increasing intraspecific body size and water content of calcium and chloride (Camargo, 2003). In terrestrial animals increased concentrations of fluoride results in significant depression of feed intake, body weight gain, feed efficiency and egg quality.

2.3 Effects of Fluoride on Vegetation

In plants fluoride is taken up both by roots and leaves through fluoride contaminated water and aerial deposition. Excessive fluoride in plants affects most of their physiological and biochemical functions leading to decreased yield and growth. Signs of inorganic fluoride phytotoxicity such as chlorosis, necrosis and decreased growth rates, are most likely to occur in the young, expanding tissues of broad leaf plants and elongating needles of conifers (Pal et al., 2012). Various morphological features affected are plant height, length of shoot and root, number and size of leaves, fruit, flower and seed germination rate. The biochemical processes affected are photosynthesis, transpiration rate (stomatal conductance and gas exchange), respiration, change in pigments activity such as chlorophyll, secondary metabolites like sugars, ascorbic acid, amino acids, proteins, lipids and antioxidant enzyme activity.

2.4 Toxicity of Fluoride to Microorganisms

At high concentrations, fluoride also has toxic effects on bacteria and fungi. More than a century ago, fluoride was used as an antimicrobial agent. The antimicrobial effects of this ion have continued to be elucidated over the last decades, with an emphasis on the bacteria that cause dental caries. Fluoride is also toxic to eukaryotic organisms. At the beginning of the 20th century, it was noted that baker's yeast, *Saccharomyces cerevisiae*, was killed in 1% sodium fluoride. Inhibitory growth effects have also been observed for other species of fungi, including several pathogens (Treshow, 1965).

Although fluoride is often present as a wastewater contaminant, published data on its inhibitory effect to microbial populations present in wastewater treatment systems are very scarce. Fluoride is inhibitory towards microbial populations involved in various metabolic steps in anaerobic digestion processes, i.e., mesophilic and thermophilic acetoclastic methanogens, as well as propionate- and butyrate-degraders, at concentrations lower than those found in some fluoride-containing industrial effluents. In contrast, very high concentrations of soluble fluoride (>500 mg/L) can be tolerated by microbial communities involved in the aerobic activated sludge and in denitrification processes without significant inhibitory impact. Nitrification processes are somewhat more sensitive but they appear to acclimate rapidly to fluoride (Ochoa-Herrera et al., 2009).

2.5 Molecular Mechanism of Fluoride Toxicity

The mechanisms by which fluoride is toxic to these different species are complex and incompletely understood. Several reports have suggested that the toxic action of fluoride resides in the fact that this ion acts as an enzymatic poison, inhibiting enzyme activity and also sometimes as enzyme stimulator and, ultimately, interrupting metabolic processes. In humans, fluoride interacts with vital processes which are widely involved in signalling pathways such as gene expression, metabolism, ion transport, secretion, proliferation and migration, apoptosis, endocytosis, oxidative stress. In plants also, fluoride accumulation leads to changed enzyme activity involved in the inhibition of protein synthesis and secretion, alteration in gene expression and generation of reactive oxygen radicals (Barbier et al., 2010).

3 NEED FOR FLUORIDE REMEDIATION

Owing to rapid increase in fluoride level concentration in ecosystem and its harmful effects to animals, plants and humans, there is a need for development of long lasting and eco-friendly technologies for fluoride remediation from environment. What makes the task of arresting the spread of fluorosis, the dreaded disease caused by excessive intake of fluoride through drinking water, is the fact that there is no cure available anywhere in the world for it. Prevention thus is the only way to control it. But an effective prevention strategy can hardly be drawn up without first assessing the extent of the problem and understanding it in its entirety.

3.1 Conventional Methods of Fluoride Remediation

Many physical and chemical methods are available for defluoridation of drinking water. Fluoride from water or wastewater can be removed by membrane filtration, precipitation, nanofiltration, ion-exchange, electro-coagulation, reverse osmosis and adsorption. But these techniques have high operational and maintenance cost, low fluoride removal capacities, lack of selectivity for fluoride, undesirable effects on water quality and generation of large volumes of sludge. In recent years, considerable attention has been focused on the study of fluoride removal using natural, synthetic and biomass materials such as activated alumina, fly ash, alum sludge, chitosan beads, red mud, zeolite, calcite, hydrated cement, attapulgite, and acid-treated spent bleaching earth (Rao and Metre 2012).

3.2 Bioremediation of Fluoride

Bioremediation is one of the most cost-effective and eco-friendly technologies for managing the polluted environment and recover contaminated soil and water. The biological agents include both plants and microorganisms.

The use of plants (phytoremediation) for remediation of fluoride contaminated soil and water has been reported in many studies. Many plant species such as *Prosopis juliflora, Camellia japonica, Pittosporum tobira,* and *Saccharum officinarum,* some aquatic plants such as *Ipomoea aquatica, Spirodela polyrrhiza, Hydrilla verticillata* are known to be fluoride hyperaccumulators. These hyperacccumulators have the ability to accumulate and tolerate high fluoride concentration without showing any symptoms of injuries and thus have the potential to remediate fluoride contaminated soil and water. The mechanism of fluoride hyperaccumulation owes to the fact that tolerant species may be able to deactivate fluoride better than sensitive ones.

A large number of microorganisms like bacteria, fungi, algae are used for bioremediation. Microorganisms can be isolated from almost any environmental conditions. Microbes have the ability to adapt and grow at extreme conditions of temperature, heat, desert conditions, in water, under anaerobic or aerobic conditions, in the presence of hazardous compounds or on any waste. The main requirements are an energy source and a carbon source. Because of the adaptability of microbes and other biological systems, these can be used to degrade or remediate environmental pollutants. Depending upon the nature of polluting agent different types of organisms are selected and utilized for bioremediation as they only survive within a limited range of chemical contaminants (Latha and Reddy 2013).

3.3 Fluoride Tolerant Microorganisms for Bioremediation

Because of the ubiquitous presence of fluoride in the environment and its toxic effects, organisms must have evolved mechanisms of resistance to fluoride. However, little is known about these strategies and pathways. Microorganisms have acquired a variety of mechanisms for adaptation to the presence of toxic elements. Many studies done on bacterial diversity in heavy metal contaminated sites have demonstrated that there are indigenous organisms that have not only adapted to the new environments but have also flourished under them. To deal with high concentration of these toxic pollutants, tolerant microorganisms possess a variety of mechanisms such as sorption, mineralization, uptake and accumulation, extracellular precipitation, complexation, enzymatic oxidation or reduction to a less toxic form, efflux or to use them as terminal electron acceptors in anaerobic respiration (Haq and Shakoori 2010).

Most mechanism involves the efflux of metal ions outside the cell, and genes for tolerance. These mechanisms are sometime encoded in plasmid genes facilitating the transfer of toxic metal resistance from one cell to another. Microbial species, such as *Pseudomonas aeroginosa, Klebsiella pneumoniae, Proteus mirabilis, Staphylococcus* and *Bacillus subtilis* have been shown to be relatively efficient in bioaccumulation of uranium, mercury, copper, lead, zinc and other metal ions from polluted effluents (Hookoom and Puchooa 2013; Latha and Reddy 2013). Research on fluoride remediation has suggested that many plants and bacteria secrete high affinity anion binding compounds called ionophores. The ionophores bind specific chemical forms of anions (Chouhan et al., 2012).

This anion ionophore complex is then absorbed back into the organism for utilization. Such organisms may provide the opportunity to make fluoride less available and less dangerous.

In a study conducted to isolate, identify and characterize fluoride-resistant bacteria that may play a role in bioremediation of excessive fluorides from environments; samples of water and soil were collected from different sites of Gwalior and Morena regions of Madhya Pradesh, India. Five bacterial isolates were found to be fluoride resistant, viz., *Micrococcus luteus, Aeromonas hydrophyla, Micrococcus varians, Pseudomonas aerugenosa* and *Escherichia coli*. The protein profile study revealed the expression of one extra band in *Micrococcus luteus* and three extra bands in *Pseudomonas aerugenosa,* these extra bands expressed in response to fluoride may act as ionophores. Also both isolates were able to reduce the fluoride concentration to significant level of 19.8% and 22% respectively after 12 days of treatment. These preliminary results support the hypothesis that the isolates expressing ionophores within the cytosol are found to be efficient in removing fluoride from the media (Chouhan et al., 2012).

In another study, the fluoride tolerance of the uranium bioleaching microorganisms was studied. Five typical bioleaching strains including *Acidithiobacillus ferrooxidans* ATCC 23270, *Leptospirillum ferriphilum YSK, Sulfobacillus thermo sulfidooxidans* ST, *Acidithiobacillus thiooxidans* A01, *Acidithiobacillus caldus* S1 were selected to compare their growth fluctuation and their oxidation rate of iron or sulphur under different concentrations of fluoride. The growth curve results at increasing fluoride concentration showed that *A. ferrooxidans* ATCC 23270 held the best fluoride tolerance. In order to explore the mechanism of microbial resistance to fluoride, whole-genome array (WGA) was used to analyze the genome-wide expression profiling of *A. ferrooxidans* ATCC 23270 at 4.8 mM fluoride stress. Study on the gene expression profiles revealed that fluoride stress mainly involved cell membranes, energy metabolism, transport and binding proteins, DNA metabolism, cell processing, protein synthesis and transport, and the synthesis of biological factors (Liyuan et al., 2013).

In yet another study on *A. ferrooxidans* ATCC 23270, the mechanism of bacterial tolerance to fluoride was domesticated and studied under the conditions of different fluorine concentrations and pH values with or without treatment by Proteinase K. The bacteria was inoculated in 9K medium containing different concentrations of fluoride, 0, 5, 10, 20 and 50 mg/L, respectively. The activity of bacteria was determined by measuring solution redox potential. For domestication of higher tolerant strain, it was inoculated and incubated in a series of fresh media containing 0, 10, 20, 30, 40, 50 and 60 mg/L fluoride, respectively. The redox potential of these solutions were measured every day. When the solution potential reached about 600 mV (vs Ag/AgCl), the ferrous ions in the medium were thought to be almost completely oxidized. Then, the culture was inoculated and incubated again in the solution containing higher concentration of fluoride. Finally, a new higher fluoride-resistant strain was obtained, and named as *Frs*. Due to adaptation for a long time in fluoride containing medium, *Frs* strain could resist up to 40 mg/L fluoride. The results of intracellular fluoride concentration with and without proteinase K treatment showed that, fluoride traverses the membrane likely by assistant diffusion which needed some proteins to help to accomplish the transport course. The treatment of Proteinase K would destroy part of outer membrane proteins involved in free fluoride ion's exclusion at neutral conditions. Hence, when some kinds of proteins were damaged, intracellular fluoride increased, which suggests that some proteins of cell wall or cell membrane are involved with the bacterial fluoride tolerance (Peng et al., 2013).

In Eubacteria and Archaea, a class of regulatory RNAs, or riboswitches were discovered that bind to fluoride and regulate the expression of genes in response to fluoride anion. Riboswitches are metabolite or ion-sensing structured RNA motifs typically located in the non-coding regions of certain mRNAs. These control the expression of adjoining protein-coding regions through several different mechanisms including transcription termination, translation blocking, and alternative splicing. One of the genes most commonly associated with fluoride-sensing riboswitches is the crcB gene. This gene was originally identified in *Escherichia coli* and encodes a small protein (127 amino acids) that has four predicted transmembrane domains. Due to its association with the fluoride-binding riboswitch, it was hypothesized that *crcB* may be involved in fluoride resistance, specifically fluoride efflux. An *E. coli* strain lacking *crcB* was 200-fold more sensitive to fluoride than the Wild type (WT) strain (Baker et al., 2012).

Li et al. (2012) used genetic approaches involving three organisms: the filamentous fungus *Neurospora crassa,* the budding yeast *Saccharomyces cerevisiae,* and the pathogen *Candida albicans* to demonstrate that CrcB proteins play a role in fluoride resistance in eukaryotes and are involved in removing fluoride from the cell. This is the only known example of a gene identified to function in fluoride efflux in eukaryotes. Based on these results, these genes were renamed as *FEX,* for fluoride exporters. The protein FEX is required for the rapid expulsion of cytoplasmic fluoride, indicating that many eukaryotic species that carry *FEX* genes likely avoid fluoride toxicity by purging cellular fluoride.

4 CONCLUSION

Bioremediation is a technique for cleaning up pollution by enhancing the natural biodegradation processes. The results of case studies suggest that bacteria and fungi can actively reduce fluoride contamination that may provide opportunity to develop a new bioremediation technique for extracting excess of fluoride in a simple way. Further research should be focussed on identification of a few more fluoride tolerant or resistant microorganism that can survive under extreme conditions. These microorganisms should be given prime consideration in bioremediation due to alarming situation raised by fluoride contamination.

References

Baker JL, Sudarsan N, Weinberg Z, Roth A, Stockbridge RB and Breaker RR (2012). Widespread genetic switches and toxicity resistance proteins for fluoride. *Science*. 335(6065): 233–235.

Barbier O, Arreola-Mendoza L and DelRazo LM (2010). Molecular mechanisms of fluoride toxicity. *Chemico Biological Interactions* 188: 319–33.

Camargo JA (2003), Fluoride toxicity to aquatic organisms: a review. *Chemosphere* 50(3): 251–64.

Chouhan S, Tuteja U and Flora SJS (2012), Isolation, identification and characterization of fluoride resistant bacteria: Possible role in bioremediation. *Applied Biochemistry and Microbiology* 48(1): 43–50.

Haq R and Shakoori AR (2000). Microorganisms resistant to heavy metals and toxic chemicals as indicators of environmental pollution and their use in bioremediation. *Folia Biologica* 48: 143–147.

Hookoom M and Puchooa D (2013). Isolation and identification of heavy metals tolerant bacteria from industrial and agricultural areas in Mauritius. *Current Research in Microbiology and Biotechnology* 1(3): 119–123.

Janardhan RN, Deay S and Das K (2009). Fluoride contamination in groundwater's of Sonbhadra District, Uttar Pradesh, India. *Current Science* 96(7): 979–985.

Latha PA and Reddy SS (2013). Review on bioremediation - Potential tool for removing environmental pollution. *International Journal of Basic and Applied Chemical Sciences* 3(3): 21–33.

Li S, Smith KD, Davis JH, Gordon PB, Breaker RR and Strobel SA (2013). Eukaryotic resistance to fluoride toxicity mediated by a widespread family of fluoride export proteins. *Proc Natl Acad Sci USA*, 110: 19018–19023.

Liyuan Ma, Qian Li , Yunhua Xiao , Qingliang Wang , Huaqun Yin , Yili Liang, Guanzhou Qiu and Xueduan Liu (2013). Comparative study of Fluoride-tolerance of five typical bioleaching microorganisms. *Advanced Materials Research* 825: 214–218.

Ochoa-Herrera V, Ochoa-Herrera V1, Banihani Q, León G, Khatri C, Field JA and Sierra-Alvarez R (2009). Toxicity of fluoride to microorganisms in biological wastewater treatment systems. *Water Res* 43(13): 3177–3186.

Pal KC, Mondal NK, Bhaumik R, Banerjee A and Datta JK (2012). Incorporation of Fluoride in vegetation and associated biochemical changes due to Fluoride contamination in water and soil: A Comparative Field Study. *Annals of Environmental Science* 6: 123–139.

Peng Z Yu R, Qiu G , Gu G, Wang Q, Li Q and Liu X (2013). Really active form of fluorine toxicity affecting *Acidithio bacillus ferrooxidans* activity in bioleaching uranium. *Transcation of Non-ferrous Metals Society of China* 23: 812–817.

Rao MK and Metre M (2012). Effective Low Cost Adsorbents for Removal of Fluoride from Water: A Review. *International Journal of Science and Research* 3(6): 120–124.

Sharma MR and Gupta V (2014). Fluoride and its ecological effects in water: A Review. *Global Journal for Research Analysis* 3(8): 178–180.

Treshow M (1965). Response of some pathogenic fungi to sodium fluoride. *Mycologia* 57: 216–221.

7

PGPR Bio-inoculants for Sustainable Crop Production

Manoj Kaushal* and Rajesh Kaushal
Department of Basic Science, Dr. Y. S. Parmar University of Horticulture and Forestry, Nauni-173230, Solan, Himachal Pradesh
**Email: kaushal.mbg@gmail.com*

1 INTRODUCTION

Plant growth promoting rhizobacteria are groups of bacteria that are found in rhizosphere (at root surfaces) and endorhizosphere (inside the roots) which improve the extent and quality of plant growth directly and/or indirectly. They directly promote the growth of plants usually by the production of phytohormones and/or by increasing the available nutrients through production of secondary metabolites to solubilize native P content, siderophore producers and/or as biocontrol agent that are able to protect plant from infections by phytopathogenic organisms through production of antibiotics, HCN and cell wall degrading enzymes. PGPR have been applied to various crops to enhance seed emergence, growth and crop yield, but only a few isolate have been commercialized (Dey et al., 2004; Herman et al., 2008).

Cauliflower (*Brassica oleracea* var. *botrytis* L.) is a member of family cruciferae grown as commercial crop of the state for both table purpose and seed crop. It is grown throughout the year in different agro-climatic zones occupying an area of 1400 ha with annual production of 24,390 million tonnes (Anonymous 2007). Since there is no commercial PGPR/ biofertilizers formulation for this particular crop therefore, there is an urgent need to develop an effective inoculum of plant growth promoting rhizobacteria (PGPR) with multifarious plant growth promoting traits such as phosphorous solubilizers, asymbiotic nitrogen fixers and production of growth hormones besides acting as biocontrol agents (Cleyet-Marcel et al., 2001). For sustainable production of cauliflower and less dependence on chemical inputs, it is important to have commercial inocula of PGPR with multifarious traits. Therefore, the present investigations were undertaken to screen the PGPR isolates from its natural growing zones with multifarious activities, at various levels of N and P with other recommended package of practices for commercial cultivation of cauliflower.

2 MATERIALS AND METHODS

2.1 Isolation of PGPR

Soil and root samples were collected from the rhizosphere of cauliflower plants from Bilaspur, Hamirpur and Kangra districts of Himachal Pradesh. The samples were stored at 4°C in the Soil Microbiology Laboratory, Dr. Y.S. Parmar University of Horticulture and Forestry, Nauni, Solan, India. One gram of the rhizosphere soil was placed in 9 ml of sterilized distilled water under aseptic conditions. The soil suspension was diluted in 10-fold series and the microbial count was determined by the standard pour plate technique (Subba Rao 1999). For endophytic microorganisms root sample was surface sterilized by 0.2 per cent mercuric chloride ($HgCl_2$) for two minutes followed by repeated washing in sterilized distilled water. The surface sterility of roots were cross-checked by incubating the surface sterilized roots in sterilized nutrient agar medium for 24 hrs at 35± 2°C. One gram of surface sterilized root sample was placed in 9 ml of sterilized distilled water and was crushed to produce slurry using pestle and mortar under aseptic conditions. Then serially diluted suspension of roots were spread on pre-poured nutrient agar medium. After incubation of 24–48 hrs. Modified Replica plate technique was used for isolation.

2.2 Screening and Characterization of Bacterial Isolates Under Controlled Conditions

Screening of the bacterial isolates for the various plant growth promoting activities like P-solubilization, siderophore formation, HCN concentration, growth on N-free medium, auxin production and antagonism against *Fusarium* spp., *Rhizoctonia solani* and *Pythium* spp. were performed by adopting the standard methods as given in Table 1. The optimization of growth conditions (physical, chemical and nutritional) of selected bacterial isolates were standardized by conducting separate experiments as per standard methods. Phosphate solubilization in liquid PVK medium containing TCP (tri calcium phosphate) by culture supernatant was determined by Bray and Kurtz (1945).

Table 1 Methods adopted to check multifarious plant growth promoting activities of the bacterial isolates

Activity	*Method Adopted*
Phosphate solubilizing	Pikovskaya (1948)
Nitrogen fixing ability	Jensen (1987)
Siderophore production	Schwyn and Neilands (1987)
HCN production	Bakker and Schippers (1987)
IAA production	Gorden and Paleg (1957)
Antagonistic activity	Vincent (1947)

2.2.1 Field experimentation

Three bacterial isolates (MK_5, MK_7 and MK_9) were selected on the basis of their multifarious plant growth promoting activities during *in vitro* and net house conditions. Bacterial cell suspension (O.D. 1.00 at 540 nm) of 72 h old culture grown in nutrient broth at the rate of 10 per cent was used as inoculum for field experimentation. The seeds were treated with bacterial inoculum for 8 hours. Untreated seeds were treated in sterilized distilled water for same time and designated as control. The seeds were sown in nursery and one-month-old seedlings were transplanted in the field. A booster dose of bacterial culture was added at one month interval till harvesting. The experiment was conducted at RHRS, Bajaura (Kullu) situated at 32° N latitude and 77°E longitude. The treatments combinations viz.: T1 (Control), T2 (MK_5 + 50% NP), T3 (MK_5 + 75% NP), T4 (MK_5 + 100% NP), T5 (MK_7 + 50% NP), T6 (MK_7 + 75% NP), T7 (MK_7 + 100% NP), T8 (MK_9 + 50% NP), T9 (MK_9 + 75% NP) and T10 (MK_9 + 100% NP) were arranged in RBD design. Soil samples were collected before the start of experiment and at termination of the experiment and analyzed for physico-chemical parameters like pH, electrical conductivity, organic carbon, available nitrogen, available phosphorus and available potassium as per the methods described by Jackson, 1973. The observations were recorded at different time intervals on different quantitative characters of cauliflower, viz. (number of non-wrapper leaves, curd diameter, curd depth and curd weight and curd yield). The data recorded were statistically analyzed as per the design suggested by Gomez and Gomez (1984).

3 RESULTS

3.1 Morphological Characteristics of PGPR Isolates

Five bacterial isolates (MK_2, MK_4, MK_5, MK_7 and MK_9) were successfully isolated from the rhizosphere and endorhizosphere of cauliflower. The isolates varied in their morphological characteristics, isolates were found to be fast growers, round shaped, raised colonies having rough surface with undulate to erose margin. All isolates were motile and Gram-positive in reaction.

3.2 Screening and Characterization of Bacterial Isolates

The selected bacterial isolates were screened for their ability of performing multifarious activities, i.e., P-solubilization, growth on nitrogen-free medium, siderophore, auxin, HCN production and antagonism against *Fusarium* spp., *Pythium* spp. and *Rhizoctonia solani.* The performance of various isolates is summarized in Table 2.

Then these isolates (MK_2, MK_4 MK_5, MK_7 and MK_9) were further characterized and the activities were also quantified. The results

Table 2 Screening of selected bacterial isolates for multifarious plant growth promoting activities

Isolates	*P-Solubilization*	*N-free medium*	*Auxin production*	*Siderophore production*	*HCN production*	*Antagonism against*		
						Fusarium sp.	*Pythium sp.*	*R. solani*
MK_2	+	+	+	+	–	–	–	–
MK_4	+	++	+	+	–	–	–	–
MK_5	+++	+++	++	++	++	+++	+	++
MK_7	++	++	+	+	+	++	+	++
MK_9	++	++	+	+	+	+	++	+

+ Very good activity, ++ good activity, +++ fair activity, – no activity represented in Table 3 reveal that MK_5 had highest phosphate solubilizing efficiency (PSE, i.e., 172.21%) which was statistically at par with rest of the four isolates. In liquid medium maximum phosphate was solubilized by MK_5 (664.30 μg/ml), which was statistically at par with MK_4, MK_7 and MK_9. Growth inhibition shown by various bacterial isolates against *Fusarium* spp., *Rhizoctonia solani* and *Pythium* spp. are also summarized in Table 3. Also the isolate MK_5 produced a significantly higher concentration of IAA (29.67 μg/ml) after 72 hours of incubation as compared to other isolates. All the five bacterial isolates produced a bright zone with yellowish colour around the bacterial colony on Chrome-azurol-S medium. Quantitative estimation of siderophore using Chrome-azurol-S (CAS) liquid assay revealed that bacterial isolate MK_5 produced maximum (33.02% siderophore unit) at 72 hours of incubation.

3.3 Soil Properties and Nutrient Status

The physico-chemical properties of soil were recorded at the start and termination of the experiment. The data on initial soil parameters are pH (6.71), EC (0.34 dSm^{-1}), bulk density (1.50 mgm^{-3}) and organic matter (1.03%). The initial available N (313.60 kg/ha) and available K (191.70 kg/ha) was medium, however available P (86.00 kg/ha) was in high range. There was no significant change in basic physico-chemical properties of soil, i.e., pH, EC, organic matter and bulk density after the termination of experiment. However, the available nutrient contents N, P and K were increased by 2.43-36.24%, 0.18-45.75% and 1.83-19.38%, respectively over control as given in Table 4.

3.4 Plant Parameters

The application PGPR isolates at different levels of N and P significantly increased number of non-wrapper leaves, curd diameter, curd weight, curd depth and curd yield/ha (Table 5) over uninoculated control. The highest yield (355.56 q/ha) was recorded in MK_5 isolate with 100% NP fertilizers (T_4 treatment) which was statistically at par with T7 and T10, i.e., with the isolates MK_7 (315.80 q/ha) and MK_9 (343.21 q/ha) at same doses of N and P fertilizers respectively.

4 DISCUSSION

PGPR colonizes plant roots and exert beneficial effects on plant growth and development by a wide variety of mechanisms. The exact mechanism by which PGPR stimulate plant growth is not clearly established, although several hypothesis such as production of phytohormones, suppression of deleterious organisms, activation of phosphate solubilization, and promotion of the mineral nutrient uptake are usually believed to be

growth differences between mycorrhizal and nonmycorrhizal were less than in perlite. However, root colonization rate was not reduced during the NFT culture period. In this system, high amounts of fungal biomass were produced. This would allow the determination of metal and other nutrient concentrations in fungal hyphae. Furthermore, they found large amounts of external fungal hyphae surrounding the root surface. As much as 130 mg fungal biomass was collected per culture plate (three plants). Therefore, they suggested that this modified NFT culture system would be suitable for fungal biomass production on a large scale with a view to additional aeration by intermittent nutrient supply, optimum P supply, and a use of glass beads as support materials. Furthmore, bulk inoculum composition with a mixture of spores, colonized roots and hyphae grown in soilless media by the modified NFT system might be useful way to mass produce mycorrhizal crops and inoculums for commercial horticultural purposes.

Besides variation in techniques used in aseptic inoculation of AM fungi to the host plant, the major concern in the NFT culture system is the concentration of nutrients. The preferred values for various nutrient element varies from one mycorrhizal system to the other depending on the size and other features of the plant. Another important factor is the compromise between plant growth and Mycorrhizal infection since water-logged conditions affect mycorrhizal growth adversely (Tarafdar 1995). It is therefore, necessary to maintain the nutrients in the form of thin film. Form of nutrient elements also affects AM infection and therefore, it is necessary to use balanced and proper composition.

3.2.2 Aeroponic

Aeroponics is the process of growing plants in an air and mist environment without the use of soil or an aggregate medium. The aeroponic is derived from the Greek meanings of "aero" (air) and "ponos" (labour). Aeroponic culture differs from both conventional hydroponics and *in vitro* (plant tissue culture) growing. Unlike hydroponic, which use water as growing medium and essential minerals to sustain plant growth, aeroponics is conducted without a growing medium. This system was first used for the study of legume-rhizobia interaction by Zobel et al. (1976) and then for AM fungi by Sylvia and Hubbell (1986). Aeroponic is a highly erased environment than traditional hydroponics and has proven to be an efficient system for growing AM inoculum without physical substrate (Hung and Slyvia 1988). AM culture has been successfully established on maize, sweet potato, leek, adzuki bean and bahia grass using this culture system (Sylvia and Hubbell 1986; Hung and Sulvia 1988; Weathers and Zobel 1992: Wang and Tschen 1994; Mohammad et al., 2000). There are three basic methods for producing atomised nutrient solution: (a) an impeller system makes use of an atomising disc, (b) pressurised spray through nozzles, or (c) ultrasonically generated fog (Weathers and Zobel 1992).

Studies conducted showed that in the aeroponic system, colonization and sporulation on the host plant by VAM fungi were superior to those reported for soil-based pot cultures. Uniform colonization (75%) of *Paspalum notatum* roots by *Glomus mosseae* and abundant sporulation (5 chlamydospores per cm colonized root length) were obtained after 8 weeks in aeroponic cultures. Colonization of roots by *Glomus intraradices* reached 50% after 12 weeks in an aeroponic culture and a mean of 8 chlamydospores per cm of colonized root were observed. This method should allow more efficient and uniform cultures of AM (Sylvia and Hubbell 1986).

Three AM fungal species (*Glomus deserticola, G. intraradices, G. mosseae*) established successfully in aeroponic culture with bahia grass (*Paspalum notatum*) and industrial sweet potato (*Ipomoea batatas*) when precolonized seedlings and a modified Hoagland's nutrient solution with a low phosphorus (1 mM) level were used. These fungi not only colonized the roots but also sporulated profusely. All colonized roots and spores of *G. mosseae* produced aeroponically were viable and could be used as effective inocula (Hung and Sylvia 1987). In another study on bahia grass (*Paspalum notatum*) and industrial sweet potato (*Ipomoea batatas*) inoculated with *Glomus deserticola, G. etunicatum* and *G. intraradices* and grown in aeroponic cultures showed colonization after 12–14 weeks, by the inoculated AM fungi. Abundant vesicles and arbuscules were formed in the roots and profuse sporulation was detected within and outside roots. The mean percent root colonization and sporulation per cm of colonized roots generally increased with time, although with some treatments, colonization declined by week 14. Spore production ranged from 4 spores per cm of colonized roots for *G. etunicatum* to 51 spores per cm of colonized root by *G. intraradices*. Infectivity trials with root inocula resulted in a mean of 38, 45, and 28% of *P. notatum* roots colonized by *G. deserticola, G. etunicatum*, and *G. intraradices*, respectively. The germination rates *G. etunicatum* spores produced in soil were significantly higher than those produced in aeroponic cultures (64% versus 46%) after incubation at 28°C in 2 weeks. However, infectivity studies comparing *G. etunicatum* spores from soil and aeroponic cultures indicated no biological differences between the spore sources (Hung and Sylvia 1988).

Spore production by aeroponic cultures was significantly higher than that by nutrient flow culture. Spore production in the culture of 0.5 strength Hoagland's solution containing ¼ amount of phosphorus was better than that of 0.5 strength Hoagland's solution containing the normal amount of phosphate or Taifei nutrient solution. Among the three species of VAM fungi used in the study, the spore production of *G. etunicatum* and

G. fasciculatum was more abundant than that of *G. mosseae*. VAM fungal chlamydospores were produced on the surface of adzuki bean roots in erlenmeyer flasks, 60 days after incolulation with VAM fungi (Wang and Tschen 1994).

AM inoculum was produced using an aeroponic system consisting of a tank with 200 litres of nutrient solution pulverizing the root plants through micro-irrigation nozzles. During the first 28 days of cultivation, the plants suffered an adaptation stress with a significant reduction of the root colonization rate and spore number. However, after this adaptation period, the AMF colonization rate increased significantly, stabilizing at 56 to 72 days in sweet potato plants inoculated with *Entrophospora colombiana* with 72% of colonization. The number of spores recovered in the root system was 5082 and 156336 per g of root dry matter at 70 and 98 days, respectively. However, with *Gigaspora margarita* and *Glomus etunicatum*, the values were lower than expected (De Souza et al., 1996).

Mohammad et al. (2000) compared conventional atomizing disc aeroponic technology with ultrasonic nebulizer technology for the production of *Glomus intraradices* inocula. The piezo-ceramic element technology used in the ultrasonic nebulizer employs high frequency sound waves to nebulize nutrient solution into micro droplets, 1 micrometer in diameter. The growth of precolonized AM roots of sudan grass was achieved in both chambers used but both root growth and mycorrhization were significantly faster and more extensive in the ultrasonic nebulizer system than in the atomizing disc system. Shearing of the roots infected by AM fungi in both the systems did not reduce inoculum viability as was evident from the MPN data. However, sheared roots from the ultrasonic nebulizer system had significantly more infective propagules than those produced in the atomizing disc system. Thus, the latest ultrasonic nebulizer aeroponic technology appears to be superior and an alternative to the conventional atomizing disc or spray nozzle system for the production of high quality AMF inocula. These inocula can be used in small doses to produce a large response which is a prerequisite for the commercialization of AMF technology.

The aeroponically grown fungal infected roots can be sheared to produce a fine powdered AM inocula consisting of spores, hyphae and roots with numerous intraradical vesicles, which can be formulated as alginate beads to give sheared-root inocula a more versatile from root fragments of AM fungi to regenerate in alginate beads and infected roots (Strullu and Romand 1987; Strullu and Plenchette 1991). AM fungal infected roots produced in aeroponic system allows better extraction of viable propagules such as spores, hyphae and roots and can be used as a slurry or dried root fragments (Jarstfer and Sylvia 1992). Periodic nutrient change can reduce the population of deleterious organisms and the concentration of toxins that may occur in the nutrient solution. Frequent root pruning can reduce interplant root contact and could aid in sanitation. Being relatively a closed system, many environmental parameters can be easily controlled. Hydrogen ion and P-concentrations are most obvious, but temperature, atmospheric gases, plant growth regulators, and the microbial community can more easily manipulated than in culture systems that employ physical substrates. The inocula produced by this technique costs very less and allows a grower to economically sell plants colonized by mycorrhizal fungi (Jarstfer and Sylvia 1992).

3.2.3 Axenic culture

Propagation of AM fungi under axenic condition continues to be a challenging goal of modern plant biology. Empirical tests of many chemical substances and physiological conditions failed to provide sustained hyphal growth from germinated spores in single organ culture. In recent years it has been possible to grow AM fungi in sterile conditions in root-organ culture (Fortin et al., 2002).

The establishment of *in vitro* root organ culture has great potential in understanding the arbuscular mycorrhizal symbiosis and monoxenic inoculum production (Chandran and Potty 2009). Root organ culture was first developed by White and co-workers (White 1943; Butcher and Street 1964; Butcher 1980). These authors used excised roots on synthetic mineral media supplemented with vitamins and a carbohydrate source. Pioneering work by Mosse and Hepper (1975) used root cultures obtained from tomato and red clover to establish *in vitro* mycorrhiza with *Glomus mosse*. The authors demonstrated for the first time that spores of an AM fungus could be successfully used to colonize excised roots growing on a mineral based medium. Later, Strullu and Romand (1986, 1987) showed that it was also possible to establish mycorrhiza on excised roots of strawberry, onion and tomato using the intraradical phase (i.e., vesicles or entire Mycorrhizal root pieces) of several species of *Glomus* as inoculums. Low mineral media minimal (M) and modified Strullu–Romand (MSR) media were also successfully used to obtain mycorrhiza and fungal sporulation using nontransformed tomato root cultures (Chabot et al., 1992a; Diop et al., 1994a, 1994b; Bago et al., 1996). *In vitro* grown plantlet of *Ziziphus mummularia* inoculated with spores of *Glomus deserticola* and after 40 days maintenance in flasks under controlled conditions, the plantlets were transferred to sand in pots and maintained in a green house. Spores of *G. deserticola* proliferated vigorously under *in vitro* conditions and produced external and internal hyphae, vesicles and arbuscules (Mathur and Vyas, 1995).

A natural genetic transformation of plants by ubiquitous soil bacterium *Agrobacterium rhizogenes* (Riker et al., 1930) produces a condition known as hairy roots. This stable transformation (Tepfer 1989) produces Ri T-DNA transformed plant tissues that are morphogenetically programmed to develop as root. The bacterium, *Agrobacterium rhizogenes* inserts in transformed roots copies of T-DNA which occur in a large plamid of *A. rhizogenes* (Chilton et al., 1982). Therefore, transformed roots have a quick, vigorous and homogenous growth in relatively poor substrates without supplementation of hormonal substances. The negative geotropism of transformed root facilitates contacts with hyphae of AM fungi (Becard and Fortin 1988; Mungnier and Mosse 1987; Genre et al., 2005). Diop (1990) observed less contamination in culture of transformed carrot roots due to negative geotropism and Tepfer (1984) indicated that they can survive for a long time without subculture.

Daucus carota L. (carrot) and *Convolvulus sepium* L. (bindweed) were among the earliest species to be transformed using *Agrobacterium rhizogenes* Conn. (Tepfer and Tempé 1981). These Ri T-DNA transformed roots have since served in a wide range of fundamental and applied studies. One of the most important of these has been the study of the AM symbiosis. The first culture of hairy roots colonized by an AM fungus was achieved by Mugnier and Mosse (1987). These authors successfully colonized *Convolvulus sepium* hairy roots using spores of *G. mosseae* but, as was the case with nontransformed clover root-organ cultures (Hepper and Mosse 1975), no sporulation occurred. The first *in vitro* sporulation of an AM fungus was obtained by Bécard and Fortin (1988) using carrot hairy roots colonized by *Glomus intraradices.* Spore production followed reductions in the concentration of certain nutrients in the culture medium (see Bécard and Piché 1992) that allowed mycorrhizal inhibition to be avoided, but did not affect root growth and development. This led to the production of reproducible monoxenic cultures of *G. intraradices* that were characterized by large quantities of mycelium and spores (Diop et al., 1992). In recent years Ri T-DNA transformed hairy roots from tomatoes (de Souza and Berbara 1999), clover (Bago et al., 1998), *Manihot esculenta* (Potty and Chandran 2001), *Ipomoea batatas* (Chandran and Potty 2009) and *Vigna vexillata* (Chandran and Potty 2010) were used successfully for co-cultivation AM fungi. These plant species could provide alternative for mass production of AM Fungi axenically.

Many AM fungal species have been successfully grown in Ri-TDNA transformed roots. They included *Gigaspora margarita* (Diop et al., 1992; Verma and Adholeya 1995; Gadkar and Adholeya 2000); *Glomus etunicatum* (Schreiner and Koide 1993; Pawlowska et al., 1999), *Gigaspora gigantea* (Gadkar et al., 1997), *Gigaspora rosea* (Bago et al., 1998), *Acaulospora rehmii* (Dalpe and Declerck 2002), *Glomus intraradices* (Chabot et al., 1992; St-Arnaud et al., 1996; Jolicoeur et al., 1999), *Glomus versiforme* (Diop et al., 1994; Declerck et al., 1996), *Glomus caledonium* (Karandashov et al. 2000), *Glomus fasciculatum* and *Glomus macrocarpum* (Declerck et al., 1998), *Glomus proliferum* (Declerck et al., 2000), *Glomus deserticola* (Mathur and Vyas 1995), *Glomus fistulosum* (Gryndler et al., 1998), *Glomus microcarpum* (Potty and Chandran 2001; Chandran and Potty 2010), *Glomus sinuosum* (Bi et al. 2004), *Gigaspora gigantia* (Douds and Becard 1993), *Glomus vesiforme* (Declerck et al., 1996). Tiwari et al., (2002) reported for the first time co-culture of two AM isolates *Gigospora margarita* and *Glomus intraradices* on RiT-DNA transformed carrot roots. The co-culture model under the *in vitro* system appears as model approach towards mass production of multiple mycorrhizal fungal isolates.

All types of AM propagules (spores, vesicles, mycelia, sheared mycorrhizal roots) can be used to initiate *in vitro* cultures. Chlamydospores of Glomus species (Mosse and Hepper 1975; Mugnier and Mosse 1987; Budi et al., 1999) and non-sporocarpic azygospores of *Gigaspora margarita* (Becard and Fortin 1988; Becard and Piche 1989a; Diop et al., 1992) are the preferred inoculums starter. However, the inabilities of some AM fungi to produce spores (Johnson 1977; McGee 1989) use of intraradical forms seem to be a good alternative to establish AM symbiosis. As intravesicles mycorrhizal root acts as reserves and propagules, they have a high inoculums potential than other AM propoagules such as spores and hyphae (Magrou 1946; Bierman and Liderman 1983; Strullu and Romand 1986). Several authors have used intraradical forms to achieve mass production of *Glomus* sp. (Diop 1995; Diop et al., 1994; Strullu et al., 1996; Decklerk et al., 1996a,1998). Before being used as *in vitro* inoculums, AM propagules must be properly surface sterilized. This step is critical because success depends upon elimination of all contaminants. Spores are often surface sterilized using two-step procedure of Metz et al., (1979) as modified by Becard and Fortin (1988) and AM sheared inocula according to Diop et al., (1994) method. A solution containing strong oxidising agent, chloramine T (2%) solution with traces of surfactant (Tween 20/80) and antibiotic such as streptomycin or gentamycin is widely used. Sterilized AM propagules must be stored at 4°C until use.

Pure culture of AM fungi is still a big challenge. The pre-symbiotique growth of AM is characterized by the formation of so called running hyphae. After some weeks without additional host partner, growth of germinated AM propagule stops, the hyphae septate and the cytoplasm retract. Several factors such as chemical treatments (Gianinazzi–Pearson et al., 1989; Becard and Piche 1989a), genetical factors (Burggraaf and Beringer, 1989; Becard and Pfeffer 1993; Bianciotto and Bonfante 1993) and nutrition (Hepper 1983) have been studied to explain the lack of pure growth of the extraradical phase of AM fungi. Low amounts of natural or synthesized flavonoids positively stimulated growth (Morandi 1989, Chabot et al., 1992b). Otherwise, certain amino acids (Hepper and Jacobsen 1983), volatile compounds, particularly carbon dioxide and root exudations, promote elongation (Carr et al., 1985; Elias and Safir, Becard and Piche 1989b).

Hildebrant et al. (2002) observed the presence of slime forming bacteria *Paenibacillus validus* on the surface sterilized spores of *Glomus intraradices*. These bacteria stimulate the growth of this fungus up to spore formation in the absence of any plant tissue. Similar results on presymbiotic sporulation have been reported with other *Glomus* sp (Diop 2003). The authors explained this unusual sporulation by possible utilization by the fungi of a chemical component (or components) secreted by the bacteria. Buee et al. (2000) isolated a semi-purified fraction called "branched factor" from 8 mycotrophic plant species that are very active in stimulation of germination and nuclear division of *Gigaspora* sp.

4 CONCLUSION

AM fungi are ubiquitous soil inhabitants forming symbioses with most naturally growing terrestrial and most aquatic plants. Their potential to enhance plant growth is well documented and recognized but not fully exploited. They are rarely found in nurseries due to the use of composted soil-less mixes, high levels of fertilizer and regular application of fungicidal drenches. The potential advantages of inoculation of nursery plants with AM fungi in horticulture, agriculture, floriculture and forestry is not perceived by these industries as a significant issue. This is partially due to inadequate methods for large-scale inoculum production. Thus, the exploitation of AM fungi still presents a great challenge. Pot culture in pasteurized soils has been the most widely used method for producing AMF inocula but it is time consuming, bulky, and often not pathogen free. To overcome these problems, various soil-free methods, namely, soil-less growth media aeroponic system, hydroponic and axenic culture of AM fungi with transformed or non-transformed living roots of various hosts have been used successfully to produce AMF-colonized root inocula. But efforts should be made to produce AM inoculum in the field using indigenous isolates, if it supports better growth of the desired crop species. Production of AM fungus inoculum on the farm potentially offers several important advantages. In future, plant growth promoting rhizobacteria (PGPR), mycorrhization helper bacteria (MHB), etc., may be incorporated in AM production strategies to develop multi-agent inoculum so that plants achieve maximum benefits. This area merits greater attention.

References

Addy HD, Miller MH and Peterson RL (1997). Infectivity of the propagules associated with extraradical mycelium of two fungi following winter freezing. *New Phytol* 135: 745–753.

Bago B, Pfeffer PE and Shachar-Hill Y (2000). Carbon metabolism and transport in arbuscular mycorrhizas. *Plant Physiol* 124: 949–957.

Bago B, Vierheilig H, Piche Y and Azcon-Aguilar C (1996). Nitrate depletion and pH changes induced by the extraradical mycelium of the arbuscular mycorrhizal fungus, *Glomus intraraduces* grown in monoxenic culture. *New phytol* 133: 273–280.

Bago B, Azcon-Aguilar C, Goulet A and Piche Y (1998). Branched absorbing structures (BAS): a feature of the extra radical mycelium of symbiotic arbuscular mycorrhizal fungi. *New Phytol* 139: 375–388.

Bago B, Azcon-Aguilar C and Piche Y (1998). Architecture and developmental dynamics of the external mycelium of the arbuscular mycorrhizal fungus *Glomus intraradices* grown under monoxenic conditions. *Mycologia* 90: 52–62.

Bagyaraj DJ (1990). *Glomus fasciculatum* inoculums: its shelf life and the possibility of introducing rhizobium in it. In: Allen MF (ed) Innovation and Hierarchical Integration University of Wyoming, *Wyoming* 16.

Baltruschat H (1987a). Field inoculation of maize with vesicular arbuscular mycorrhizal fungi by using expanded clay as carrier material for Mycorrhiza. *J Plant dis Protect* 94: 419–430.

Baltruschat H (1987b). Evaluation of the suitability of expanded clay as carrier material for VA Mycorrhiza spores in field inoculation of maize. *Angew Botanic* 61: 163–169.

Becard G and Fortin JA (1988). Early events of vesicular arbuscular mycorrhiza formation on Ri T-DNA transformed roots. *New Phytol* 108: 211–21.

Becard G and Piche Y (1989a). Fungal growth stimulation by CO_2 and roots exudates in vesicular-arbuscular Mycorrhizal symbiosis. *Appl Environ Microbial* 55: 2320–2325.

Becard G and Piche Y (1989b). New aspects of the acquisition of biotrophic status by a vesicular-arbuscular mycorrhizal fungus *Gigospora margarita*. *New Phytol* 112: 77–83.

Biermann B and Linderman RG (1983). Use of vesicular-arbuscular mycorrhiza roots, intraradicle vesicles and extraradicle vesicles as inoculums. *New Phytol* 95: 97–105.

Bil Y, Li X, Wang H and Christie P (2002). Establishment of monoxenic culture between the arbuscular Mycorrhizal fungi *Glomus sinousum* and Ri T-DNA-transformed carrot roots. *Plant and Soil* 261: 239–244.

Brechelt A (1988). Einflub verschiedener organischer Dungemittel auf die Effizienz der VA Mykorrhiza und einger N_2 fixierender Bakterien im Boden bei tropischen und sibtropischen Pflanzen. Diss Gottinger Beitrage zur Land-und Forstwirtschen in den Tropen und Subtropen, Hoft 32. Margraf Gaimersheim.

Buee M, Rossignol M, Jauneau A, Ranjeva R and Becard G (2000). The pre-symbiotic growth of arbuscular Mycorrhizal fungi is reduced by a branching factor partially purified from plant root exudates. *Am Phytopathol Soc* 6: 693–698.

Burggraaf AJP and Beringer JE (1989). Abence of nuclear DNA synthesis in vesicular-arbuscular mycorrhizal fungi during *in vitro* development. *New Phytol* 11: 25–33.

Butcher DN (1980). The culture of isolated roots. In: Ingram DS and Helgelson JP (eds) Tissue culture methods for plant pathologists. Blackwell Scientific, Oxford. 13–17.

Butcher DN and Street HE (1964). Excised root culture. *Bot. Rev.* 30: 513–586.

Cabello M, Irrazabal G, Bucsinszky AM, Saparrat M and Schalamuk S (2005). Effect of an arbuscular mycorrhizal fungus, *Glomus mosseae*, and a rock phosphate-solubilizing fungus *Penicillium thornii*, on *Mentha piperita* growth in a soilless medium. Journal of Basic Microbiology 45: 182–189. doi: 10.1002/jobm.200410409.

Caar GR, Hinkley A, Le Tacon F, Hepper CM, Jones MGK and Thomas E (1985). Improved hyphal growth of two species of vesicular-arbuscular Mycorrhizal fungi in the presence of suspension-cultured plants cells. *New Phytol* 101: 417–426.

Chabot S, Becard G and Piche Y (1992). Life cycle of *Glomus intraradix* in root organ culture. *Mycologia* 84: 315–321.

Chambers CA, Smith SE and Smith FA (1980). Effects of ammonium and nitrate ions on mycorrhizal infection, nodulation and growth of *Trifolium subterraneum*. *New Phytol*. 85: 47–62.

Chandran PR and Potty VP (2009). Arbuscular mycorrhizal fungal inoculums production using *Ipomoea batatas* hairy roots in bioreactor. *IUP Journal of Biotechnology* 3: 356–360.

Chandran PR and Potty VP (2010). *In vitro* co-culture of VAM fungi *Glomus microcarpum* in Ri T-DNA transformed hairy roots of *Vigna vexillata*. *IJAEB* 3: 333–338.

Chilton MD, Tepfer DA, Petit A, David C, Casse-Delbart F and Tempe J (1982). *Agrobacterium rhizogenes* inserts T-DNA into the genomes of the host plant root cells. *Nature* 295: 432–434.

Cooper AJ (1979). The ABC of NFT. *Grower book, London* 2–40.

Cooper AJ (1985). Crop production in recirculating nutrient solution. *Sci Hortic* 3: 251–258.

Coltman RR, Warerer DR and Huang RS (1988). A simple method for production of *Glomus aggregatum* inoculums using controlled released fertilizer. *Hortic Sci* 23: 213–215.

Dalpe Y and Declerck S (2002). Development of *Acaulospora rehmii* spore and hyphal swellings under root-organ culture. *Mycologia* 94: 850–855.

Dehne HW and Backhaus GF (1986). The Use of vesicular arbusccular Mycorrhizal fungi in plant production. I. Inoculums production. *J Plant Dis Protect* 93: 415–424.

De Souza ES, Burirty HA, Santo ACDE and daSiva MLRB (1996). Alternative for arbuscular Mycorrhizal fungi inoculum productionin aeroponic culture. Pesqu. *Agropecu. Bras.* 31: 153–158.

Declerck S, Strullu DG and Plenchette C (1996a). *In vitro* mass production of the arbuscular mycorrhizal fungus *Glomus versiforme,* associated with Ri T-DNA transformed carrot roots. *Mycological Research* 100: 1237–1242.

Declerck S, Strullu DG, Plenchette C and Guillemette T (1996b). Entrapment of *in vitro* produced spores of *Glomus versiforme* in alginate beads: *in vitro* and *in vivo* inoculum potentials. *J Biotechnol* 48: 51–57.

Declerck S, Strullu DG and Plenchette C (1998). Monoxenic culture or the intra radical forms of *Glomus* sp. isolated from a tropical ecosystem. A proposed methodology for germplasm collection. *Mycologia* 90: 579–565.

Declerck S, Cranenbrouck S, Dalpe Y, Seguin S, Grandmougin–Ferjani A, Fontaine J, and Sancholle M (2000). *Glomus proliferum* sp. *nov.,* A description based on morphological, biochemical, molecular and monoxenic cultivation data. *Mycologia* 92: 1178–1187.

Diop TA (1995). Ecologie des champignons mycorhiziens a vesicules et a arbuscules associes a *Acacia albida* (del) dans les zones saheliennes et soudano-guineennes du Senegal. These de docteur en Biologie et Physiologie vegetales, Angers, France.

Diop TA (2003). *In vitro* culture of arbuscular mycorrhizal fungi: advances and future prospects. *Afr J Biotechnol* 2: 692–697.

Diop TA, Becard G and Piche Y (1992). Long term *in vitro* culture of an endomycorrhizal fungus, *Gigaspora margarita* on Ri T-DNA transformed root of carrot. *Symbiosis* 12: 249–259.

Diop TA, Plenchette C and Strullu DG (1994a). Dual axenic culture of sheared-root inocula of vesicular arbuscular mycorrhizal fungi associated with tomato roots. *Mycorrhiza* 5: 17–22.

Diop TA, Plenchette C and Strullu DG (1994b). *In vitro* culture of sheared mycorrhizal roots. *Symbiosis* 17: 217–227.

Douds D and Becard G (1993). Competitive interaction between *Gigaspora margarita and Gigaspora* gigantia *in vitro.* In: Proceedings of the ninth North America symposium on Mycorrhizae. Guelph, ON, Canada, 8–12.

Douds Jr. DD, Nagahashi G, Pfeffer PE, Kayser WH and Reider C (2005). On-farm production and utilization of mycorrhizal fungus inoculum. *Can J Plant Sci* 85: 15–21.

Douds Jr. DD, Nagahashi G, Pfeffer PE, Reider C and Kayser WH (2006). On-farm production of AM fungus inoculums in mixtures of compost and vermiculite. *Bioresour Technol* 97: 809–818.

Douds Jr. DD, Nagahashi G, Reider C and Hepperly PR (2007). Inoculation with arbuscular mycorrhizal fungi increases the yield of potatoes in high P soil. *Biol Agric Hortic* 25: 67–78.

Douds Jr. DD, Nagahashi G, Reider C and Hepperly PR (2008a). Choosing a mixture ratio for the on-farm production of AM fungus inoculum in mixture of compost and vermiculite. *Compost Sci Util* 16: 52–60.

Douds Jr. DD, Nagahashi G, Shenk JE and Demchak K (2008b). Inoculation of strawberries with AM fungi produced on-farm increased yield. *Biol Agric Hortic* 25: 67–78.

Douds Jr. DD, Nagahashi G and Hepperly PR (2010). On-farm production of inoculums of indigenous arbuscular mycorrhizal fungi and assessment of diluents of compost for inoculums production. *Bioresour Technol* 101: 2326–2330.

Douds Jr. DD and Schenck NC (1990a). Increased sporulation of vesicular arbuscular mycorrhizal fungi in plant production. *J Plant Dis Protect* 93: 415–418.

Douds Jr. DD and Schenck NC (1990b). Cryopreservation of spores of vesicular-arbuscular mycorrhizal fungi. *New Phytol* 115: 667–674.

Dugassa DG, Grunewaldt-Stocker G and Schonbeck F (1995), Growth of *Glomus intrardices* and its effect on linseed (*Linum usitatissimum* L) in hydroponic culture. *Mycorrhiza* 5: 279–282.

Elmes RP and Mosse B (1994). Vesicular arbuscular Endomycorrhiza production II. Experiments with maize (*Zea mays*) and other hosts in nutrient flow culture. *Can J Bot* 62: 1531–1536.

Elmes RP, Hepper CM, Hayman DS and O'shea J (1984). The use of vesicular arbuscular mycorrhizal roots by nutrient film technique as inoculum for field sites. *Ann Appl Biol* 104: 437–441.

Elias KS and Safir GR (1987). Hyphal elongation of *Glomus fasiculatus* in response to root exudates. *Appl Environ Microbial* 53: 1928–1933.

Ferguson JJ and Menge JA (1982). Factors that affect production of endomycorrhizal inoculum. *Proc Fla State Hortic Soc* 95: 37–39.

Ferguson JJ and Woodhead SH (1982). Increase and maintenance of vesicular-arbuscular mycorrhizal fungi. In: Schenck NC (ed) Methods and principles of mycoohizal research. *American Phtopathological Society* 47–54.

Fortin JA, Becard G, Declerck S, Dalpe Y, St Arnaud M, Coughlan AP and Piche Y (2002). Arbuscular Mycorrhiza on root organ culture. *Can J. Bot* 80: 1–20.

Gaur A (1997). Inoculum production technology development of Vesicular Arbuscular Mycorrhizae. Ph.D. Thesis. University of Delhi, Dehli, India

Gaur A and Adholeya A (2000). Effect of the particle size of soil-less substrate upon AM fungus inoculum production. *Mycorrhiza* 10: 43–48.

Gaur A and Adholeya A (2002). Arbuscualr Mycorrhizal inoculation of five tropical fodder crops and inoculum production in marginal soil amended with organic matter. *Biol Fert Soils* 35: 214–218.

Govindarajulu M, Pfeffer PE, Jin H, Abubaker J, Douds DD, Allen JW, Bu¨cking H, Lammers PJ and Shachar-Hill Y (2005). Nitrogen transfer in the arbuscular mycorrhizal symbiosis. *Nature* 435: 819–823.

Gadkar V, Adholeya A and Satyanarayana T (1997). Randomly amplified polymorphic dna using the m13 core sequence of the vesicular-arbuscular Mycorrhizal fungi *Gigaspora margarita* and *Gigospora gigantean*. *Can J Microbiol* 43: 795–798.

Gryndler M, HrseLova H, Chvatalova L and Vosatka M (1998). *In vitro* proliferation of *Glomus fistulosum* intraradical hyphae from mycorrhizal root segments of maize. *Mycol Res* 102: 1067–1073.

Guttay AJR (1983). The interaction of fertilizers and vesicular-arbuscular mycorrhizae in composted plant residues. *J Amer Soc Hort Sci* 108: 222–224

Hayman DS (1987). VA mycorrhizas in field crop systems. In: Safir G (ed) Ecophysiology of VA mycorrhizal plants, CRC Press, Inc., Boca Raton, Florida.

Harrison MJ, Dewbre GR and Liu JY (2002). A phosphate transporter from *Medicago truncatula* involved in the acquisiton of phosphate released by arbuscular mycorrhizal fungi. *Plant Cell* 14: 2413–2429.

Hildebrandt U, Janetta K and Bothe H (2002). Towards growth of arbuscular mycorrhizal fungi independent of a plant host. *Appl Environ Microbial* 68: 1919–1924.

Hung LL and Sylvia DM (1987). VAM inoculum production by an aeroponic culture. In: Sylvia DM, Hung LL and Graham JH (eds) proceedings of the Seventh North American Conference on Mycorrhiza. University of Florida. Florida 364.

Hung LL and Sylvia DM (1988). Production of vesicular arbuscular mycorrhizal fungi in aeroponic culture. *Appl Environ Microbiol* 54: 353–357.

Jarstfer AG and Sylvia DM (1992). Inoculum production and inoculums strategies for vesicular arbuscular mycorrhizal fungi. In: Meeting B (ed) Soil microbial technologies: applications in agriculture, forestry and environmental management. *Marcel (New York: Decker)* 349–377.

Jarstfer AG and Sylvia DM (1994). Aeroponic culture of VAM fungi. In: Varma A and Hock B (eds) Mycorrhiza: structure, function, molecular biology and biotechnology. Springer-Verlag, Berlin. 427–441.

Johnson PN (1977). Mycorrhizal endogonaceae in a New Zealand forest. *New Phytol* 78: 161–170.

Karandashov V, Kuzovkina I, Hawkins HJ and George E (2000). Growth and sporulation of the arbuscular mycorrhizal fungus *Glomus caledonium* in dual culture with transformed carrot roots. *Mycorrhiza* 10: 23–28.

LEE Yun-Jeong and George Eckhard (2005). Germplasm Acquisition from Conceptions to products. Workshop, providence, Fodhe Island, ETATS-UNIS (03/10/2003) 40: 378–380.

Liyanage HD (1989). Effects of phosphorus nutrition and host Species on root colonization and sporulation by Vesicular-Arbuscular (VA) mycorrhizal fungi in sand-vermiculite medium. M.S. thesis, University of Florida, Gainesville.

Magrou J (1946). Sur la culture de quelques champignons de mycorhizes a arbuscules et al., vesicules. *Rev Gen Bot* 53: 49–77.

Martin F, Perotto S and Bonfante P (2000). Mycorrhizal fungi: A fungal community at the interface between soil and roots. In: Pinton R, Varanini Z and Nannipieri P (eds) The Rhizosphere: Biochemistry and organic substances at the soil-plant interface. New York/Basel: Marcel Dekker 263-/296.

Mathew J and Johri BN (1987). Vesicular arbuscular mycorrhizal inoculum through nutrient film technique (NFT) grown moong (*Phaseolus mungo*). In: Sylvia DM, Hung LL and Graham JH (eds) proceedings of the seventh North American Conference on Mycorrhiza, Florida, University of Florida. 364.

Mathew J and Johri BN (1988). Propagation of vesicular-arbuscular mycorrhizal fungi in moong (*Phaseolus radiata* L.) through nutrient film technique (NFT). *Curr Sci* 57: 156–158.

Mathur N and Vyas A (1995). *In vitro* production of *Glomus deserticola* in association with *Ziziphus nummularia. Plant Cell Rep* 14: 735–737.

McGree PA (1989). Variation in propagules numbers of vesicular-arbucular mycorrhizal fungi in semi arid soil. *Mycol Res* 92: 28-33

Menge JA (1982). Effect of soil fumigants and fungicides on vesicular-arbuscular fungi. *Phytopathology* 72: 1125–1131.

Menge JA (1983). Utilization of vesicular arbuscular mycorrhizal fungi in agriculture. *Can J Bot* 61: 1015–1024.

Menge JA (1984). Inoculums production. In: Powell CL and Bagyaraj DJ (eds), VA Mycorrhiza. CRC press, Inc., Boca Raton, Florida 187–203.

Menge JA (1984). The feasibility of using VA mycorrhizal fungi to improve crop production in fumigated and non-fumigated soils. In: Fergusson JJ (ed) Application of mycorrhizal fungi in crop production. University of Florida, Gainesville, FL, 24–34.

Mertz SM, Heithaus lll JJ and Bush RL (1979). Mass productionof axenic spore of the endomycorrhizal fungus *Gigaspora margarita*. *Trans Br Mycol Soc* 72: 167–169.

Miller PD and Kitt DG (1992). The Beltsville method for soilless production of vesicular arbuscular mycorrhizal fungi. *Mycorrhiza* 2: 9–15.

Mohammad A, Khan AG and Kuek C (2000). Improved aeroponic culture of inocula of arbucular Mycorrhizal fungi. *Mycorrhiza* 9: 337–339.

Morton JB, Bentivenga SP and Wheeler WW (1993). Germplasm in the International collection of Arbuscular Mycorrhizal Fungi. (INVAM) and production for culture development, documentation and storage. *Mycotaxon* 48: 491–528.

Mosse B (1981). Vesicular arbuscular mycorrhiza research for tropical agriculture. Research Bulletin 194, Hawaii Institue of Tropical Agriculture and Human Resources, University of Hawaii.

Mosse B (1988). Some studies relating to "independent" growth of vesicular-arbuscular endophytes. *Can J Bot* 66: 2533–2540.

Mosse B And Hepper C (1975), Vesicular arbuscular Mycorrhizal infections in root organ cultures. *Physiol Plant Pathol* 5: 215–223.

Mosse B, Stribley DP and Tacon F (1981). Ecology of mycorrhizae and mycorrhizal fungi. *Advances in Microbial Ecology* 5: 137–210.

Mosse B and Thompson JP (1984). Vesicular arbuscular endomycorrhizal inoculums production. I. Exploratory experiments with beans (*Phaseolus vulgaris)* in nutrient flow culture. *Can J Bot* 62: 1523–1530.

Mugnier J and Mosse B (1987). Vesicular arbuscular Mycorrhizal infection in transformed root-inducing T-DNA roots grown axenically. *Phytopathology* 77: 1045–1050.

Nemec S (1987). VA myvorrhizae in horticulture systems. In: Ecophysiology of VA Mycorrhizal plants. Safir GR (ed) CRC Press, Inc., Boca Raton, 193–211.

Nelsen CE and Safir GR (1982). Increased drought tolerance of Mycorrhizal onion plants caused by improved phosphorus nutrition. *Plants* 154: 407–413.

Oseni TO, Shongwe NS and Masarirambi MT (2010). Effect of arbuscular mycorrhiza (AM) inoculation on the performance of tomato nursery seedlings in vermiculite. *Int J Agric Biol* 12: 789–792.

Pawlowakai TE, Doud Jr DD and Charvat I (1999). *In vitro* propagation and life cycle of the arbuscular mycorrhizal fungus *Glomus etunicatum*. *Mycol Res* 103: 1549–1556.

Potty VP and Chandran PR (2001). Growth and establishment of AM fungi *Glomus microcarpum var. microcarpum* in genetically transformed roots of cassava *(Manihot esculenta* Crantz). Proceedings of the 5th international meeting of cassava biotechnology network. November 4–9, St. Louis Mo, USA.

Potty VP and Chandran PR (2010). *In vitro* culture of VAM Fungi *Glomus microcarpum* in Ri T-DNA transformed hairy roots of *Vigna vexillata.* IJAEB 3: 333–338.

Remy W, Taylor TN, Hass H and Kerp H (1994). Four hundred million year old vesicular arbuscular mycorrhizae. *Proceedings of the National Academy of Sciences,* USA 91: 11841–11843.

Resh HM (1978). Hydroponics food production. Woodbridge Press Publication 73.

Riker AJ, Banfield WM, Wright WH, Keitt GW and Sagen HE (1930). Studies on infectious hairy root of nursery apple trees. *J Agric Res* 41: 507–540.

Saif SR (1986). Vesicular-arbuscular mycorrhizae in tropical forage species as influenced by season, soil texture, fertilizers, host species and ecotypes. *Angew Botanik* 60: 125–139.

Same BI, Robson A and Abbott LK (1983). Phosphorus, soluble carbohydrates and endomycorrhizal infections. Soil Bio Biochem 15: 593–597.

Sieverding E (1987). A VA mycorrhizal fungus, *Glomus glomerulatum* sp. nov., with two hyphal attachments and spore formed only in sporocarps. *Mycotaxon* 29: 73–79.

Sieverding E (1991). Vesicular-arbuscular mycorrhiza management in tropical agrosystems. Deutsche Gesellschaft fur Technische Zusammansabeit (GT2) GmbH. Eschbon, Germany.

Simon L, Bousquet J, Levesque RC and Lalonde M (1993). Origin and diversification of endomycorrhizal fungi and coincidence with vesicular land plants. *Nature* 363: 67–69.

Schreiner RP and Koide R (1993). Stimulation of vesicular arbuscular mycorrhizal fungi by mycotropic and non-mycotrophic plant root systems. *Appl Environ Microbiol* 59: 2750–2752.

Schuessler A, Schwarzott D and Walker C (2001). A new fungal phylum, the Glomeromycota: *Phylogeny and evolution. Mycol Res* 105: 1413–1421.

Srinivasan MN and Bagyaraj DJ (1987). Selection of suitable substrate for mass multiplication of *Glomus fasiculatum*. In: Mycorrhiza Round Table: Proceeding of a workshop (Eds Verma AK, Oka AK, Mukerji KG, Tilak KVBR, and Janki Raj), New Delhi, 592–599.

Srinivasan MN and Bagyaraj DJ (1988). Optimum frequency of addition of nutrient solution for mass production of *Glomus fasiculatum* inoculum. *Current Science* 57: 665–667.

Srinivasan MN and Bagyaraj DJ (1989). Use of pesticides for mass production of vesicular arbuscular mycorrhizal inoculum. *Plant Soil* 119: 127–132.

Srinivasan MN and Bagyaraj DJ (1990). Suitable source and level of nitrogen for mass production of the VA Mycorrhizal fungus *Glomus fasiculatum.* In: Jalali BL and Chand H (eds) Current trends in Mycorrhizal Research. HAU, Hisar, Harayana. 35–36.

St-Arnaud M, Hamel C, Vimard B, Caron M and Fortin JA (1996). Enhanced hyphal growth and spore production of the arbuscular mycorrhizal fungus *Glomus intraradices* in an *in vitro* system in the absence of host roots. *Mycological Research* 100: 328–332.

Streiner AA (1976). Nomenculture with Hydroponics. Proc. Int. Con. on soilless culture (IWUSU) 4(1): 19–20.

Strullu DG and Plentchette C (1991). The entrapment of *Glomus* sp in alginate beads and their use as root inoculum. *Mycol Res* 95: 1194–1196.

Strullu DG and Romand C (1986). Methode d'obtention d'endomycorhizes an vesicules et a abuscles en conditions axeniques C.R. Acad. *Sci. Ser III Sci. Vie* 303: 245–250.

Strullu DG and Romand C (1987). Culture axenique de vesicules isolees a partir d'endomycorhizes et re-association *in vitro* a des raciness de tomate. *C.R. Acad. Sci Paris Ser.* III, 305: 15–19.

Suriyapperuma SP and Koske RE (1995). Attraction of germ tubes and germination of spores of the arbuscular Mycorrhizal fungus, *Gigaspora gigentea* in the presence of roots of maize exposed to different concentrations of phosphorus. *Mycologia* 87: 772–778.

Sylvia DM and Hubbell DH (1986). Growth and sporulation of vesicular arbuscular Mycorrhizal fungi in aeroponic and membrane system. *Symbiosis* 1: 259–267.

Sylvia DM and Jarstfer AG (1994). Sheared root inocula of vesicular arbuscular Mycorrhizal fungi. *Applied Environ Microbiol* 58: 229–232.

Sylvia DM and Schenck NC (1983). Application of super phosphate to Mycorrhizal plants stimulates sporulation of phosphorus tolerant vesicular arbuscular Mycorrhizal fungi. *New Phytol* 95: 655–661.

Talukdar NC and Germida JJ (1993). Propagation and storage of vesicular-arbuscular mycorrhizal fungi isolated from Saskatchewan agriculture soil. *Can J Bot* 71: 1328–1335.

Tarbell TJ and Koske RE (2007). Evaluation of commercial arbuscular Mycorrhizal inocula in a sand/peat medium. *Mycorrhiza* 18: 51–56.

Tarafdar JC (1995). Role of a VA Mycorrhizal fungus on growth and water relation in wheat in presence of organic and inorganic phosphorus. *J Ind Soc Soil Sci* 43: 200–204.

Tepfer D (1989). Ri T-DNA from *Agrobacterium rhizogenes*: a source of genes having applications in rhizosphere biology and plant development, ecology and evolution. In: Kosuge T and Nester EW (eds) Plant–microbe interactions. McGraw-Hill Publishing, New York. 294–342.

Tepfer DA and Tempé J (1981). Production d'agropine par des racines formées sous l'action d'*Agrobacterium rhizogenes*, souche A4. *C.R. Acad. Sci. Paris*, 292: 153–156.

Thompson JP (1986). Soilless cultures of vesicular arbuscular Mycorrhizae of cereals: Effects of nutrient concentration and nitrogen source. *Can J Bot* 64: 2282–2294.

Tiwari P and Adholeya A (2002). *In vitro* co-culture of two AM isolates *Gigospora margarita* and *Glomus intraradicles* on RiT-DNA transformed roots. *FEMS Microbiol lett* 206: 39–43.

Trappe JM, Molina R and Castellano MA (1984). Reactions of mycorrhizal fungi and mycorrhizal formation to pesticides. *Ann Rev Phytopathol* 22: 331–359.

Van der Heijden MGA, Klironomos JN, Ursic M, Moutoglis P, Streitwolf-Engel R, Boller T, Wiemken A and Sanders IR (1998). Mycorrhizal fungal diversity determines plant biodiversity, ecosystem variability and productivity. *Nature* 396: 69–72.

Van der Heijden MGA and Sanders IR (2002). Mycorrhizal ecology: Synthesis and perspectives. In: Van der Heijden MGA and Sanders IR (eds) Mycorrhizal ecology. Heidelberg: Springer-Verlag. 441-/456.

Vestberg M and Kukkonen S (2009). Microbiologically improved peat based media for nursery production by addition of Arbuscular Mycorrhizal fungi. Acta Hort. (ISHS) 819: 403-410 (http://www.actahort org/books/819_49:htm).

Vilarino A, Arines J and Schuepp H (1992). Propagule production by VA fungi in red clove plants subjected to periodic removal of the aerial parts. In: Read DJ, Lewis DH, Fitter AJ and Alexander IJ (eds) Mycorrhizas in Ecosystem. Wallingford, CAB International.

Warner A (1985). Mycorrhizal seed pellets. United States Patent 4, 551,165

Wang CL and Tschen JSM (1994), Production of mycorrhizal inoculum by soilless culture. Journal of the Agricultural Association of China 167: 50–60.

Weathers PJ and Zobel RW (1992). Aeroponics for the culture of organisms, tissue and cells. *Biotech Adv* 10: 93–115.

White PR (1943). A handbook of plant tissue culture. J. Cattel, Lancaster, Pa.

Zobel RW, Del Tredici P and Torrey JG (1976). Method for growing plants aeroponically. *Plant Physiol* 57: 344–346.

9

Physicochemical and Bacteriological Studies of Wetland Soils of Keoladeo National Park, Bharatpur, Rajasthan (India)

Narendra Kumar*[1&3], Nivedita Sharma[1] and Surendra Singh[2]

[1]School of Studies in Microbiology, Jiwaji University, Gwalior-474011, (M.P.) India

[2]Department of Botany, Banaras Hindu University, Varanasi-221005, (U.P.) India

[3]Department of Biotechnology, IMS Engineering College, Ghaziabad (U.P.) India

**Email: nkrathore1@gmail.com*

1 INTRODUCTION

Keoladeo National Park is a 29 km^2 area situated on the extreme western edge of the Gangetic basin that was once confluence of rivers Gambhir and Banganga in Bharatpur district in the state of Rajasthan (Fig. 1). It has a unique mosaic of habitats that include wetlands, woodlands, scrub forests, grasslands that supports an amazing diversity of both plant and animal species. The project site Keoladeo National Park is a man-made marshy wetland and is a designated Ramsar site at Bharatpur (Rajasthan). The soil of the wetland area is mainly thick alluvium with some patches of saline soil in the terrestrial areas. Wetlands are integral part of river basins and are considered one of the most productive ecosystems of the earth. They may include a wide range of habitats constituting about six percent of earth's surface. According to Ramsar convention, wetlands are areas of marsh, fen, peat land or water bodies, whether natural or artificial, permanent or temporary, static or flowing, fresh, brackish or salt including areas of marine water, the depth of which at low tide does not exceed six metres. The characteristics of wetlands are soils that remain waterlogged or are submerged underwater for whole or part of the year and the wetland biota depend upon and are adapted to this waterlogging or submergence during at least a part of their life cycle. Wetlands, as vital ecosystems, have influenced the development of human societies since historical times. They have been recognized as peat bogs, grass and sedge marshes, floodplains consisting of recent alluvial deposits bordering rivers, inland valleys, shallow ponds, mudflats and littoral areas of larger bodies of water which can be grouped into naturally occurring and man-made or anthropogenic wetlands (Gopal et al., 1982).

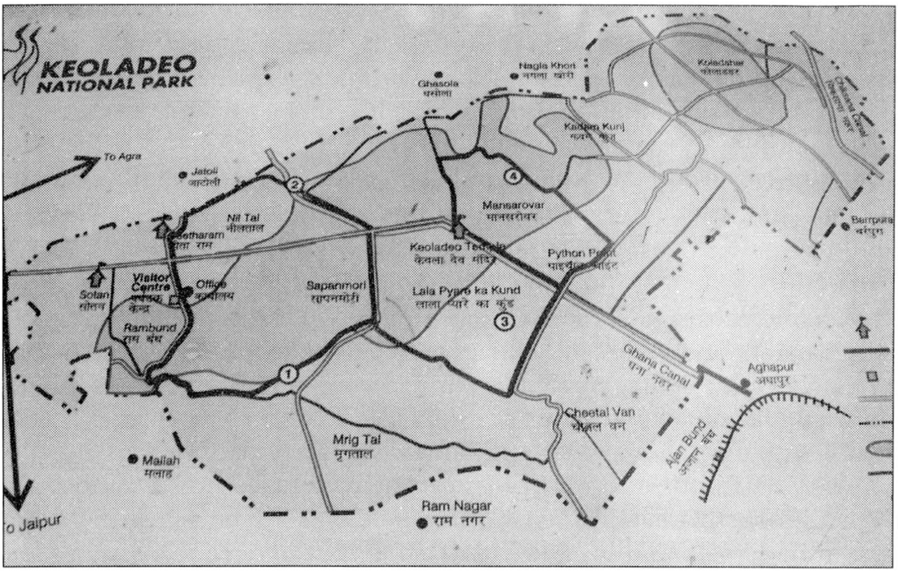

Fig. 1 Map showing the Keoladeo National Park

International efforts for protection of wetlands emerged in the background of environmental movements. A shift from the view that 'wetlands are wastelands' and hence to be developed, to the growing awareness of its intrinsic values and functions have resulted in several international efforts to conserve, protect and restore wetlands in recent decades (World Resources 2001). Considering the ecological values and benefits of wetlands, international scientific review boards and political leaders have promoted these benefits and called for their protection. 'Ramsar convention in 1971 and Rio Summit in 1992 were the milestones in the history of conservation of the wetlands (Denny 1994). This has resulted in re-designating some of the wetlands as 'wetlands of international importance', 'hemispheric reserves for shore birds' or 'conservation wetlands' (Wigham et al., 1993).

Microorganisms play an important role in the functioning of wetland ecosystem. They contribute towards circulation of matter/nutrient, by biogeochemical cycling, which ultimately reflects upon the eco-functional status of the ecosystem. Wetlands have been attributed for several ecological functions and their sustainability is of great importance as they contain rich biodiversity. Increase in natural and anthropogenic activities is, however, resulting in degradation of wetlands and as a consequence the microorganisms are continuously depleting from the wetlands. Keeping in mind the importance of microorganisms in wetland ecosystem, the present study was undertaken.

Annual rainfall, the amount of water received from Ajan bund, the time of its release and time of rainfall are some of the important factors which have an impact on the flora, fauna and other ecological parameters of Keoladeo National Park. Vijayan (1991) in his report has included quite a bit on the chemistry of water, flora, fauna and primary productivity of the Keoladeo National Park. Nevertheless, a systematic study on microbial flora of the park is needed. The poor rainfall and inadequate water supply from the Ajan bund led to a drought condition inside the park and therefore the whole aquatic area of the park has become completely dried and appeared as open grassland during the period of visit and study (August 2006 – November 2007).

Wetland ecosystems are the sites of rapid biogeochemical cycling due to the interactions between the surfaces soil (oxic) and deeper soils (anoxic) (Weiss et al., 2003). The understanding of carbon and nutrient cycling is important for regional scale restoration, as well as for global scale climate modeling efforts and protection of wetland systems (Gutknecht et al., 2006). Long lasting waterlogging results in reduced soil conditions in wetlands. The diffusion of gases in water is about 10,000 times slower than in air (Colmer 2003), and also the oxygen is depleted rapidly by roots, microorganisms, and soil reductants. The processes of oxygen depletion include denitrification, reduction of manganese, iron and sulphate, as well as changes in soil pH (Pezeshki 2001). Hence, wetland plants are adapted to survive anoxic conditions in oxygen limited soils or sediments by the formation of aerenchymatous tissue (Visser et al., 2000) in their roots, stems and leaves (Armstrong 1964).

Wetlands, both artificial and natural, have been used to remove potential toxic metals from contaminated drainage (Hallberg and Johnson 2005). Iron plaques have been observed on a number of wetland plants, including those growing in contaminated and polluted areas, which shows that these iron plaques act as deposits of heavy metals (Batty et al., 2000). In artificially constructed wetlands, bacteria have also been exploited for wastewater treatment (Hallberg and Johnson 2005; Nicomrat et al., 2006; Nicorarat et al., 2008; Faulwetter et al., 2009). Several researchers have studied the bacterial communities of these low-cost systems. Some based on specific bacterial groups, such as *Enterobacteriaceae* (Vacca et al. 2005), or methanotrophic bacterial populations (DeJournett et al., 2007). Other researches have focused on communities from constructed wetlands that treat acid coal mine drainage (Nicomrat et al., 2008) or wastewater with differing phosphorus loadings (Ahn et al., 2007). Nevertheless, it remains difficult to find works that make reference to the bacterial communities.

There has been paucity of information on the bacteriological and physicochemical characteristics of this wetland in particular. The purpose of this work was to characterize and physiochemical properties and the bacterial communities of the soil in Keoladeo National Park wetland which provide some baseline data for sustainable management of Keoladeo National Park wetland soils.

2 MATERIALS AND METHODS

2.1 Study Area

The project site, Keoladeo National Park is highly diverse in nature as far as its habitat is concerned. Looking its habitat diversity, the Park has been divided into six experimental sites for the periodic collection of soil samples. These sites represent the entire habitat of the park. The sites selected for the periodic collection of soil samples are named as follows: A block, B block, D block, E block, K block and L block.

2.2 Sample Collection

Soil samples were collected up to 0–6 cm depth from the surface with the help of sterilized iron borer, in sterilized polyethene bags previously marked with adhesive stickers following the method of Johnson and Curl (1972), and taken in ice-packed coolers to the laboratory for bacteriological and physicochemical analysis.

2.3 Soil pH

Calomel electrode method was used for determining the soil pH (Piper 1966). Soil was dispersed in the distilled water in a ratio of 1:5 (w/v) and was shaken vigorously for one hour on an electric shaker. The pH of the soil suspension was recorded by using pH meter.

2.4 Physicochemical Studies of Wetland Soils

Organic carbon in soil samples was analyzed following the method of Datta et al. (1962). Inorganic phosphorous was analyzed following the method of Fiske and Subbarow (1925). Sodium, calcium and potassium was analyzed by using flame photometer, total nitrogen content was estimated by Kjeldhal digestion method in soil samples.

2.5 Bacteriological Analysis of Wetland Soils

The soil sample was mixed, and a suspension of 1 g (dry weight equivalent) in 10 ml of sterile water was prepared. One ml of the soil suspension was then diluted serially (tenfold) and used for the isolation of aerobic heterotrophic bacterial populations by standard spread-plate dilution method described by Seeley and VanDemark (1981), in triplicate. Nutrient agar containing 0.015% (w/v) nystatin (to inhibit fungi growth) was used for bacteria isolation and incubation was at 35°C for five days. Pure isolates of representative communities were maintained on agar slant at 4°C. Identification of isolates was based on cultural, microscopic, and biochemical characteristics with reference to Bergey's manual of determinative bacteriology (1989).

3 RESULTS

3.1 Soil pH

The pH values ranged from 6.60 – 8.40. The soil pH in D and K blocks was higher than A, B, E and L blocks as shown in Table 1. However, differences in the soil pH values of the different sampling locations were not observed to be statistically significant.

Table 1 pH of soil samples collected from different blocks of Keoladeo National Park

S. No.	*Sampling Sites*	*pH of soil samples*		
		Ist Sampling	*IInd Sampling*	*IIIrd Sampling*
1.	A block	6.6	7.0	7.1
2.	B block	6.7	7.2	7.2
3.	D block	8.4	7.9	7.3
4.	E block	7.7	7.3	7.3
5.	K block	8.1	8.0	7.1
6.	L block	7.3	7.4	7.3

3.2 Levels of Organic Carbon in Soil Samples of Different Blocks

It is evident from the data of Table 2 that the concentration of organic carbon ranged from 30 mg/g to 128 mg/g in the soil samples collected from different sites (blocks) during August 2006 (Ist sampling). However, the concentration of organic carbon ranged from 41 mg/g to 132 mg/g during April 2007 (IInd sampling). While, during Nov. 2007 (IIIrd sampling) the concentration of organic carbon ranged found from 32 mg/g to 130 mg/g. The data of Table 2 indicate that highest concentration of organic carbon in A block and the lowest concentration of organic carbon in D block were recorded.

Table 2 Organic carbon levels in soil samples collected from different blocks of Keoladeo National Park

S. No.	*Sampling Sites*	*Organic carbon (mg/g)*		
		Ist Sampling	*IInd Sampling*	*IIIrd Sampling*
1.	A block	128	132	130
2.	B block	62	71	66
3.	D block	30	41	32
4.	E block	51	60	55
5.	K block	52	63	55
6.	L block	49	55	51

3.3 Levels of Total Nitrogen Soil Samples of Different Blocks

It is evident from the data of Table 3 that total nitrogen level ranged from 0.323 to 0.542% in the soil samples collected from different sampling sites (blocks) during August 2006 (I[st] sampling). However, the concentration of total nitrogen ranged from 0.256 to 0.723 % in the soil samples collected from different sampling sites (blocks) during April 2007 (IInd sampling). While, the concentration of total nitrogen ranged from 0.168 to 0.828% in the soil samples collected from different sampling sites during November 2007. During first sampling, the highest (0.542%) and the lowest (0.323%) concentration of nitrogen was recorded in A and L blocks, respectively. In contrast, during the second and third sampling, the highest (0.723 and 0.828%) and the lowest (0.256 and 0.168%) levels of nitrogen were recorded in A and B blocks, respectively.

Table 3 Total nitrogen levels in soil samples collected from different blocks of Keoladeo National Park

S. No.	*Sampling Sites*	*Total Nitrogen (%)*		
		I[st] Sampling	*IInd Sampling*	*IIIrd Sampling*
1.	A block	0.542	0.723	0.828
2.	B block	0.378	0.256	0.168
3.	D block	0.523	0.612	0.758
4.	E block	0.311	0.325	0.378
5.	K block	0.451	0.675	0.655
6.	L block	0.323	0.400	0.423

3.4 Levels of Phosphorus in Soil Samples of Different Blocks

It is evident from the data of Table 4 that the concentration of phosphorus ranged from 0.8 µg/gm to 2.8 µg/gm in the soil samples collected from different sampling sites (blocks) during August 2006 (I[st] sampling). The concentration of phosphorus ranged from 1.3 µg/gm to 3.1 µg/gm in the soil samples collected from different sampling sites (blocks) during April 2007 (IInd sampling). However, the concentration of phosphorus ranged from 1.0 µg/gm to 2.9 µg/gm during November 2007 (IIIrd sampling).

Table 4 Phosphorus levels in soil samples collected from different blocks of Keoladeo National Park

S. No.	*Sampling Sites*	*Concentration of PO_4^{2-} (µg/g)*		
		I[st] Sampling	*IInd Sampling*	*IIIrd Sampling*
1.	A block	0.8	1.3	1.0
2.	B block	1.7	2.1	1.9

Contd...

Contd...

3.	D block	2.6	2.9	2.7
4.	E block	2.8	3.1	2.9
5.	K block	2.0	2.6	2.2
6.	L block	1.5	1.9	1.7

3.5 Levels of Calcium in Soil Samples of Different Blocks

It is evident from the data of Table 5 that the concentration of calcium ranged from 361.52 mg/100 g to 822.12 mg/100 g in the soil samples collected from different sampling sites (blocks) during August 2006 (Ist sampling). The concentration of calcium ranged from 144.29 mg/100 g to 490.58 mg/100 g in the soil samples collected from different sampling sites (blocks) during April 2007 (IInd sampling). However, the concentration of calcium ranged from 116.59 mg/100 g to 373.23 mg/100 g during November 2007 (IIIrd sampling).

Table 5 Calcium levels in soil samples collected from different blocks of Keoladeo National Park

S. No.	*Sampling Sites*	*Concentration of Ca^{2+} (mg/100g)*		
		Ist Sampling	*IInd Sampling*	*IIIrd Sampling*
1.	A block	822.12	490.58	373.23
2.	B block	383.16	296.59	234.53
3.	D block	361.52	144.29	116.59
4.	E block	538.67	388.77	251.34
5.	K block	649.20	291.78	154.44
6.	L block	516.23	248.49	134.21

3.6 Levels of Potassium in Soil Samples of Different Blocks

It is evident from the data of Table 6 that potassium concentration ranged from 1320 ppm/g to 3645 ppm/g in the soil samples collected from different sampling sites (blocks) during August 2006 (Ist sampling). However, the concentration of potassium ranged from 530 ppm/g to 3100 ppm/g in the soil samples collected from different sampling sites (blocks) during April 2007 (IInd sampling). The concentration of potassium ranged from 2100 ppm/g to 3200 ppm/g in the soil samples during November 2007 (IIIrd sampling). During first sampling, the highest (3645 ppm/g) and lowest (1320 ppm/g) concentration of potassium was recorded in K and A blocks, respectively. In contrast, during the second sampling, the highest (3100 ppm/g) and lowest (530 ppm/g) concentration of potassium was recorded in A block and K block, respectively. However, during the third sampling, the highest (3200 ppm/g) and lowest (2100 ppm/g) concentration of potassium in soil was recorded in E and K blocks, respectively. However, the concentration of sodium was increasing in the soil collected from all the blocks from previous sampling.

Table 6 Potassium levels in soil samples collected from different blocks of Keoladeo National Park

S. No.	Sampling Sites	K Concentration (ppm/g)		
		Ist Sampling	IInd Sampling	IIIrd Sampling
1.	A block	1320	3100	3125
2.	B block	2200	2210	2230
3.	D block	2820	1580	2321
4.	E block	3600	2487.5	3200
5.	K block	3645	530	2100
6.	L block	2980	2750	3000

3.7 Levels of Sodium in Soil Samples of Different Blocks

It is evident from the data of Table 7 that sodium concentration ranged from 27.5 ppm/g to 137.50 ppm/g in the soil samples collected from different sampling sites (blocks) during August 2006 (I^{st} sampling). However, the concentration of sodium ranged from 100 ppm/g to 160 ppm/g in the soil samples collected from different sampling sites (blocks) during April 2007 (IInd sampling). The concentration of sodium 105 ppm/g to 162.3 ppm/g were recorded in the soil samples collected from different sampling sites during November 2007. During first sampling, the highest (137.50 ppm/g) and lowest (27.5 ppm/g) concentration of sodium was recorded in D and A blocks, respectively. In contrast, during second and third sampling, the highest (160 ppm/g and 162.3 ppm/g) and lowest (100 ppm/g and 105 ppm/g) concentration of sodium was recorded in D and L blocks, respectively.

Table 7 Sodium levels in soil samples collected from different blocks of Keoladeo National Park

S. No.	Sampling Sites	Na Concentration (ppm/g)		
		I^{st} Sampling	IInd Sampling	IIIrd Sampling
1.	A block	27.5	105	108
2.	B block	110	132.5	136
3.	D block	137.50	160	162.3
4.	E block	102.50	142.5	145
5.	K block	82.5	110	112.4
6.	L block	65.0	100	105

3.8 Variations in Bacteria Types in Different Blocks during I[st] Sampling

Throughout the sampling locations, a total of eleven (11) distinct cultivable isolates of heterotrophic bacteria were recorded during I[st] sampling with *Bacillus* and *Staphylococcus* genera being the most dominant. The least genera were *Flavobacterium, Proteus* and *Shigella.* The total bacteria cultural types in A and E blocks were seven (7), D block were five (5) and B, K and L blocks were eight (8) genera (Table 8). The total number of bacteria cultural types recorded was significant, i.e., between D block vs A, B and L block, K block vs A, B and L block.

Table 8 Occurrence of some genera of aerobic heterotrophic bacteria during Ist Sampling in different blocks of Keoladeo National Park, Bharatpur

Bacteria	*Occurrence*					
	A block	*B block*	*D block*	*E block*	*K block*	*L block*
Bacillus	+	+	+	+	+	+
Diplococcus	+	+	+	–	+	–
Corynebacterium	–	+	+	+	+	+
Flavobacterium	–	–	–	+	–	–
Micrococcus	+	+	–	+	–	+
Pseudomonas	+	–	+	+	–	+
Proteus	+	–	–	–	–	+
Shigella	–	+	–	–	+	–
Staphylococcus	+	+	+	+	+	+
Yersinia	+	+	–	+	+	+
Enterobacter	–	+	–	+	+	+

3.9 Variations in Bacteria Types in Different Blocks During IInd Sampling

It is evident from the data Table 9 that *Alcaligens, Clostridium, Enterobacter, Flavobacterium, Pseudomonas* and *Proteus* were found to be the common bacteria present in the blocks or sampling sites during IInd sampling. Whereas *Corynebacterium* and *Nitrosomonas* were found only in K and B blocks and *Salmonella* and *Mycobaterium* only in D and B blocks. While *Azotobacter* was found only in K and A blocks, *Agrobacterium* occurred only in D and E blocks. *E. coli* only in K and D, *Shigella* only in D and A, *Staphylococcus* was found only in A and L blocks. Whereas *Arthrobacter* and *Micrococcus* were isolated from L block, *Cellulomonas* was isolated from D block, *Erwinia, Bacillus* and *Yersinia* were isolated from A block.

However, *Thiobacillus* and *Streptococcus* were isolated from B block and from E block respectively. Whereas, the soil sample of K block of the park supported large number of bacterial spp.

Table 9 Occurrence of genera of aerobic heterotrophic bacteria during IInd Sampling in different blocks of Keoladeo National Park, Bharatpur

Bacteria	*Occurrence*					
	A block	*B block*	*D block*	*E block*	*K block*	*L block*
Azotobacter	+	+	–	+	+	–
Bacillus	+	+	+	+	+	+
Corynebacterium	–	+	+	–	+	–
Flavobacterium	+	+	–	–	+	–
Micrococcus	–	+	–	–	–	+
Pseudomonas	+	–	+	+	+	–
Proteus	–	–	+	+	+	+
Shigella	+	+	+	–	–	–
Staphylococcus	+	+	+	+	–	+
Yersinia	–	–	+	+	–	+
Enterobacter	+	+	+	+	–	+
Erwinia	+	–	+	–	+	–
Alcaligens	–	+	–	+	+	–
Nitrosomanas	+	+	+	–	+	–
Clostridium	–	+	–	+	+	+
E. coli	+	+	–	–	–	+
Thiobacillus	–	+	+	–	–	+
Mycobacterium	–	+	–	–	–	–

3.10 Variations in Bacteria Types in Different Blocks during IIIrd Sampling

It is evident from the data of Table 10 that *Bacillus* spp., *E. coli, Enterobater* spp., *Micrococcus* spp., *Pseudomonas* spp., and *Staphylococcus* spp., were found to be the most common bacteria present in approximately all the blocks or sampling sites during IIIrd sampling. Whereas *Corynebacterium* spp., was found only in K and B blocks, *Diplococcus* spp., was, however, found in D and L blocks only. *Proteus* spp., *Salmonella* spp., and *Yersinia* spp., were found only in A and L, D and L and A and E blocks, respectively. Whereas *Agrobacterium* spp., and *Streptococcus* spp., were isolated from E block only, *Flavobacterium* spp., was, however, isolated from A block only. Over all the soil sample of K and A blocks of the park supported large number of bacterial species.

Table 10 Occurrence of some genera of aerobic heterotrophic bacteria during IIIrd Sampling in different blocks of Keoladeo National Park, Bharatpur

Bacteria	Occurrence					
	A block	B block	D block	E block	K block	L block
Bacillus	+	+	+	+	+	+
Diplococcus	–	–	+	–	–	+
Corynebacterium	–	+	–	–	+	–
Flavobacterium	+	+	–	–	–	–
Micrococcus	+	+	–	–	+	–
Pseudomonas	+	–	–	+	+	–
Proteus	+	–	–	–	–	+
Shigella	+	–	+	–	–	+
Staphylococcus	+	–	+	+	+	+
Yersinia	+	–	–	+	–	–
Enterobacter	+	+	–	+	+	+
E. coli	+	+	+	–	+	+

4 DISCUSSION

Wetland ecosystems play an important role on earth for all kinds of organisms, as well as a helpful in removal of pollutants. These are among the most productive ecosystems on the earth providing crucial ecosystem services (Costanza et al., 1997). They are the transition between the terrestrial and aquatic realms and function as major nutrient transforming ecosystems (Mitsch and Gosselink 2000). The key biogeochemical transformations affecting the mobility and bioavailability of nutrients take place within soils and sediments of wetland (Laverman et al., 2007). A characteristic of wetland soils and sediments is the close juxtaposition of oxic and anoxic conditions that enable intense cycling of carbon, nutrients and metals (Roden and Wetzel 1996). The elemental redox cycles are driven by O_2 entering the anoxic zone, not only at the sediment surface, but also at greater depths due to O_2 input via aerenchymatous roots of wetland plants and macrofaunal burrows (Armstrong 1964; Koretsky et al., 2005). Redox conditions in wetland soils and sediments are highly dynamic, because of variations in primary productivity, tidal forcing, temperature, sediment deposition and groundwater inputs, among others. These environmental characteristics promote the establishment of metabolically versatile microbial communities that are able to rapidly respond to changes in their chemical environment (Koretsky et al., 2005; Thullner et al., 2005).

The soil is a highly heterogeneous environment and may exhibit high spatial variations (Franklin and Mills 2003; Baker et al., 2009; Ferreira et al., 2010). The microbial community structure may depend on the spatial scale of observation, in the range of centimeters to a few hundred meters (Ettema and Wardle 2002). Wetlands are essential ecosystems which play vital roles in purification of wastewater, preventing soil erosion and flooding, harboring high biodiversity and also providing sites for recreational activities. They are also sites of rapid biogeochemical cycling due to the presence of redox gradients. The redox cycles of various elements are directly related to the biological and geochemical functions of wetlands (Borch et al., 2010).

Ramsar Convention Bureau (2002) attributes major values for the wetland. Flood control is said to be one of the most important function and value provided by the wetlands. Several countries in the tropics as well as in the temperate regions experience great economic loss due to floods every year. Reclamation of wetlands, destruction of a forest, or conversion of a specific land area anchors on the economic values or immediate monetary gains of the respective resource. It is relevant to determine the priceless values and the silent services rendered by the natural or artificial ecosystems like the wetland to the humanity. Determination of the economic values of the wetland may also be important and useful in the context of conservation and management of the system. One of the major difficulties in economic valuation of a wetland is the insufficient information available on important ecological and hydrological processes, on the basis of which such values are generated by wetlands (Barbier et al., 1997).

The characterization of bacterial communities is one of the major objectives in microbial ecology. Species diversity consists of species richness, total number of species, species evenness and species distribution (Ovreas 2000). Many different methods and approaches have been applied to gain access to microorganisms residing in soil and to assess the microbial diversity. However, approximately 99% of bacteria observed under a microscope are not cultivable under standard conditions (Torsvik et al., 1990; Atlas and Bartha 1998; Hill et al., 2000).

Soil bacterial communities play a critical role in regulating the cycling, retention, and release of major nutrients and soil carbon in freshwater wetlands, with demonstrably large effects on water quality (Richardson and Marshall 1996) and global carbon cycling (Roulet 2000). All the soil pH in this study was near neutral ranges, which favours microbial growth. The composition and diversity of culturable, heterotrophic bacteria observed in this study were similar for the different sampling locations. This corroborates with the finding of Okoh et al. (1999). However, the variation of species differed from one sample to another. Most of the bacteria isolated in this study have been reported by other workers (Amir and Pineau 1998; Okoh et al., 1999). Amir and Pineau (1998) reported that, among the topsoil they investigated, species of Actinomycetes were the most dominant.

5 CONCLUSION

The outcome of this study shows that diverse types of bacteria were isolated from Keoladeo National Park, Bharatpur, Rajasthan. Although samples were collected from different locations, the distinct types of bacteria isolated were generally similar, though occurring at different locations during the period of the study. The abundance of bacteria in this study was typical of environment with high species richness and functional diversity. Despite the fact that it is possible that a number of bacteria may be missed in this study, the isolates could be readily assigned dominant (e.g. *Bacillus* sp.) or transient/successional roles in the isolation of organisms from different sampling locations, which form the basis of this study. Although the results of this study would not be considered to be exhaustive, as it was done within the limits of facilities available in the laboratory, an insight into the population dynamics and distribution of culturable aerobic bacterial diversity has been elucidated. This is without prejudice to the possible influence which a substantial proportion of bacteria are not culturable *in vitro* could have on the overall picture of event. It would require more modern technology (nuclei acid probes) to obtain such detailed overview of microbial diversity. This should be a subject of extension of this investigation in future.

Acknowledgements

Financial assistance provided by Ministry of Environments and Forests, New Delhi, India Grant No. (F.No. J-22012/44/2005-CS (W) is highly acknowledged.

References

Ahn C, Gillevet PM and Sikaroodi M (2007). Molecular characterization of microbial communities in treatment microcosm wetland as influenced by macrophytes and phosphorus loading. *Ecol Indic* 7: 852–863.

Amir H and Pineau R (1998). Influence of plants and cropping on microbiological characteris tics of some new Caledonian Ultramafic soils. *Aust J soil Res* 36(3): 457–470.

Armstrong W (1964). Oxygen diffusion from roots of some British bog plants. *Nature* 204: 801–802.

Atlas RM and Bartha R (1998). Microbial ecology: fundamentals and applications. Benjamin and Cummings, Redwood City, CA. 694–699.

Baker KL, Langenheder S, Nicol GW, Ricketts D, Killham K, Campbell CD and Prosser JI (2009). Environmental and spatial characterisation of bacterial community composition in soil to inform sampling strategies. *Soil Biol Biochem* 41: 2292–2298.

Barbier EB, Mike Acreman M and Knowler D (1997). Economic Valuation of Wetlands: A Guide for Policy Makers and Planners. Gland, Switzerland: Ramsar Convention Bureau.

Batty LC, Baker AJM, Wheeler BD and Curtis CD (2000). The effect of pH and plaque on the uptake of Cu and Mn in *Phragmites australis* (Cav.) Trin ex. Steudel. *Ann Bot.* 86: 647–653.

Bergey (1989). Bergey's manual of systematic bacteriology. Sterley JT (ed). Vol. 3, Williams and Eilkins, Baltimore. pp. 450.

Borch T, Kretzschmar R, Kappler A, Van Cappellen P, Ginder-Vogel M, Voegelin A and Campbell K (2010). Biogeochemical redox processes and their impact on contaminant dynamics. *Environ. Sci. Technol.* 44: 15–23.

Colmer TD (2003). Long-distance transport of gases in plants: A perspective on internal aeration and radial oxygen loss from roots. *Plant Cell Environ* 26: 17–36.

Costanza R, dArge R, deGroot R, Farber S, Grasso M and Hannon B (1997). The value of the world's ecosystem services and natural capital. *Nature* 387: 253–260.

Datta NP, Khera MS and Saini TR (1962). A rapid calorimetric procedure for the determination of the organic carbon in soils. *J Ind Soc Soil Sci* 10: 67–74.

DeJournett TD, Arnold WA and LaPara TM (2007). The characterization and quantification of methanotrophic bacterial populations in constructed wetland sediments using PCR targeting 16S rRNA gene fragments. *Appl Soil Ecol* 35, 648–659.

Denny P (1994). Biodiversity and wetlands. *Wetland Ecol Managem* 3: 55–61.

Ettema CH and Wardle DA (2002). Spatial soil ecology. *Trends Ecol Evol* 17: 177–183.

Faulwetter JL, Gagnon V, Sundberg C, Chazarenc F, Burr MD and Brisson J (2009). Microbial processes influencing performance of treatment wetlands: A review. *Ecol Eng* 35: 987–1004.

Ferreira TO, Otero XL, de Souza VS, Vidal-Torrado P, Macias F and Firme LP (2010). Spatial patterns of soil attributes and components in a mangrove system in Southeast Brazil (So Paulo). *J Soils Sedim* 10: 995–1006.

Fiske CH and Subbarow Y (1925). The colorimetric determination of phosphorus. *J Biol Chem* 66: 375–385.

Franklin RB and Mills AL (2003). Multi-scale variation in spatial heterogeneity for microbial community structure in an eastern Virginia agricultural field. *FEMS Microbiol Ecol* 44: 335–346.

Gopal B, Turner RE, Wetzer RG and DF Whigham (1982). Wetlands: Ecology and Management. India: National Institute of Ecology and International Scientific Publications, pp. 514.

Gutknecht JLM, Goodman RM and Balser TC (2006). Linking soil process and microbial ecology in freshwater wetland ecosystems. *Plant and Soil* 289: 17–34.

Hallberg KB and Johnson DB (2005). Microbiology of a wetland ecosystem constructed to remediate mine drainage from a heavy metal mine. *Sci. Total Environ.* 338: 53–66.

Hill GT, Mitkowski NA, Aldrich-Wolfe L, Emele LR, Jurkonie DD, Ficke A, Maldonado-Ramirez S, Lynch ST and Nelson SB (2000). Methods for assessing the composition and diversity of soil microbial communities. *Appl. Soil Ecol.* 15: 25–36.

Johnson LF and Curl EA (1972). Methods for research on the ecology of soil-borne plant pathogens. Minneapolis, Burges Publishing Company.

Koretsky CM, Van Cappellen P, DiChristina TJ, Kostka JE, Lowe KL and Moore CM (2005). Salt marsh pore water geochemistry does not correlate with microbial community structure. *Estu. Coas. Shelf Sci.* 62: 233–251.

Laverman, AM, Meile C, Van Cappellen P and Wieringa EBA (2007). Vertical distribution of denitrification in an estuarine sediment: Integrating sediment flow through reactor experiments and micro-profiling via reactive transport modeling. *Appl Environ Microbiol* 73: 40–47.

Mitsch WJ and Gosselink JG (2000). The value of wetlands: importance of scale and landscape setting. *Ecolog Econo* 35: 25–33.

Nicomrat D, Dick WA, Dopson M and Tuovinen OH (2008). Bacterial phylogenetic diversity in a constructed wetland system treating acid coal mine drainage. *Soil Biol Biochem* 40: 312–321.

Nicomrat D, Dick WA and Tuovinen OH (2006). Microbial populations identified by fluorescence *in situ* hybridization in a constructed wetland treating acid coal mine drainage. *J Environ Qual* 35: 1329–1337.

Nicorarat D, Dick WA, Dopson M and Tuovinen OH (2008). Bacterial phylogenetic diversity in a constructed wetland system treating acid coal mine drainage. *Soil Biol Biochem* 40: 312–321.

Okoh LA, Badejo MA, Nathanies IT and Tian G (1999). Studies on the bacteria, fungi and springtails (collembola) of an agroforestry arboretum in Nigeria. *Pedobio* 43: 18–27.

Ovreas L (2000). Population and community level approaches for analyzing microbial diversity in natural environments. *Ecol Lett* 3: 236–251.

Pezeshki SR (2001). Wetland plant responses to soil flooding. *Environ. Exper. Bot.* 46: 299–312.

Piper CS (1966). Soil and Plant analysis. Hans Publishers, Bombay.

Richardson CJ and Marshall PE (1986). Processes controlling the movement, storage, and export of phosphorus in a fen peatland. *Ecol Monograph* 56: 279–302.

Roden EE and Wetzel RG (1996). Organic carbon oxidation and suppression of methane production by microbial Fe (III) oxide reduction in vegetated and unvegetated freshwater wetland sediments. *Limnol Oceano* 41: 1733–1748.

Roulet NT (2000). Peatlands, carbon storage, greenhouse gases and the Kyoto Protocol: Prospects and significance for Canada. *Wetlands* 20: 605–615.

Seeley HW and Van Demark PJ (1981). Microbes in action, A laboratory manual of microbiology. 3rd Edition W.H Freeman and Company USA pp. 350.

Thullner M, Van Cappellen P and Regnier P (2005). Modeling the impact of microbial activity on redox dynamics in porous media. Geoch. et Cosmoch. Acta 69: 5005–5019.

Torsvik V, Goksoyr J and Daae F (1990). High diversity of in DNA of soil bacteria. *Appl Environ Microbiol* 56: 782–787.

Vacca V, Wand H, Nokolausz M, Kuschk P and Kästner M (2005). Effect of plants and filter materials on bacteria removal in pilot-scale wetlands. *Water Res* 39: 1361–1373.

Vijayan VS (1991). Keoladeo National Park ecology study. Bombay Natural History Society, Mumbai. pp. 23.

Visser EJW, Bogemann GM, Van de Steeg HM, Pierik R and Blom CWPM (2000). Flooding tolerance of carex species in relation to field distribution and aerenchyma formation. *New Phytol* 148: 93–103.

Weiss JV, Emerson D, Backer SM and Megonigal JP (2003). Enumeration of Fe(II)-oxidizing and Fe(III)-reducing bacteria in the root zone of wetland plants: Implications for a rhizosphere iron cycle. *Biogeochemistry* 64: 77–96.

Whigham DF, Dykjova D and Hejny S (1993). Wetlands of the world I: Inventory, Ecology and Management. Handbook of Vegetation Science, Kluwer Academic Publishers, Dordrecht, Netherlands.

10

Production of Bacterial Proteases Using Agro-industrial Wastes by Statistical Experimental Designs

Rajni Singh*, **Sonali Gupta and Rajshree Saxena**

Amity Institute of Microbial Biotechnology, Amity University Uttar Pradesh Noida-201303, India

**Email: rsingh3@amity.edu*

1 INTRODUCTION

The rapid development of nations across the world in terms of industrial, agricultural and scientific growth has led to extreme exploitation of natural resources accompanied with generation of enormous amount of waste. With global population predicted to hit 9 billion people by 2050, agriculture will need to produce more to serve basic nutritional needs. In order to achieve a global average food consumption of 3130 kcal per person per day by 2050, an additional billion tonnes of agricultural and farm food would need to be produced annually (source: http://www.farmingfirst.org/2011/02/improving-nutrition-by-reducing-waste). This high production of food would be associated with increased generation of waste which would have its environmental impact if not handled and managed properly. The best utilization of this waste is as alternative substrates in bioprocesses and this would subsequently help to solve pollution problem. Biotechnological processes, especially the solid state fermentation technique have contributed enormously to such utilization (Soares et al., 2000; Soccol and Vandenberghe 2003). Solid-state fermentation (SSF) has emerged as a potential technology for the production of different enzymes with the help of microorganisms. SSF process is believed to mimic natural environment and encourage the microorganism to work at its best for the production of the product (Pandey 1994) whereby high titers of product are often obtained (Prakasham et al., 2006). In this process an insoluble substrate is fermented with sufficient moisture but without free water. Agricultural wastes contain high moisture content and can be converted to useful agricultural and industrial products by microbes with the supplementation of nitrogen sources (Yang 2009) hence are considered to be the best substrate for SSF processes and enzyme production. Several factors contribute to the success of a biotechnological process in solid state fermentation as the substrate used in the process, moisture level, time of production, and presence of trace elements. All these factors are required to be used in their optimum concentration and ratio so that their interaction leads to the maximum enzyme production.

The classical methods of optimization involve changing one independent variable while maintaining all other factors constant. This method is however more time consuming and expensive when a large number of variables are evaluated. To overcome this difficulty, experimental factorial design and response surface methodology can be employed to optimize the medium components. RSM (response surface methodolgy) is a collection of statistical and mathematical techniques useful for the modeling and analysis of problems in which a response of interest is influenced by several variables, with the objective being to optimize this response (Montgomery 2001). The primary screening of variables involve 2 factorial methods as Placket-Burman or Taguchi designs. Second-order models like Central Composite Box–Behnken and Doehlert designs are widely used in RSM as they can take on a wide variety of functional forms, and this flexibility allows them to more closely approximate the true response surface (Carvalho et al., 1997, Adinarayan and Elliah 2002, Li et al., 2007, Xiao et al., 2007).

In the present work solid state fermentation was carried out with protease producing *Bacillus* strain isolated from soil. Various agro-industrial wastes as wheat bran, rice bran and mustard oil seed cake were used as substrate. Among these, Mustard oil seed cake was found to be the best substrate for the protease production by the selected strain of bacteria. The conditions for the best enzyme production using mustard oilseed cake were optimized using statistical design experiments Placket-Burman and central composite design in response surface methodology.

2 MATERIALS AND METHODS

2.1 Qualitative Screening of *Bacillus* Strain for Protease Production

Bacillus strain used in the study was selected from the laboratory culture collection for production of protease by milk agar plate method. The culture was streaked on milk agar plates containing 1% skimmed milk powder with 2% agar. After incubation at 37°C overnight, a clear large zone around the streak indicated the protease activity.

2.1.1 Inoculum preparation

The selected bacterial strain was inoculated in nutrient broth [consisting of (g L^{-1}): peptone-5; beef extract- 3; NaCl-5] followed by incubation at 37°C for 24 h to get a standardized inoculum (0.5 OD at 600 nm).

2.1.2 Substrate

Four different types of agro-industrial wastes to be used as substrate, viz., wheat bran, gram husk, and mustard oilcake were procured from local market of Delhi and processed to obtain substrate of size 2-4 mm.

2.1.3 SSF technique

SSF experiments were conducted in 100 ml Erlenmeyer flasks containing substrate impregnated with 10 ml of sterile liquid nutrient medium containing (%): [KH_2PO_4-0.1, NaCl-0.25, $MgSO_4.7H_2O$-0.01, $CaCl_2$-0.01]. The substrate was thoroughly mixed with the liquid medium. The flasks were autoclaved at 121°C for 15 min at 15 lbs, inoculated with the prepared mother culture of the strain and incubated at 37°C for the desired time period. Samples were withdrawn and centrifuged at 10,000 rpm for 10 min. The supernatant was taken as the crude enzyme and assayed for protease activity.

2.2 Enzyme Assay

Protease activity was measured using casein as substrate. Aliquots of enzyme suitably diluted, were mixed with 50 mM glycine–NaOH buffer (pH 9) to make 1 ml volume. 1 ml of 1% casein was added and incubated for 10 min at 60°C. The reaction was stopped by addition of 4 ml TCA (5%, w/v). The mixture was allowed to stand at room temperature for 30 min to 1h and then filtered to remove the precipitate. 1 ml of the filtrate was mixed with 5 ml of 0.5 M Na_2CO_3 solution. 0.5 ml of Folin & Ciocalteu (phenol reagent) reagent was added and the mixture was incubated in dark to develop the blue colour. The blue coloured solution was then was estimated spectrophotometrically at 660 nm. A standard curve was generated using solutions of 0–100 μg/l tyrosine. One unit of protease activity was defined as the amount of enzyme required to liberate 1 μg tyrosine per milliliter in 1 min under the experimental conditions used. The experiments were carried out in triplicates and standard error was calculated.

2.2.1 Optimization studies for enzyme production

2.2.1.1 Selection of solid substrate

For different agro-industrial waste (gram husk, rice bran, wheat bran and mustard oil seed) were used as substrate in the SSF experiments following one at a time method. The substrate that gave the best enzyme production was selected for further optimization along with other production factors.

2.2.1.2 Statistical optimization of media

Two-step processes was selected for the optimization of media for enzyme production by utilizing the selected substrate mustard oilseed cake. The first step was Plackett-Burman design where the screening and selection of different variables required for enzyme production was done. The second level optimization was done by step central composite design in response surface methodology. Design Expert® 8.0.2.0 (Stat-Ease, Inc., Minneapolis, MN, USA) was used to design and analyze both the experiments.

2.2.1.3 Placket-Burman design

Plackett–Burman design (Plackett and Burman, 1946) was employed for the initial screening of factors an effective method for screening of factors required for enzyme production from agro-industrial waste. It allows the evaluation of N variables in the N + 1 experiments; each variable is examined at two levels: −1 for a low level and +1 for a high level (Plackett and Burman 1946; Ghanem et al., 2000). Eight factors were involved the production of protease enzyme in the present study. These included physical factors as substrate concentration, inoculums size, moisture content in the media, time of production and chemical factors as K_2HPO_4, NaCl, $MgSO_4$ and $CaCl_2$. These variables screened with a twelve-run Plackett–Burman design are shown in Table 1. The effect of each variable was determined by the following equation:

$$E(xi) = 2(\Sigma Mi + -Mi -) / N \quad (1)$$

where $E(xi)$ is the concentration effect of the tested variable, Mi+ and Mi– are the total production from the trials where the measured variable (xi) was present at high (+) and (-) low concentrations, respectively; and N is the number of trials.

2.2.1.4 Response surface methodology

Response surface methodology consists of a group of empirical techniques for evaluation of relations existing between a cluster of controlled experimental factors and the measured responses, according to one or more selected criteria. Central composite design (CCD) for three variables selected by Placket-Burman design was used to obtain the combination of values that optimizes the maximum enzyme production as response (Y1). Each numeric factor is varied over 5 levels — alpha points (−1.682, +1.682), 1 factor (−1, +1) and one centre point resulting in a total of 20 experiments. The design permits the estimation of all main and interaction effects, while the purpose of the centre points is to estimate the pure error and curvature. A second order polynomial equation (1) can be used to represent the function in the range of interest.

$$Y = \hat{a}_o + \hat{a}_1X_1 + \hat{a}_2X_2 + \hat{a}_3X_3 + \hat{a}_{11}X_1^2 + \hat{a}_{22}X_2^2 + \hat{a}_{33}X_3^2 + \hat{a}_{12}X_1X_2 + \hat{a}_{13}X_1X_3 + \hat{a}_{23}X_2X_3 \quad (2)$$

where Y is the measured response, $\hat{a}_0$ is the intercept term, $\hat{a}_1, \hat{a}_2, \hat{a}_3$ are linear coefficient, $\hat{a}_{11}, \hat{a}_{22}, \hat{a}_{33}$ are quadratic coefficient, $\hat{a}_{12}, \hat{a}_{13}, \hat{a}_{23}$ are interaction coefficient and $X_1, X_2 X_3$ are coded independent variables.

Table 1 Placket-Burman design

	Factor 1	*Factor 2*	*Factor 3*	*Factor 4*	*Factor 5*	*Factor 6*	*Factor 7*	*Factor 8*	*Response 1*
Std	*A: Mustard oilseed cake*	*B: Inoculum Size*	*C: Moisture*	*D: Time*	*E: KH_2PO_4*	*F: NaCl*	*G: $MgSO_4$*	*H: $CaCl_2$*	*Protease Activity*
	%	*%*	*Ratio*	*h*	*%*	*%*	*%*	*%*	*U/g*
1	5	20	2	72	0.1	0.25	0.01	0.01	2259
2	2	20	5	24	0.1	0.25	0.1	0.01	889
3	5	5	5	72	0.01	0.25	0.1	0.05	1903
4	2	20	2	72	0.1	0.1	0.1	0.05	1008
5	2	5	5	24	0.1	0.25	0.01	0.05	998.2
6	2	5	2	72	0.01	0.25	0.1	0.01	1378
7	5	5	2	24	0.1	0.1	0.1	0.05	956
8	5	20	2	24	0.01	0.25	0.01	0.05	1129.24
9	5	20	5	24	0.01	0.1	0.1	0.01	1024
10	2	20	5	72	0.01	0.1	0.01	0.05	1278
11	5	5	5	72	0.1	0.1	0.01	0.01	2185.62
12	2	5	2	24	0.01	0.1	0.01	0.01	806

A total of 20 experiments were designed by RSM in CCD with three significant variables, namely, substrate, inoculum size and moisture ratio. The runs were performed in duplicates. The design is shown in Table 3 (a and b).

3 RESULTS AND DISCUSSION

3.1 Selection of the Solid Substrate

The initial SSF experiments were performed with all the four substrates used one at a time. The maximum enzyme production 1931 U/g was observed with mustard oilseed cake followed by rice bran (1168 U/g) while wheat bran and gram husk produced 637 and 297 U/g of enzyme respectively in similar conditions (Fig. 1). Hence, mustard oilseed cake was selected for further optimization studies.

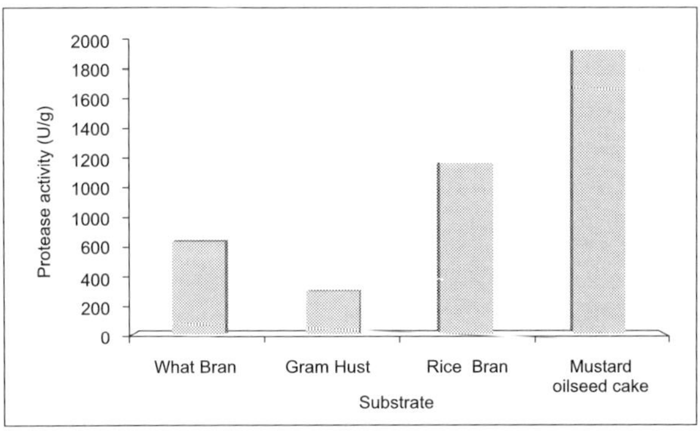

Fig. 1 Protease production with different substrates

3.2 Statistical Optimization of Media

3.2.1 Placket-Burman design

A total of eight variables were analyzed with regard to their effects on protease production using the Plackett–Burman design. The design matrix selected for the screening of significant variables for protease production and the corresponding responses are shown in Table 1. Table 2a represents the $E(x_i)$ value and ranking of the variables investigated in the Plackett–Burman design. A large $E(x_i)$ coefficient, either positive or negative, indicates a large impact on response; while a coefficient close to zero indicates little or no effect. Substrate, moisture ratio, time, K_2HPO_4 and NaCl exhibited positive effect, while inoculum size, $MgSO_4$ and $CaCl_2$ negatively affected the production. Out of all these substrate, moisture ratio and inoculum size were chosen as the variables for the second level optimization. Factors that had positive effect as time, NaCl and $MgSO_4$ were taken at their maximum limit while factors negatively effecting

Table 2(a) Ranking of the variables investigated in the Plackett-Burman design

Variable	*Component*	M_i^+–	M_i^-)	*Difference*	$E(x_i)$	*Absolute* $E(x_i)$	*Ranking*
A	MOC	9456.86	6357.2	3099.66	516.61	516.61	2
B	Inoculum Size	7587.24	8226.82	- 639.58	- 106.6	106.597	8
C	Moisture Ratio	8277.82	7536.24	741.58	123.597	123.597	7
D	Time	10011.6	5802.44	4209.18	701.53	701.53	1
E	KH_2PO_4	8295.82	7518.24	777.58	129.597	- 129.6	6
F	NaCl	8556.44	7257.62	1298.82	216.47	216.47	4
G	$MgSO_4$	7158	8656.06	- 1498.1	- 249.68	- 249.68	3
H	$CaCl_2$	7272.44	8541.62	- 1269.2	- 211.53	211.53	5

the enzyme production were taken in their lower limits for the further experiments.

The adequacy of the model was calculated via ANOVA (Table 2b). The Model F-value of 76.22 implies the model is significant. There is only a 0.22% chance that a "Model F-Value" this large could occur due to noise. Values of "Prob > F" less than 0.0500 indicate model terms are significant.In this case A, D, E, F, G, H are significant model terms. The "Pred R-Squared" of 0.9217 is in reasonable agreement with the "Adj R-Squared" of 0.9820."Adeq Precision" measures the signal to noise ratio. A ratio greater than 4 is desirable. Ratio of 24.768 of the model indicates an adequate signal. This model can be used to navigate the design space.

Table 2(b) ANOVA for Placket-Burman

	Sum of		*Mean*	*F*	*p-value*	
Source	Squares	df	Square	Value	Prob > F	
Model	2869221	8	358652.7	76.22223	0.0022	Significant
A-Mustard oil seed cake	800657.7	1	800657.7	170.1588	0.0010	
B-Inoculum Size	34088.55	1	34088.55	7.244628	0.0743	
C-Moisture	45828.41	1	45828.41	9.739627	0.0524	
D-Time	1476433	1	1476433	313.7772	0.0004	
E-KH_2PO_4	50385.89	1	50385.89	10.7082	0.0467	
F-NaCl	140577.8	1	140577.8	29.87612	0.0120	
G-$MgSO_4$	187015.3	1	187015.3	39.7452	0.0081	
H-$CaCl_2$	134234.8	1	134234.8	28.52809	0.0128	
Residual	14116.07	3	4705.355			
Cor Total	2883338	11				

Std. Dev: 68.60, R-Squared: 0.9951, Adj R-Squared: 0..9820,
Pred R-Squared: 0.9217, Adeq Precision: 24.768

3.2.2 Response surface methodology

Central composite design of 20 runs with three significant variables was used in RSM (Table 3a). The design model and the corresponding responses of RSM experiments to determine the effects of three independent variables (substrate, inoculums size and moisture ratio), along with the mean predicted values and the residual value are shown in Table 3b.

The ANOVA analysis of the optimization study is given in Table 4. The Model F-value of 3150.36 implies the model is significant. There is only a 0.01% chance that a "Model F-Value" this large could occur due to noise. Values of "Prob > F" less than 0.0500 indicate model terms are significant. In this case A, B, C, AB, BC, A^2, B^2, C^2 are significant model terms. Values

Table 3(a) Design summary CCD

Study Type			Response Surface			Runs			20	
Design Type			Central Composite			Blocks			No Blocks	
Design Model			Quadratic			Build Time (ms)			4.86	
Factor	Name	Units	Type	Subtype	Minimum	Maximum	-1 Actual	+1 Actual	Mean	Std. Dev.
A	Substrate	%	Numeric	Continuous	1.3	9.70	3	8	5.50	2.07
B	Moisture	Ratio	Numeric	Continuous	0.64	7.36	2	6	4.00	1.65
C	Inoculum Size	%	Numeric	Continuous	6.59	23.41	010.1	20	15.00	4.13
Response Name	Units	Obs	Analysis	Minimum	Maximum	Mean	Std. Dev.	Ratio	Trans	Model
Protease Activity	U/g	20	Polynomial	878	3620	2582.55	968.084	4.12301	None	Quadratic

Table 3(b) Central composite design matrix, response and predicted responses for protease activity

		Factor 1	Factor 2	Factor 3	Response 1		
Std	*Run*	*A: Substrate*	*B: Moisture*	*C: Inoculum Size*	*Protease activity*	*Predicted*	*Residual*
		%	*Ratio*	%	*U/g*	*Value*	*Residual*
1	16	3	2	10	946	926.65319	19.3468
2	4	8	2	10	1109	1101.9694	7.0306
3	1	3	6	10	2571	2529.4148	41.5852
4	7	8	6	10	2842	2855.731	- 13.731
5	12	3	2	20	2182	2149.3904	32.6096
6	18	8	2	20	2291	2313.7066	- 22.707
7	10	3	6	20	2367	2355.152	11.848
8	15	8	6	20	2670	2670.4683	- 0.4683
9	8	1.295518	4	15	878	931.56363	- 53.564
10	14	9.704482	4	15	1371	1344.1347	26.8653
11	6	5.5	0.636414	15	1643	1655.4711	- 12.471

Contd...

Contd...

12	13	5.5	7.363586	15	3289	3303.2272	-14.227
13	11	5.5	4	6.591036	2516	2539.1449	– 23.145
14	9	5.5	4	23.40896	3408	3411.5534	-3.5534
15	2	5.5	4	15	3584	3593.9032	– 9.9032
16	5	5.5	4	15	3600	3593.9032	6.09678
17	20	5.5	4	15	3617	3593.9032	23.0968
18	17	5.5	4	15	3589	3593.9032	– 4.9032
19	19	5.5	4	15	3620	3593.9032	26.0968
20	3	5.5	4	15	3558	3593.9032	– 35.903

Table 4 Analysis of variance table (ANOVA for Response Surface Quadratic Model- CCD)

Source	*Sum of* Squares	Df	*Mean* Square	*F* Value	*p-value* Prob > F	
Model	17795330	9	1977259	1762.096	< 0.0001	Significant
A-Substrate	205467.5	1	205467.5	183.1088	< 0.0001	
B-Moisture	3277416	1	3277416	2920.772	< 0.0001	
C-Inoculum Size	918724.9	1	918724.9	818.7506	< 0.0001	
AB	11400.5	1	11400.5	10.15991	0.0097	
AC	60.5	1	60.5	0.053916	0.8211	
BC	975804.5	1	975804.5	869.6189	< 0.0001	
A^2	10866476	1	10866476	9684.002	< 0.0001	
B^2	2237769	1	2237769	1994.258	< 0.0001	
C^2	689236.4	1	689236.4	614.2347	< 0.0001	
Residual	11221.06	10	1122.106			
Lack of Fit	8561.726	5	1712.345	3.219501	0.1126	Not significant
Pure Error	2659.333	5	531.8667			
Cor Total	17806551	19				

Std. Dev: 33.50; R-Squared: 0.9994; Adj R-Squared: 0.9988;
Pred R-Squared: 0.9961; Adeq Precision: 112.606

greater than 0.1000 indicate the model terms are not significant. The "Lack of Fit F-value" of 1.33 implies the Lack of Fit is not significant relative to the pure error. There is a 37.97% chance that a "Lack of Fit F-value" this large could occur due to noise. Non-significant lack of fit is good.

The regression equation coefficients were calculated and the data was fitted to a second-order polynomial equation. Thus, the response (Y), (in terms of coded factors) protease production by the selected *Bacillus* sp. can be expressed in terms of the following regression equation:

$$\text{Protease Activity} = +3593.90 + 122.66A + 489.88B + 259.37C + 37.75AB - 2.75AC - 349.25BC - 868.35A^2 - 394.05B2 - 218.69C^2$$

where A is substrate, B is moisture and C is inoculum size.

The regression equation obtained from the ANOVA showed that the R^2 (multiple correlation coefficient) was 0.9994 (a value >0.75 indicates fitness of the model). The "Pred R-Squared" of 0.9988 is in reasonable agreement with the "Adj R-Squared" of 0.9961."Adeq Precision" measures the signal to noise ratio. A ratio greater than 4 is desirable. In our study the ratio of 112.606 indicates an adequate signal. This model can be used to navigate the design space.

Three-dimensional response surface contour graphs were constructed by plotting the response (protease production) on the Z-axis against any two independent variables, while maintaining one variable at its optimal level. Figure 2(a) shows an increase in protease production was observed when moisture ratio and inoculum size increase but further increase in these two factors resulted in decrease of the response, when substrate concentration was maintained at its optimum. Similarly, the enzyme production increased by increasing the substrate concentration and inoculums size (Fig. 2b) and substrate and moisture ratio (Fig. 1c) but in the both the cases the response decreased after an optimal level of conditions was reached. All the three figures show a fairly strong degree of curvature of 3D surface where the optimum can easily be determined.

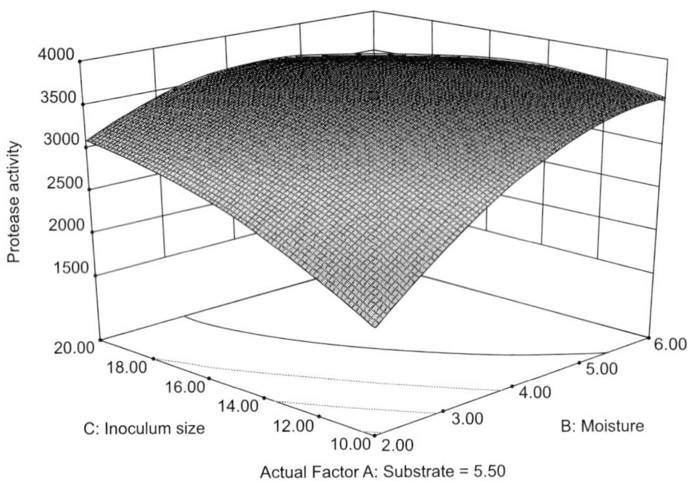

(a)

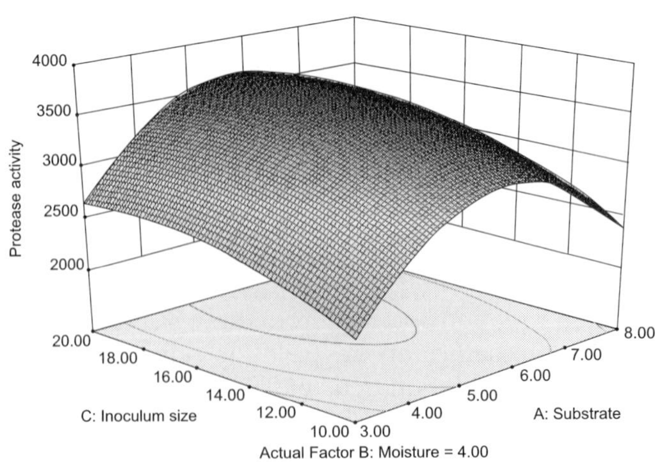

(b)

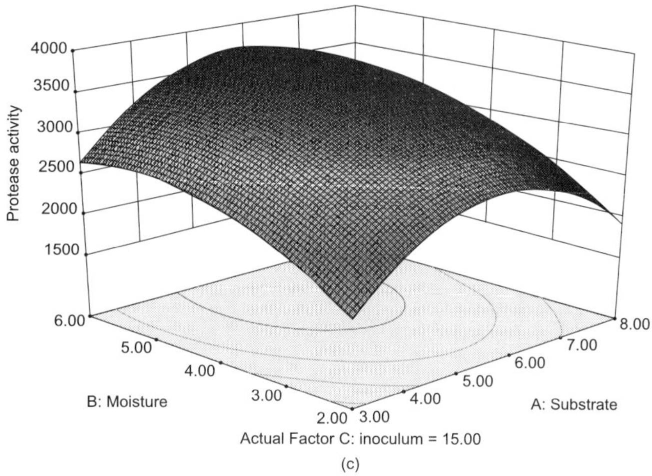

Fig. 2 Contour plots of enzyme activity as a function of the interactions of two variables by keeping the other at centre level: interactions of inoculums size and moisture ratio with substrate at 5.5% (a), interactions of substrate and inoculums size with moisture at 1 : 4 ratio (b), and interactions of substrate and moisture with inoculums size at 15% (c).

Thus, the maximum protease production was 3600 U/g after 72h when the substrate concentration was 5.5%, moisture ratio was 1 : 4 and inoculums size 15%.

4 CONCLUSION

Mustard oil seed cake was utilized in solid state fermentation for protease production. Production conditions were optimized through Plackett–Burman and Central Composite designs in response surface methodology. The optimization resulted in about twofold increase in the production of the enzyme by the selected bacterial strain. An increase in 2 to 3-fold of product yield is always expected with SSF when compared to that of submerged fermentation (Raimbault 1998). The results show that selected strain possesses a high biotechnological potential for the production of enzyme by utilizing agro industrial residues.

References

Adinarayana K and Ellaiah P (2002). Response surface optimization of the critical medium components for the production of alkaline proteases by a newly isolated *Bacillus* sp. *J Pharm Pharmaceut Sci* 5(3): 272–278.

Carvalho CML, Serralheiro MLM, Cabral JMS and Airebarros MR (1997). Application of factorial design to the study of transesterification reactions using cutinase in AOT-reversed micelles. *Enzyme Microb Technol* 21: 117–123.

Improving Nutrition by Reducing Waste (2011) http://www.farmingfirst.org/2011/02/improving-nutrition-by-reducing-waste.

Montgomery DC (2001). Design and analysis of experiments. New York: John Wiley & Sons.

Pandey A (1994). Solid state fermentation: an overview. In: Pandey A (ed) Solid State Fermentation. New Delhi, Wiley Eastern 3–10.

Plackett RL and Burman JP (1946). The design of optimum multifactorial experiments. *Biometrika* 33: 305–325.

Prakasham RS, Subba Rao Ch. and Sharma PN (2006). Green gram husk- an inexpensive substrate for alkaline protease production by *Bacillus* sp. in solid-state fermentation. *Biores Technol* 97: 1449–54.

Raimbault M (1998). General and microbiological aspects of solid substrate fermentation. *Elect J Biotechnol* 1: 1–15.

Soares M, Christen P, Pandey A and Soccol CR (2000). Fruity flavour production by *Ceratocystis fimbriata* grown on coffee husk in solid state fermentation. *Process Biochem* 35: 857-861

Soccol CR and Vandenberghe LPS (2003). Overview of applied and solid state fermentation in Brazil. *Biochem Eng* J 13: 205–218.

Yang SS (2009). Biomass Conversion Technology with solid state fermentation. ICESCO, *Science and Technology Vision* 5(7): 7–14.

Xiao ZJ, Liu PH, Qin JY and Xu P (2007). Statistical optimization of medium components for enhanced acetoin production from molasses and soybean meal hydrolyzate. *Appl Microbiol Biotechnol* 74: 61–68.

11

Adverse Effect of Indiscriminate use of Chemical Pesticides and Its Alternative Ecofriendly Means for Management of Diseases and Pests Infecting Agricultural Crops

Rajesh Kumar Pandey[1*] and Ram Prasad[2,3]

[1]*Department of Botany, Bundelkhand University, Jhansi-284128, U.P., India*

[2]*Amity Institute of Microbial Technology, Amity University Uttar Pradesh Noida-201303, U.P., India*

[3]*Department of Mechanical Engineering, Johns Hopkins University, Baltimore-21218 Maryland, USA*

**Email: rkp_vam@rediffmail.com*

1 INTRODUCTION

Advances in plant protection have contributed considerably to increasing yields and ensuring regular production. Easy to obtain and apply, and rather inexpensive, chemical control products have proved to be extremely efficient and reliable in a very large number of cases, on large surface areas. More than in many other countries, Indian farming has developed production systems based on using these products; it is currently highly dependent on pesticides and India now in Asia after China and ranks twelfth globally in worldwide pesticide consumption. However, today, the systematic use of pesticides is being called into question, with the increasing awareness of their negative impacts, the demonstration of undesirable adverse effects on ecosystems, on non-targeted useful or domestic species and on human health. During the last three and half decades of the twentieth century, the productivity of the Indian agriculture increased dramatically. Yield of the main food crops, rice and wheat, almost doubled, and the yield of other food as well as non-food crops too increased considerably. This remarkable increase resulted from an intensive use of high yielding seeds, fertilizers, pesticides, and water. The yield-augmenting technological change led to the self-sufficiency in the food production. Yet, the concerns of food security have not vanished altogether. By 2030, demand for food grains is expected to increase to 260–265 million tonnes. In other words, food grain production in India has to increase at least 2 million tonnes a year to meet the food security needs of the growing population. The incremental production to meet this demand would have to come from the substantial increase in the agricultural productivity, as the scope for bringing additional land under cultivation is limited. Concurrently, a number of biotic and abiotic factors also continue to constrain the growth in the agricultural productivity.

Insect pests, diseases and weeds are the main yield-limiting factors in most parts of the world. A significant proportion of the potential agricultural production is lost due to pests. Reliable estimates of crop loss, however, are scarce. Amongst the best documented information, Oerke et al. (1995) estimate 42 per cent loss in the global output due to insect pests, diseases and weeds, despite use of various plant protection tactics. The loss could have increased to 70 per cent if the pests were left uncontrolled. Similar observations have been made by Pimental (1993b) who estimates 40 per cent pre-harvest production loss; the post-harvest loss adds another 10–20 per cent. The loss is higher in the Asian countries (Oerke et al., 1995). In India, the pre-harvest production loss due to insect pests is estimated to be 25 per cent for rice, 5 to 10 per cent for wheat and 30 per cent for pulses (Dhaliwal and Arora 1993). For cotton, rapeseed-mustard and sugarcane, the loss is estimated at 50, 35 and 20 per cent, respectively.

2 HISTORY OF PESTICIDE

Humans have used pesticides for a long time to prevent damage to their crops. Common traditional pesticides include plant and animal derivatives such as neem, chrysanthemums, and rotenone derivates or common chemicals such as sulphur, mercury and arsenic. However, most were either ineffective or toxic to both humans and animals. Thus, only a small set of these were used. Traditional farmers used other (non-chemical) means to control pests. The story of modern synthetic pesticides starts with World War I and the introduction of DDT. During the 19th century, active research into science had yielded several chemicals. Accidental contact with some of these indicated they have some effect on humans. Thus, the idea that these chemicals could be used to affect living organisms, took shape. Large-scale use of synthetic chemicals as weapons against humans first occurred in World War I by Germany, Britain, France and other major participants. Chemicals such as mustard gas, cyanides and chlorine were observed to cause burns, sicknesses and death in soldiers. World War I is often called "the chemists' war", both for the extensive use of poison gas and the importance of nitrates and advanced high explosives. Their effects evoked interest among people and many scientists started studying them closely. Research into these and similar chemicals was funded by many companies and governments and continued well after the war.

2.1 What is a Pesticide?

The pest can be insects, plant pathogens, weeds, mollusks, birds, mammals, fish, nematodes (roundworms) and microbes that compete with humans for food, destroy property, spread or help carry diseases or are seen as a nuisance. A pesticide may be a:

- Chemical substance,
- Biological agent (such as a virus or bacteria),
- Antimicrobial,
- Disinfectant or
- Device

Many chemical pesticides are poisonous to humans also. Pesticides are used both in farms and within our homes and gardens. In this presentation, we will look at pesticide use in farming and how it affects many entities including us.

Types of Pesticides

Pesticides come in many forms:

Bactericides for the control of bacteria

Fungicides for the control of fungi and oomycetes

Herbicides for the control of weeds

Insecticides for the control of insects – these can be Ovicides (for eggs), Larvicides (for larva or baby insects) or Adulticides (for mature insects)

Miticides for the control of mites

Molluscicides for the control of slugs and snails

Nematicides for the control of nematodes

Rodenticides for the control of rodents

Virucides for the control of viruses

Some pesticides are absorbed by the plants and thus become part of the plants themselves. Such pesticides are called **systemic** pesticides. The poison then appears in all parts of the plants such as leaves and flowers causing many organisms that consume parts of the plant to be affected, including humans. Often useful insects such as bees which pollinate plants die from this.

Production and Usage of pesticide in India. The production of pesticides started in India in 1952 with the establishment of a plant for the production of BHC near Calcutta, and India is now the second largest manufacturer of pesticides in Asia after China and ranks twelfth globally. There has been a steady growth in the production of technical grade pesticides in India, 90,000 MT are used annually to control pests and plant diseases, which is about 3% of the total world market. The pesticides are classified as insecticides, fungicides, weedicides, herbicides, nematodicides and rodenticides; of which insecticides constitutes 77% of the total pesticides used in different agricultural and animal husbandry practices and in public health operations. The pattern of pesticide usage in India is different from that for the world in general. As can be seen from Figure below, in following India 76% of the pesticide used is insecticide, as against 44% globally. The use of herbicides and fungicides is correspondingly less. The main use of pesticides in India is for cotton crops (45%), followed by paddy and wheat. Although per hectare pesticide use in India is about 250 g, pesticides are used indiscriminately (Dhaliwal and Arora, 1996). About half of the total pesticides used in agriculture goes towards controlling insect pests and diseases of cotton, which occupies only 5 per cent of the total cultivated area. Cotton receives as many as 15–20 rounds of insecticide sprays right from the vegetative stage till its maturity.

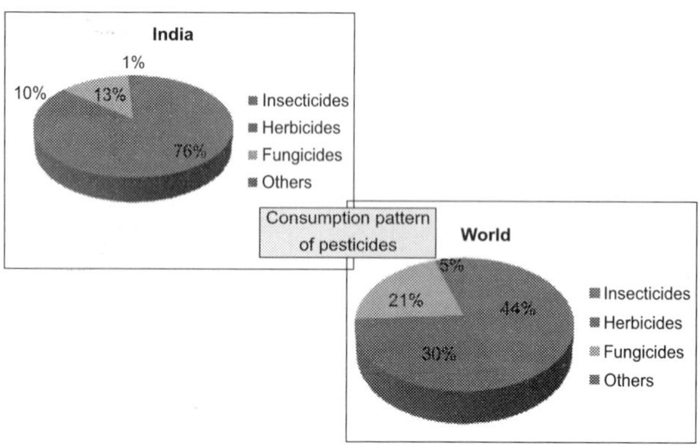

2.2 How Pesticides Work

Some pesticides are **contact** pesticides. To be effective these need to be absorbed through the external body surface or the exposed plant tissue where the pests reside; for example, tetramethrin used in household fly sprays, and bipyridillium herbicides such as paraquat. Contact pesticides have to reach their target directly to be effective. Other pesticides are **systemic** in action. Systemic pesticides are absorbed by a plant or animal and moved (translocated) from the site of application to another site within the plant or animal where they become effective; for example, insecticides that are absorbed by foliage and translocated throughout the plant where they kill chewing or sucking insects; or nematicides that are applied to the leaves of plants and are transferred to the roots of the plant to kill worms or caterpillars that are attacking the plant's roots. Similarly, blood anticoagulant rodenticides in baits take effect once they have been transferred from the digestive system to the bloodstream of rats or mice. In modern concept several pesticides are available as strong weapon in mixture form of **contact and systemic** which has ability to manage of epidemic of diseases in quick time. Pesticides are most effective on eggs, larvae or seedlings. Thus, they must also be applied at the right time - early in the life of the pest. Any delay in application would render them ineffective.

Once it reaches the target pest, the chemical may act in different ways:

2.2.1 Blocking the cellular processes of target organisms in a purely mechanical way

By this, the pesticide physically prevents a basic cellular function even without any chemical reactions. Example of this would be spray oils that clog the respiratory mechanism of insects, petroleum oils that dissolve protective waxes on some insects and plants, or the bypyridylium herbicides that destroy the membranes of plant cells leading to their desiccation.

2.2.2 Destroy or alter the pest's metabolism

Metabolism is the transfer of energy within the cells of organisms, which is essential for the growth and survival of all living things. There are many inhibitory pesticides in this category, such as by obstructing the respiratory, germination mechanisms. Sometimes, the pesticide will undergo transformation to another harmful chemical during the pest's metabolism and thus affect it. Example of these are rotenone and cyanide which disrupt respiratory functions in animals and sodium monfluoroacetate which gets converted to fluorocitrate inside the cell which is then poisonous to many animals.

2.2.3 Disrupt enzyme processes or denature proteins

It is a very important part of every living cell. Examples include inorganic copper compounds, dithiocarbamate fungicides, phosphono amino acid herbicides such as glyphosate, and organophosphate insecticides.

2.2.4 Simulate or otherwise interfere with hormones

Which are messenger chemicals that activate needed biological functions such as reproduction, to disrupt these cycles. Examples are the phenoxy herbicides that interfere with plant growth hormones and insect growth regulators that interfere with cuticle formation in insects during moulting.

2.2.5 Nervous system disruptors

Affect mainly animal groups such as insects, nematodes and rodents. Some are narcotics such as some fumigant pesticides. Others disrupt the movement of nerve impulses, such as the organophosphate, carbamate and pyrethroid pesticides.

2.2.6 Disrupt photosynthesis

Prevent the plant (weeds) from producing or storing energy and ultimately kill the plant. Examples include the triazine, substituted urea and uracil herbicides.

Some pesticides are residual in action and continue to be effective for days, weeks or months after their application. Examples are the triazine herbicides that persist in the soil and kill emerging weeds over the lifetime of a crop and insecticides that remain active for several years as a barrier to termites entering buildings.

Many modern pesticides do not persist for long in the environment. They act quickly and are then degraded to non-toxic substances by chemical or microbial processes. This helps prevent their build-up in crops or other organisms. How quickly a pesticide breaks down depends on its chemical properties, how much is applied and how it is distributed, as well as environmental factors such as temperature, moisture, soil pH and the availability of microorganisms.

2.3 Hazards of Pesticides

2.3.1 Direct impact on human beings

If the credits of pesticides include enhanced economic potential in terms of increased production of food and fiber, and amelioration of vector-borne diseases, then their debits have resulted in serious health implications to man and his environment. There is now overwhelming evidence that some of these chemicals do pose potential risk to humans and other life forms and unwanted side effects to the environment. No segment of the population is completely protected against exposure to pesticides and the potentially serious health effects, though a disproportionate burden is shouldered by the people of developing countries and by high risk groups in each country. The worldwide deaths and chronic illnesses due to pesticide poisoning number about 1 million per year. The high risk groups exposed to pesticides include the production workers, formulators, sprayers, mixers, loaders and agricultural farm workers. During manufacture and formulation, the possibility of hazards may be more because the processes involved are not risk free. In industrial settings, the workers are at increased risk since they handle various toxic chemicals including pesticides, raw materials, toxic solvents and inert carriers. In India, the first report of poisoning due to pesticides was from Kerala in 1958, where over 100 people died after consuming wheat flour contaminated with parathion. This prompted the Special Committee on Harmful Effects of Pesticides constituted by the ICAR to focus attention on the problem. Further, Carlson in 1962 warned that OC compounds could pollute the tissues of virtually every life form on the earth, the air, the lakes and the oceans, the fishes that live in them and the birds that feed on the fishes. Later, the US National Academy of Sciences stated that the DDT metabolite, DDE causes eggshell thinning and that the bald eagle population in the United States declined primarily because of exposure to DDT and its metabolites. Certain environmental chemicals including pesticides, known as endocrine disruptors, are known to elicit their adverse effects by mimicking or antagonising natural hormones in the body and it has been postulated that their long-term, low-dose exposure are increasingly linked to human health effects such as immunosuppression, hormone disruption, diminished intelligence, reproductive abnormalities and cancer.

A study on workers (N = 356) in four units manufacturing HCH revealed neurological symptoms (21%) which were related to the intensity of exposure. The magnitude of the toxicity risk involved in the spraying of methomyl, a carbamate insecticide, in field conditions was assessed by the National Institute of Occupational Health (NIOH).

Significant changes were noticed in the ECG and the levels of serum LDH and ChE activities in the spraymen indicating the cardiotoxic effects of methomyl. Observations confined to health surveillance in male formulators engaged in production of dust and liquid formulations of various pesticides (malathion, methyl parathion, DDT and lindane) in industrial settings of the unorganised sector revealed a high occurrence of generalized symptoms (headache, nausea, vomiting, fatigue, irritation of skin and eyes) besides psychological, neurological, cardiorespiratory and gastrointestinal symptoms coupled with low plasma cholinesterase (ChE) activity. Data on reproductive toxicity were collected from 1106 couples when the males were associated with the spraying of pesticides (OC, OP and carbamates) in cotton fields.

A study in malaria spraymen was initiated to evaluate the effects of a short term (16 week) exposure in workers (N = 216) spraying HCH in field conditions.

2.3.2 *Impact through food commodities*

The UK Pesticide Residue Committee annual report (2002) showed that over 70% of the food in the UK contained no pesticide residues at all and only 1.09% contained residues above the statutory maximum residue levels (MRLs). It concluded that "none of these residues caused concern for people's health". Yet these very small quantities of chemicals in our food, detected at ever lower levels due to increasingly sensitive laboratory equipment, are now easy targets for the media. In India, a study revealed that 50% of the vegetable samples taken from farm gate were found contaminated with various pesticides (0.01–2.23 ppm) of which 16% were above MRL.

2.3.3 Impact on environment

Pesticides can contaminate soil, water, turf, and other vegetation. In addition to killing insects or weeds, pesticides can be toxic to a host of other organisms including birds, fish, beneficial insects, and non-target plants. Insecticides are generally the most acute toxic class of pesticides, but herbicides can also pose risks to non-target organisms.

2.3.4 Surface water contamination

Pesticides can reach surface water through runoff from treated plants and soil. Contamination of water by pesticides is widespread. The results of a comprehensive set of studies done by the U.S. Geological Survey (USGS) on major river basins across the country in the early to mid-90s yielded startling results. More than 90 per cent of water and fish samples from all streams contained one, or more often, several pesticides. Pesticides were found in all samples from major rivers with mixed agricultural and urban land use influences, and 99 per cent of samples of urban streams. The USGS also found that concentrations of insecticides in urban streams commonly exceeded guidelines for protection of aquatic life. Twenty-three pesticides were detected in waterways in the Puget Sound Basin, including 17 herbicides. According to USGS, more pesticides were detected in urban streams than in agricultural streams.

2.3.5 Groundwater contamination

Pesticides, including herbicides, can and do leach to contaminate groundwater. According to the USGS, at least 143 different pesticides and 21 transformation products have been found in the groundwater, including pesticides from every major chemical class. Over the past two decades, contamination has been found in the groundwater of more than 43 states. Contamination of groundwater is of concern because groundwater supplies 50 per cent of the U.S. population with drinking water. During one survey in India it has been found that 58% of drinking water samples drawn from various hand pumps and wells around Bhopal are contaminated with Organ Chlorine pesticides above the EPA standards. Once groundwater is polluted with toxic chemicals, it may take many years for the contamination to dissipate or be cleaned up. Cleanup may also be very costly and complex, if not impossible.

2.3.6 Soil contamination

Pesticides have various characteristics that determine how they act once in soil. Mobility refers to how much a pesticide will move around in the soil. The half life of a pesticide refers to the length of time it takes for half of the pesticide to degrade. Persistence refers to the length of time until all measurable residues of a pesticide are gone.

2.3.7 Effect on soil fertility (Beneficial Soil Microorganisms)

One spoonful of healthy soil has millions of tiny organisms including fungi, bacteria, and a host of others. These microorganisms play a key role in helping plants utilize soil nutrients needed to grow and thrive. Microorganisms also help soil store water and nutrients, regulate water flow, and filter pollutants. The heavy treatment of soil with pesticides can cause populations of beneficial soil microorganisms to decline. Sometimes pesticides have a negative impact on the available NPK from soil. According to soil scientist Dr. Elaine Ingham, "If we lose both bacteria and fungi, then the soil degrades. Overuse of chemical fertilizers and pesticides have effects on the soil organisms that are similar to human overuse of antibiotics. Indiscriminate use of chemicals might work for a few years, but after a while, there aren't enough beneficial soil organisms to hold onto the nutrients."

2.3.8 Contamination of air, soil, and non-target vegetation

Pesticide sprays can directly hit non-target vegetation, or can drift or volatilize from the treated area and contaminate air, soil, and non-target plants. Some pesticide drift occurs during every application, even from ground equipment. Drift can account for a loss of 2 to 25 % of the chemical being applied, which can spread over a distance of a few yards to several hundred miles. There are thousands of reported complaints of off target spray drift each year in the U.S. Many pesticides can volatilize (that is, they can evaporate from soil and foliage, move away from the application, and contaminate the environment.). As much as 80–90 per cent of an applied pesticide can be volatilized within a few days of application. Despite the fact that only limited research has been done on the topic, studies consistently find pesticide residues in air. According to the USGS, pesticides have been detected in the atmosphere in all areas of the USA sampled. Nearly every pesticide investigated has been detected in rain, air, fog, or snow across the nation at different times of the year. Many pesticides have been detected in air at more than half the sites sampled nationwide.

2.3.9 Non-target organisms

Pesticides are found as common contaminants in soil, air, water and on non-target organisms in our urban landscapes. Once there, they can harm plants and animals ranging from beneficial soil microorganisms and insects, non-target plants, fish, birds, and other wildlife.

3 ALTERNATIVE APPROACH FOR PROMOTING NON-CHEMICAL DISEASE AND PEST MANAGEMENT

Despite the enormous benefits that society continues to derive from the indiscriminate use of pesticide chemicals, there is widespread and increasing public concern over their potentially adverse effect on human health and the environment. The pesticides now available lack several desirable qualities, and are phytotoxic and harmful to natural enemies, animals, etc.

3.1 Integrated Pest Management – A Viable Alternative

Integrated pest management (IPM), a new concept in the field of crop protection emphasizes the need for simpler and ecologically safer measures for pest control to reduce environmental pollution and other problems caused by excessive and indiscriminate use of the pesticides. The main components of IPM are pest surveillance, use of crop varieties resistant to pest, sound cultural practices, biological control and use of ecofriendly pesticides having less mammalian toxicity. Agenda 21 of the United Nations Conference on Environment and Development (UNCED) at Rio de Janeiro in June 1992 identified IPM as one of the requirements for promoting sustainable agriculture and rural development. Considering the harmful impact of pesticides on the environment, the Government of India recognized the benefits of IPM and adopted it as the main plank of the plant protection strategy in the overall crop production programme.

IPM is a method used to manage diseases and pests in an environmentally responsible manner. By reducing our dependence on pesticides, IPM protects the environment and our health. It also saves money. IPM can be applied wherever pests are found: on and in farms, schools, homes, hospitals, restaurants, golf courses and home gardens. IPM combines different techniques to prevent pest damage without harming the environment. Pests can include insects and mites, rodents and certain birds, plant diseases, and weeds. IPM practices include monitoring, modifying pest habitat, protecting natural enemies, and, when needed, the use of pesticides.

3.1.1 IPM and its evolution

The conceptualization of IPM started with the discovery of pest resistance to pesticides during the early 1950s. IPM was first referred to as an integrated control mechanism by Stern et al. (1959) — as applied pest control, which combined and integrated biological and chemical controls. Over the years, IPM has evolved to encompass every activity that influences not only the pests but all the living beings - man, animal, plant and environment.

This prompted plant protectionists to investigate some alternative eco-friendly methods for the management of the diseases and pests. Prior to management of diseases and pests through IPM which is presently in fore-front, some investigations on management of diseases and pests through non-chemical methods are from:

3.1.1.1 Cultural practices

Cultural methods of pest control consist of regular farm operations in such a way which either destroy the pests or prevent them from causing economic loss. The various cultural practices have been grouped as under.

- Preparation of nurseries or main fields free from pest infestation by removing plant debris, trimming of bunds, treating of soil and deep summer ploughing which kills various stages of pests. Proper drainage system in field is to be adopted.
- Testing of soil for nutrients deficiencies on the basis of which fertilizers should be applied.
- Selection of clean and certified seeds and treating seeds with fungicide or biopesticides before sowing for seed-borne diseases control.
- Selection of seeds of relatively pest resistant/tolerant varieties which play a significant role in pest suppression.
- Adjustment of time of sowing and harvesting to escape peak season of pest attack.
- Rotation of crops with non-host crops. It helps in reduction of incidence of soil-borne diseases.
- Proper plant spacing which makes plants healthier and less susceptible to pests.
- Optimum use of fertilizer. Use of FYM and biofertilizers should be encouraged.
- Proper water management (alternate wetting and drying to avoid water stagnation) as high moisture in soil for prolonged periods is conducive for development of pests especially soil-borne diseases.
- Proper weed management. It is a well-known fact that most of weeds besides competing with crop for micronutrients also harbour many pests.
- Setting up yellow pan sticky traps for white flies and aphids at far above canopy height.
- Synchronized sowing. Here, community approach is required to sow the crops simultaneously in vast area so that pest may not get different staged crops suitable for its population build up. If pest appears in damaging proportion, control operation could also be applied effectively in the whole area.
- Growing trap crops on the borders or peripheries of fields. There are certain crops which are preferred more by a pest species. These are known as trap crops for that pest. By growing such crops on the border of the fields, pest population develops there which can be

either killed by using pesticides or its natural enemies are allowed to develop there for natural control.

- Root dip or seedling treatment in pest infested area.
- Inter-cropping or multiple cropping wherever possible. All the crops are not preferred by each pest species and certain crops act as repellents, thus keeping the pest species away from preferred crops resulting in reduction of pest incidence.
- Harvesting as close as to ground level. This is because certain developmental stages of insect pests/diseases remain on the plant parts which act as primary inoculum for the next crop season. Hence, harvesting crops at ground level will lessen the incidence of pests in next season.
- Before planting, nursery plants could be sprayed/dipped in copper fungicide/biopesticide solutions to protect the plants from soil-borne diseases.
- While pruning fruit trees remove crowded/dead/broken/diseased branches and destroy them. Do not pile them in the orchards which may act as source of pest infestation.
- Large pruning wounds should be covered with Bordeaux paste/paint to protect the plants from pest/disease attack.
- For excellent fruit set, pollinizer cultivars should be planted in required proportion in the orchards.
- Keeping beehives or placing flower bouquets of pollinizer cultivars facilitates better pollination and subsequent fruit set.

3.1.1.2 Mechanical practices

Removal and destruction of egg masses, larvae, pupae and adults of insect pests and disease parts of plants wherever possible.

- Installation of bamboo cage-cum-bird perches in the field and placing parasitized egg masses inside them for conservation of natural enemies and withholding of pest species wherever possible.
- Use of light traps and destruction of trapped insects.
- Use of rope for dislodging leaf feeding larvae, e.g., caseworm and leaf folders.
- Installation of bird scarer in the field where required.
- Installation of bird perches in the field for allowing birds to sit and feed on insects and their immature stages, viz., eggs, larvae and pupae.
- Use of pheromones for mating disruption, monitoring pest levels and mass trapping.

3.1.1.3 Genetic practices

Selection of comparatively pest resistant/tolerant varieties with reasonable yield levels.

3.1.1.4 Regulatory practices

In this process, regulatory rules framed by government are brought into force under which seeds and infested plant materials are not allowed to enter the country or from one part to other parts of the country. These are known as quarantine methods and are of two types, i.e., domestic and foreign quarantine.

3.1.1.5 Biological practices

Biological control of insect pests and diseases through biological means is the most important component of IPM. In broader sense, biocontrol is use of living organisms to manage crop damaging living organisms (pests). Some of the common biocontrol agents are:

Parasitoids These are the organisms which lay eggs in or on the bodies of their hosts and complete their life cycles on host bodies. As a result of this, the hosts die. A parasitoid may be of different type depending on the host developmental stage in or on which it completes its life cycle. For example, egg, larval, pupal, adult, egg-larval and larval pupal parasitoids. Examples are different species of Trichogramma, Apanteles, Bracon, Chelonus, Brachemeria, Pseudogonotopus, etc.

Predators These are free living organisms which prey upon other organisms for their food. Examples are different species of spiders, dragon flies, damsel flies, lady bird beetles, Chrysopa species, birds, etc.

Pathogens These are microorganisms which infest and cause diseases in their hosts as a result of which hosts are killed. Major groups of pathogens are fungi, viruses and bacteria. Some nematodes also cause diseases in some insect pests.

3.2 Antagonistic Biocontrol Agents Practices

Disease causing agents of pests can be mass multiplied in the laboratory at a low cost in liquid or powdered formulations. These formulations are called biopesticides and can be sprayed like ordinary chemical pesticides. The different types of biocontrol practices are grouped as under:

3.2.1 Introduction

In this process, a new species of bioagent is introduced into a locality for its establishment against its host. This is done only after thorough laboratory examination and field trials for its efficacy.

3.2.2 Augmentation

In this process, the population of natural enemies already present in the area is increased. This is done by releasing either laboratory reared or field collected bioagents. The released bioagents are of same species in such number as would require suppressing the pest population in that area.

3.2.3 Conservation

This is the most important component of biological control and plays a major role in pest suppression. In this process, natural enemies present in the nature are protected from being killed. The different practices required to protect the natural enemies are as below:

- Collection of parasitised egg masses and placing them in bamboo cage-cum-bird perches for allowing emergence of parasitoids and withholding of pest larvae.
- Awareness to differentiate pests and defenders and sparing the defenders during field sprays.
- Chemical spray should be adopted as the last resort and that too after observing pest defender ratio and economic threshold level (ETL).
- Use of broad spectrum pesticides should be avoided.
- Only selective and relatively environmental friendly (REF) pesticides should be used where necessary.
- As far as possible strip or spot application of pesticides is to be carried out.
- Adjustment of time of sowing and harvesting to avoid the peak season of pest attack.
- Growing trap crop on the borders of main fields before the actual sowing of crop to trap pest and develop natural enemies.
- Root dip/seedling treatment for gall midge prone area.
- Crop rotation and inter-cropping also help in conservation of defenders.

If pesticides are used, only recommended dose and concentration is to be used.

4 TYPES OF MICROBES USED AS BIOPESTICIDES

Microbes used as biopesticides:

1. Bacteria – *Bacillus thuringiensis* (Bt), *B. sphaericus, B. papilliae, Serratia entomophila.*
2. Viruses – Nuclear polyhedrosis viruses, granulosis viruses, non-occluded baculoviruses.

3. Fungi – *Beauveria netahizium, Entomophaga zoopthora, Paecilomyces normuraea.*
4. Protozoa – Nosema, *Thelohania vairimorpha.*

4.1 Bacteria as Biopesticides

Among the 14 registered bacterial biocontrol agents: 1. Six are based on *Bacillus.* 2. Five on *Pseudomonas.* 3. Two on *Agrobacterium* and 4. One on *Streptomyces.*

Bacteria as Biopesticides: Bacteria used as biopesticides can be divided into 2 types:

(a) Non-spore-forming—species in the *Pseudomonaeae* and the *Enterobacteriaceae*

(b) Aspore-forming bacteria—bacteria belong to the *Bacilliaceae.*

Xenorhabdus Entomopathogenic nematodes of the families Steinernematidae and Heterorhabdititae, in conjunction with bacteria of the genus Xenorhabdus, have been used successfully deployed as biopesticides. For example, BioVECTOR 4 are usually applied to control insects in cryptic and soil environments. The nematodes harbor the bacteria in their intestines. The infective third-stage larvae enter the host through natural openings and penetrate into the haemocoel. The bacteria are voided in the insect and cause septicemia, killing the insect in approximately 48 h. The nematodes feed upon the bacteria and liquefying insect; and mature into adults.

Pseudomonaeae

Pseudomonads occur commonly in the rhizosphere of plants and they are important for the control of soil-borne plant pathogens. Numerous compounds have been shown to be important for biocontrol activity. Inhibitory compounds include: Degradative enzymes such as: Protease. Cellulase. Chitinase. *B-glucanase.* Antibiotics such as: Hydrogen cyanide. Phloroglucinol. Pyoluteorin. Pyrrolnitrin. Phenazines. Wheat root-colonizing *Pseudomonas* fluorescens. Antifungal activity of *Pseudomonas* sp. and derivative strains (mutant) against *Sclerotinia sclerotiorum* (in the centre) (Berry et al. 2010). DF41 (wild type). DF41-1278 (lipopeptide synthesis) DF41-469 (gacS) DF41-469 (pUCP23-gacS).

Members of the *Bacillus* genus are often considered microbial factories for the production of a vast array of biologically active molecules potentially inhibitory for phytopathogen growth. For example, *B. subtilis,* has an average of 4–5% of its genome devoted to antibiotic synthesis and has the potential to produce more than two dozen structurally diverse antimicrobial compounds. Their spore-forming ability makes these bacteria some of the best candidates for developing efficient biopesticide.

Bacillus spores have a high level of resistance to the dryness necessary for formulation into stable products. Bacilliaceae. *Bacillus thuringiensis* (Bt) occurs naturally in the soil and on plants. It produces a crystal protein that is toxic to specific groups of insects. The insecticidal activity of Bt was first discovered in 1911; however, a product containing this bacterium as the active ingredient was not commercially manufactured until the 1950s. Several strains of Bt have been identified that infect and kill insects. Bt is used mainly to control larvae (caterpillars) of insects in the Lepidoptera order (butterflies and moths) and for the control of mosquito larvae. Bt should be applied to the underside of the leaves because most larvae feed on the underside of leaves. Bt breaks down faster in sunlight, which reduces its effectiveness.

Bt kills the caterpillars in a rather interesting way. The Bt spores must be eaten by the young caterpillar. The bacteria grows inside the caterpillar, reproduces and produces crystalline toxins.

Bt mode of action ruptures gut cell lining Bt spores.

Insect larvae can damage the plant and this allows fungal spores to enter corn tissue. These fungi may produce mycotoxins which can be fatal to some animals and are possibly related to human cancers. Use of genetically modified hybrids controls the ability of insects and diseases to attack crops.

Bt Gene for Insect Control

Bt gene for insect control Bt delta endotoxin. *Bacillus popilliae,* a naturally occurring bacteria have been in use for the control of Japanese beetle larvae in turf since the 1940s. Cause "milky disease" in the beetle larvae and establish a resident population capable of causing mortality over several seasons if soil conditions are appropriate. *Bacillus popilliae* was the first insect pathogen to be registered in the U.S. as a microbial control agent.

4.2 Fungal as Bio-pesticide

Entomopathogenic fungi are important natural regulators of insect populations and have potential as mycoinsecticide agents against diverse insect pests in agriculture. These fungi infect their hosts by penetrating through the cuticle, gaining access to the hemolymph, producing toxins, and grow by utilizing nutrients present in the haemocoel. Entomo- and nematopathogenic fungi may be applied in the form of conidia or mycelium which sporulates after application. Entomo- and nematopathogenic fungi have been used solely or in combination with various insecticides as a part of insecticide resistant management against important crop pests. Trichoderma spp. (*T. viride, T. harzianum, T. virens*), *Aspergillus niger,*

Verticillium, Beauveria bassiana. Metarhizium anisopliae. Nomuraea rileyi. Paecilomyces spp. *Lecanicillium lecanii.*

4.2.1 Indigenous technology for mass production of bio-control agents (*Trichoderma* spp) on pilot scale

Trichoderma spp. (*T. viride, T. harzianum, T. virens*) are considered potential biocontrol and growth promoting agents for many crop plants. Trichoderma populations can be established relatively easily in different types of soil and can continue to persist at detectable levels for months. Trichoderma spp. are more competitive for nutrients than many other soil microbes. *Trichoderma viride* is a biocontrol agent against soil-borne plant pathogens and it can easily colonize in plant rhizosphere and help to promote plant growth. Biocontrol activity of Trichoderma is due to combination of: Ability to serve as antagonist; Plant growth promoter; Plant defense inducer; Rhizosphere colonizer; Neutralizer of pathogen's activity favoring infection. During mycoparasitic interactions between Trichoderma and fungal pathogen, a diffusible factor released from the host before physical contact is responsible for induction of hydrolytic enzymes. During direct contact, lectins in the host's cell wall can induce coiling of the Trichoderma around the host hyphae and mycoparasite can produce appressorium-like structures to destroy the pathogen. Normal degradation of pathogen mycelia would take place after penetrating with appressorium-like structures. John et al. (2010) tested the potential of T. viride ATCC 9275 as a biocontrol of two fungal pathogens (Fusarium oxysporum f. sp. adzuki and Pythium arrhenomanes) and effect of pathogen and antagonist on growth and preliminary yield (fruit) on soybean plants. Fusarium Pythium Trichoderma.

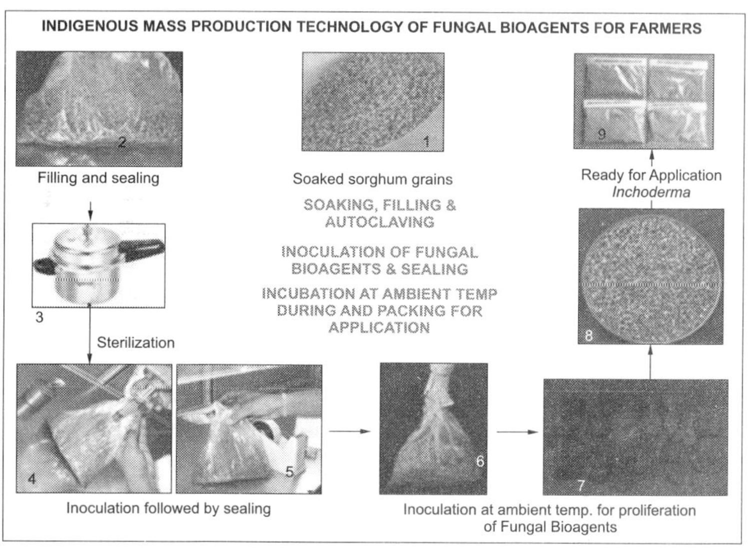

4.3 Viruses as Biopesticide

The Cydia pomonella granulovirus (famil Baculoviridae, CpGV) is used worldwide to control the codling moth. The larva has to feed on the virus during the first development stage. Since this larval stage of the insect life cycle is the most susceptible to infection with baculoviruses. Larval stage is located outside of the plant fruit enabling direct contact with the virus.

Advantage of using CpGV as a biocontrol agent:

- Absolute selectivity.
- High effectiveness towards the target organism.
- It does not harm beneficial organisms.
- It is assumed that already one CpGV particle is mortal.

Disadvantages of the untreated CpGV:

- Strong UV sensitiveness, which results in frequent reapplications during the period of growth.
- A low uptake.
- Removal of the water soluble virus through rainfall from the trees (could be avoided by the encapsulation with organic materials).

(a) **Obligate Fungal Bioagents Vesicular Arbuscular Mycorrhiza as Biopesticide:** Another friendly endoparasitic fungal bioagents Vesicular Arbuscular Mycorrhiza (VAM) represented by the genus *Glomus*, in recent years has been proved to be very useful in the management of fungal, nematode and disease complexes caused by both (Jalali and Thareja, 1981; Singh et al., 1990). *Glomus fasciculatum* and *G. etunicatum* expressed an ideal management component along with karanj and mustard oilseed cake for *M. incognita* in brinjal and tomato respectively (Singh and Goswami, 2001; Bhagwati and Goswami (2000).

(b) **Botanical antagonists:** Ever since Linford et al. (1938) observed reduced galling on incorporating chopped cabbage leaves into root-knot infested soil a long list of many weeds and neglected plants like *Andrographis paniculata, Argemone mexicana, Azadirachta indica, Calotropis gigantean, Datura metel, Eclipta alba, Eichhornea hispeda, Phyllanthus niruri,* etc. have been reported to possess nematicidal properties (Egumjobi and Afolami 1976; Goswami and Vijayalakshmi 1986; Sharma and Trivedi 1994) .

(c) **Oil-seed cakes** of Castor (*Ricinus communis*), Mahua (*Madhuca indica*). Mustard (*Brassica campestris*), Neem (*Azadirachta indica*), Karanj (*Pongamia glabra*), Undi (*Calophyllum inophyllum,* etc.) have been widely used for management of root-knot nematode infecting vegetables and pulses (Goswami and Swarup 1971; Khan et al., 1979) and also soil-borne fungi, viz., wilt, root rot, etc. (Sharma and Bedi 1980; Sen and Chakravarti). Combinations of oil-seed cakes and nematicides in reduced dose have also been demonstrated to perform better results that when either of them used alone in reducing nematode population with improvement in plant health (Bhattacharya and Goswami 1987; Goswami and Mishra 1994).

4.4 Chemical Practices

Use of chemical pesticides is the last resort when all other methods fail to keep the pest population below economic loss. Use of pesticides should be need based, judicious, based on pest surveillance and economic threshold level (ETL). This helps minimise not only the cost involved, but also to reduce associated problems. While going for chemical control, we must understand thoroughly what to spray, when to spray, where to spray and how to spray, keeping in mind the following points.

- ETL and pest defender ratio must be observed.
- Relatively safer pesticides should be selected, e.g., neem based and biopesticides.
- If pest is present in strips or isolated patches, whole field should not be sprayed.

Relevance of IPM practices is more important in vegetable and fruit crops because of their unique mode of consumption by human beings. Pesticides which are generally highly toxic and are known to have toxic residual effects should not be recommended off hand. To get more profit, farmers do not wait until waiting periods of pesticides and harvest the crop to market the same. This leads to pesticides poisoning, chronic effects, in some cases even deaths. Thus, we have to be more careful and cautious in applying pest control practices in field crops.

4.5 Judicious Use of Pesticides

Pesticides are still the most reliable and economic way of protecting crops from pests. While accepting this we need to find ways to maximize the efficacy of pesticide use, while minimizing harmful effects on the environment, and slowing down or reversing the rate of development of resistance in target pest species. Adequate knowledge of economic importance, farmer perceptions of pest losses, harmful effects of insecticides on the environment, and the potential benefits of IPM technologies for sustainable crop production is critical for setting priorities and making rational decisions on pest management.

Integrated disease and pest management Systems for foliar and soil-borne pests and diseases: The method of application of main components of IPM package for management of foliar and soil-borne pests and diseases are as following:

4.5.1 *Modules for the Application of Biopesticide in Integration with Other Safe Components*

In case of transplantable crop – tomato, brinjal, chillis, as a representative:

(A) For soil-borne diseases – wilt, root rot and root-knot nematode:

(1) At Nursery level:

(a) Soil treatment/drenching: For this, organic amendment with neem oil seed cake at 500 g/m^2 and + 200 g VAM/m^2 were done 10 days prior to sowing followed by constant watering for decomposition while at sowing, fungal bioagents at 100 g//m^2 (*Trichoderma viride* at 50 g/m^2 /acre + *Paecilomyces lilacinus* at 50 g/m^2) were applied.

(b) Foliar spray/drenching with a mixture of fungal bioagents 500 g in 5 liter/m^2 (*Trichoderma viride* at 250 g + *Beauveria bassiana* at 250 g) and neem oil seed cake at 500 g + 200 g VAM/m^2 were done with first spray after one week of germination followed 3 more sprays at intervals of 5 days.

(2) Bare root dip treatment/drenching: Transplantation was done of 20–30 days old healthy seedlings after subjecting them to bare root dip treatment for 30 minutes in a slurry containing Jaggery at 50 g + fungal bioagents (*Trichoderma* sp. at 50 g/lit. + *P. lilacinus* at 50 g/lit.) + 200 g neem oilseed cake + 200 g VAM.

(3) Transplantation of seedlings to the 'hot spots' on ridges prior to which deep ploughing followed by application of a mixture of fungal bioagents at 1 kg grown on sorghum grains or talc based + 25 kg summer solarized farm yard manure + 10 kg neem oilseed cake/acre+ 2 kg VAM/acre were + fungal bioagents at 1 kg (*Trichoderma viride* at 500 g/acre + *Paecilomyces lilacinus* at 500 g/acre) applied as soil treatment about 4-5 days before transplantation.

(B) Foliar treatment/drenching of crops for diseases (blight, Cercospora & leaf curl) and insect pests (beetle, thrips, white fly, diamond moth, etc.) was done by:

- Foliar spray/drenching with a mixture of fungal bioagents 1kg in 200 liter/acre (*Trichoderma viride* at 500 g + *Beauveria bassiana* at 500 g) and neem oil seed cake at 10 kg + 200 g VAM/acre were done with first spray after a week of germination followed 4 more sprays at intervals of 10 days.

In case of directly seeded crops – okra, cucurbits, leafy vegetables, etc: Okra as a representative,

(A) For soil-borne diseases – wilt, rot fungi, termites and root-knot nematode:

Seed treatment: For this, fungal bioagents grown on sorghum grains *T. viride* at 50 g/kg + *P. lilacinus* at 50 g/kg were dissolved in mixture of 1 litre fresh water, 50 g jaggery as a sticker and 100 g neem oilseed cake. Prior to sowing, seeds were mixed in above solution followed by shade drying.

Soil treatment/drenching: For this, organic amendment with neem oil seed cake at 50 kg/acre was done 10 days prior to sowing followed by constant watering for decomposition while at sowing, fungal bioagents (*T. viride* at 500 g/acre + *P. lilacinus* at 500 g/acre) and 2 kg VAM/acre were applied.

(B) Foliar part of crops for diseases (blight, Cercospora and mosaic) and insect pests (beetle, white fly, moth) is to be done before the time of attack of diseases on foliar part by Foliar spray with a mixture of fungal bioagents 1 kg in 200 litre/acre (*T. viride* at 500 g + *B. bassiana* at 500 g) and neem oil seed cake at 10 kg was done with first spray after a week of germination followed 4 more sprays at intervals of 10 days.

5 DISCUSSION AND CONCLUSION

The risks associated with pesticide use have been set aside in order to focus on the benefits of pesticides and redress the balance of information available for a more objective assessment of costs and benefits. In addition, there has been no attempt to quantify or rank the benefits. In considering ways to categorise and summarise the beneficial outcomes of pesticide use, it became clear that while the initial effects are usually obvious, and primary benefits follow on reasonably logically, there are secondary (and subsequent) benefits that are sometimes quite subtle and more difficult to substantiate. One such pathway illustrated below in Fig. proposes some indirect but important consequences of pesticide use for resource-poor families and ultimately the economy. However, as we try to track these events, many other factors come into play in an increasingly complicated series of interactions in which it is difficult to be sure there is cause and effect. And indeed, while the example cited above is a plausible chain of benefits, we are unlikely to find documentary evidence in individual publications that follow them right through because the chain spans such a wide range of disciplines—chemistry, economics, sociology, education. Where would it be published? So while there is plentiful evidence linking pesticide effects to primary benefits, in order to link the primary benefits to the secondary and subsequent ones and validate the complete chain of benefits, it would be necessary to review literature outside the body of publications related to pesticides, or to carry out extensive multi-disciplinary field research.

The hierarchical model of effects, primary benefits and secondary benefits helped in the unravelling and categorisation of potential benefits of using pesticides. Although there was inevitable simplification in the categorization process, it is clear that the benefits are many and diverse and can operate in all domains and dimensions, and across all countries rich and poor. There are many other technologies that parallel pesticides, such as vehicular transport and the mains electricity supply, in that they bring us enormous net benefits provided the associated risks are properly managed. However, to maximise the benefits of pesticide use at minimum human, environmental and economic cost, pesticides must be strictly regulated and used judiciously by properly trained and appropriately equipped personnel, ideally in strict integration with other complementary technologies.

References

Anderson AJ, Tari PH and Tepper CS (1988). Genetic studies on the role of an agglutinin in root colonization by *Pseudomonas putida. Appl. Environ. Microbiol.* 54: 375–380.

Anderson LM, Stockwell VO and Loper JE (2004). An extracellular protease of *Pseudomonas fluorescens* inactivates antibiotics of Pantoea agglomerans. *Phytopathology* 94: 1228–1234.

Andrews J (1992). Biological control in the phyllosphere. Annu. Rev. *Phytopathol.* 30: 603–633.

Anon (2001). FAO Desert Locust Control Guidelines—Control, second ed. Food and Agriculture Organization, Rome, Italy.

Anon (2003a). Drive and Stay Alive Inc. At: /http://www.drive and stayalive. com/info%20section/statistics/stats-usa.htmS.

Anon (2003b). Herbicide use essential to crop production. *Chem. Market Rep.* 263 (18), 4.

Anon (2004a). Water Matters for Sustainable Agriculture—A Collection of Case Studies. Crop Life International, Brussels, April 2004.

Anon (2004b). The State of Food Insecurity in the World: (2004). FAO/ Economic and Social Department. Food and Agriculture Organization, Rome, Italy.

Anon (2005b). Population and Vital Statistics Report, Series: A, No. 236/ 237, vol. LVIII, No. 1.

Anon (2005c). Conservation Technologies and the Plant Science Industry: Managing Natural Resources Sustainably. Crop Life International, Belgium, 42pp. Anon, (2006a). World Energy Council Report. At: / http://www.world energy.org/wec-geis/publications/default/tech_ papers/17th_congress/4_

Anon (2006b). World Resources Institute, (2006). Earth Trends: The Environmental Information Portal. Available at: /http://earthtrends. wri. Org S. World Resources Institute, Washington, DC.

Anon (2006c). The role of public transport to reduce greenhouse gas emissions and improve energy efficiency. European Union Position on the European Climate Change Programme and the Green Paper on Energy Efficiency. March (2006).

Anon (2004c). The World Health Organization./http://www.who.int/mediacentre/news/releases/2006/pr50/en/index.htmlS.

Anon (2005a)./http://library.thinkquest.org/C002291/high/present/stats. htmS. An End to World Hunger: Hope for the Future.

Atreya N (2006). Chemophobia—pesticide residues in food. Outlooks on Pest Management, vol. 17. December 2006, 242.

Audenaert K, Pattery T, Cornelis P, and Hofte M (2002). Induction of systemic resistance to Botrytis cinerea in tomato by Pseudomonas aeruginosa 7NSK2: role of salicylic acid, pyochelin and pyocyanin. *Mol. Plant-Microbe Interact.* 15: 1147–1156.

Austin RB (1998). Yield of wheat in the UK: recent advances and prospects. In: Annual Meeting of the Crop Science Society of America.

BK Goswami and Pandey, Rajesh Kumar (2007). Management of disease complex caused by root knot nematode and root wilt fungus on pigeonpea through soil organically enriched with Vasicular Arbuscular Mycorrhiza, karanj oilseed cake (*Pongamia pinnata*) and Farm Yard Manure. Journal of Environmental science and Health Part-B, 42 (8) 899-904 [Taylor & Francis http://www.tandf.co.uk/journals/titles/03601234.asp]

Baker KF (1987). Evolving concepts of biological control of plant pathogens. *Annu. Rev. Phytopathol.* 25: 67–85.

Bankhead SB, Landa BB, Lutton E, Weller DM and McSpadden Gardener BB (2004). Minimal changes in rhizosphere population structure following root colonization by wild type and transgenic biocontrol strains. *FEMS Microb. Ecol* 49: 307–318.

Bargabus RL, Zidack NK, Sherwood JW and Jacobsen BJ (2002). Characterization of systemic resistance in sugar beet elicited by a non-pathogenic, phyllosphere colonizing Bacillus mycoides, biological control agent. *Physiol. Mol. Plant Pathol.* 61: 289–298.

Bargabus RL, Zidack NK, Sherwood JW and Jacobsen BJ (2004). Screening for the identification of potential biological control agents that induce systemic acquired resistance in sugar beet. *Biological Contr.* 30: 342–350.

Bargabus RL Zidack NK, Sherwood JW and Jacobsen BJ (2004). Screening for the identification of potential biological control agents that induce systemic acquired resistance in sugarbeet. *Biol. Control* 30: 342–350.

Bates G, and Denton HP (2007). No-Till Establishment of Forage Crops Agricultural Extension Service. The University of Tennessee at:/ http://www.utextension.utk.edu/publications/spfiles/sp435c.pdfS.

Batson Jr, WE, Caceres J, Benson M, Cubeta MA, Elliott ML, Huber DM, Hickman MV, McLean KS, Ownley B, Newman M, Rothrock CS, Rushing KW, Kenny DS, and P Thaxton (2001). Biological seed treatment evalutations for control of the seedling disease complex of cotton, (2000). *B & C Tests* 16: F12.

Batson Jr, WE, Caceres J, Benson M, Cubeta MA, Elliott ML, Huber DM, Hickman MV, Keinath AP, Dubose V, McLean KS, Ownley B, Canaday C, Rushing KW and Kenny DS (2001). Biological seed treatment evalutations for control of the seedling disease complex of snap bean, (2000). *B & C Tests* 16: V81.

Bell J (2005). CropSaver, Summer (2005). Crop Protection Association, UK, 4.

Bellamy C (1998). The State of the World's Children. Oxford University Press, UNICEF, Oxford and New York.

Benhamou N (2004). Potential of the mycoparasite, Verticillium lecanii, to protect citrus fruit against *Penicillium digitatum,* the causal agent of green mold: A comparison with the effect of chitosan. *Phytopathology* 94: 693–705.

Benhamou N and Chet I (1997). Cellular and molecular mechanisms involved in the intersection between *Trichoderma harzianum* and *Pythium ultimum. Appl. Environ. Microbiol.* 63: 2095–2099.

Berg G, Krechel A, Ditz M, Sikora RA, Ulrich A and Hallmann J (2005). Endophytic and ectophytic potato-associated bacterial com-munities differ in structure and antagonistic function against plant pathogenic fungi. *FEMS Microbiol. Ecol.* 51: 215–229.

Berry CL (1991). Assessing environmental health risks from chemicals. *Trans. Med. Soc. Lond.* 108 (1991/1992), 72–78.

Biermann B and Linderman RG (1983). Use of vesicular-arbuscular mycorrhizal roots, intraradical vesicles and extraradical vesicles as inoculum. *New Phytol.* 95: 97–105.

Biological Control of Crop Diseases (2002). S. Gnanamanickam ed. Marcel Dekker: New York, NY Cook RJ and Baker KF (1983). The Nature and Practice of Biological Control of Plant Pathogens. American Phytopathological Society, St. Paul, MN.

Bloom B, Ehlers R, Haukeland-Salinas S, Hoddanen H, Jung K, Kuhlmann U, Ravensberg W, Strasser H, Warrior P and Wilson M (2003). Biological control agents: Safety and regulatory policy. *Bio Control* 48: 477–484.

Bridges D (Ed.) (1992). Crop Losses Due to Weeds in the United States. Weed Sci. Soc. Am., Champaign, IL, pp. 403.

Brown Ian UK Pesticides Residue Committee Report, (2004) (available online: /http://www.pesticides.gov.uk/uploadedfiles/ Web_Assets/ PRC/PRCannualreport(2004).pdfS, also available on request).

Bruns A (2003). Controlling Aflatoxin and Fumonisin in maize by crop management. *J. Toxicol. Toxin Rev.* 22 (2), 153–173.

Bull CT, Shetty KG and Subbarao KV (2002). Interactions between Myxobacteria, plant pathogenic fungi, and biocontrol agents. *Plant Dis.* 86: 889–896.

Bulla LA, Costilow RN and Sharpe ES (1978). Biology of Bacillus popilliae. Advances in Microbial Physiology 23, 1–18 [This article presents details on the biology of Bacillus popilliae].

Burges HD and Thompson EM (1971). Standardization and Assay of Microbial Insecticides. In: Microbial control on insects and mites, [Burges HD & Hussey NW, eds], Academic Press, London [This is a simulation based study showing the effect of microbial insecticides]

Catska V (1994). Interrelationship between vesicular-arbuscular mycorrhiza and rhizosphere microflora in apple replant disease. *Biologia Plant.* 36: 99–104.

Chenkin AF (1975). Economic effects of plant protection in the Russian Federated Republic. In: Papers at Sessions, VIII International Congress of Plat Protection, Moscow, USSR, 2, 27.

Chet I (1987). Trichoderma application, mode of action, and potential as a biocontrol agent of soil borne plant pathogenic fungi. In: Innovative approachs to plant disease control. Ed. By chet, John Wilay & Sons, Inc., New York. 137–159.

Chet I (1993). Biotechnology in plant disease control. Wiley Liss publication, New York. 211–235.

Chet I and Baker R (1981). Isolation and biocontrol potential of *Trichoderma hamatum* from soil naturally suppressive to *Rhizoctonia solani. Phytopathology* 71: 286–290.

Chisholm ST, Coaker G, Day B and Staskawicz BJ (2006). Host-microbe interactions: shaping the evolution of the plant immune response. Cell 124: 803–814.

Chisholm ST, Coaker G, Day B and Staskawicz BJ (2006). Host-microbe interactions: shaping the evolution of the plant immune response. Cell 124: 803-814.

Cook R (1993). Making greater use of microbial inoculants in agriculture. *Annu. Rev. Phytopathol.* 31: 53–80.

Cook RJ, Weller DM, Youssef El-Banna A, Vakoch D, and Zhang H (2002). Yield responses of direct-seeded wheat to rhizobacteria and fungicide seed treatments. *Plant Dis.* 86: 780-784.

Curtis C, Jana-Kara B, Maxwell CA (2003). Insecticide treated nets: impact on vector populations and relevance of initial intensity of transmission and pyrethroid resistance. *J. Vector Borne Dis.* 40 (1/2),

Dales MJ, Golob P (1997). The protection of maize against Prostephanus truncatus (Horn), using insecticide sprays in Tanzania. *Int. J. Pest Manage.* 43 (1), 39–43.

de Barjac H and Frachon E (1990). Classificatoin of Bacillus thuringiensis strains. Entomophaga 35,(2), 1–8 [This assignment presents data concerning the classification of different strains of *Bacillus thuringiensis*]

de Capdeville G, Wilson CL, Beer SV and Aist JR (2002). Alternative disease control agents induce resistance to blue mold in harvested 'Red Delicious' apple fruit. *Phytopathology* 92: 900–908.

De Meyer G and Hofte M (1997). Salicylic acid produced by the rhizobacterium *Pseudomonas aeruginosa* 7NSK2 induces resistance to leaf infection by *Botrytis cinerea* on bean. *Phytopathology* 87: 588–593.

Dhaliwal GS and R Arora (1993). Changing status of insect pests and their management strategies. In: Changing scenario of Indian agriculture (Eds. KS Gill, GS Dhaliwal and BS Hansara. New Delhi: Commonwealth Publishers.

Dhaliwal GS and R Arora (1996). Principles of insect management. Commonwealth Publishers, New Delhi.

Duchesne LC (1994). Role of ectomycorrhizal fungi in biocontrol. 27–45 in: Mycorrhizae and Plant Health. FL Pfleger and RG Linderman, eds. APS Press, St. Paul, MN.

Duijff BJ, Gianinazzi-Pearson V and Lemanceau P (1997). Involvement of the outer-membrane lipopolysaccharides in the endophytic colonization of tomato roots by biocontrol *Pseudomonas fluorescens* WCS417r. New Phytol. 135: 325–334.

Elad Y (1994). Biological control of grape gray mold by *Trichoderma harzianum*. Crop. Protection 13: 35–38.

Elad Y and Baker R (1985). Influence of trace amounts of cations and siderophore-producing pseudomonads on chlamydospore germination of *Fusarium oxysporum*. *Ecol. Epidemiol.* 75: 1047–1052.

Elad Y, Baker R and Chet I (1984). Parasitism of sclerotia of Sclerotium rolfsii by *Trichoderma harzianum*. *Soil Biol Biochem.* 16: 381–386.

Elad Y, Chat I and Henis Y (1982). Degradation of Plant pathogenic fungi by Trichoderma harzianum. *Canad. J. Microbiol.* 28: 719–725.

Elad Y, Chet I, Boyle P and Henis Y (1983). Parasitism of *Trichoderma harzianum* spp. on *Rhizoctonia solani* and Sclerotium rolfsii- scanning electron microscopy. *Phytopathology* 73: 85–88.

Elad Y, Hadar Y, Hadar E, Chet I and Henis Y (1981). Biological control of *Rhizoctonia solani* by *Trichoderma harzianum* in carnation. *Plant Dis.* 65: 675–677.

El-Ghaouth A, Smilanick JL, Brown GE, Ippolito A, Wisniewski M and Wilson CL (2000). Application of Candida saitoana and glycolchitosan for the control of postharvest diseases of apple and citrus fruit under semi-commercial conditions. *Plant Dis.* 84: 243-248.

Elliott ML, Jardin EAD, Batson WE, Caceres J, Brannen PM, Howell CR, Benson DM, Conway KE, Rothrock CS, Schneider RW, Ownley BH, Canaday CH, Keinath AP, Huber DM, Sumner DR, Motsenbocker CE, Thaxton PM, Cubeta MA, Adams PD, Backman PA, Fajardo J, Newman MA and Pereira RM (2001). Viability and stability of biological control agents on cotton and snap bean seeds. *Pest Mgmt. Sci.* 57: 695–706.

Environmental and economic costs of pesticide use. *Bioscience* 42, 750–760.

EPA (2000). Emmission Facts. Office of Transportation and Air Quality Average Annual Emissions and Fuel Consumption for Passenger Cars and Light Trucks EPA420-F-00-013.

Etcheverry M, Chulze S, Dalcero A, Varsavsky E, Rodriguez M, Moschetti E, Ferrero S (1987). Aflatoxins in sunflower seeds. Influence of fungicides and insecticides. *Int. J. Food Microbiol.* 5.

European Food Information Council (2006). At: /http://www.eufic.org/page/en/health-lifestyle/healthy-eating/S.

Fekete C, Weszely T and Hornok L (1996). Assignment of a PCR amplified chitinase sequence cloned from Trichoderma hamatum to resolved chromosomes of potential bio-control species of Trichoderma. EMBL: Locus THENDOCHS, Accession Z 71415.

Fischoff DA, Bowdish KS, Perlak EJ, Marrone PG, McCormic SM, Nierdermyer JG, Deam DA, Kusan-Kretzmer K, Mayer EJ, Rochester DE, Roger SG and Fraley RT (1987). Insect tolerant in transgenic tomato. Plant biotechnology 5,807 [This is a comprehensive discussion on problems of genetic engineering technology uses in insect tolerant to biopesticides]

Fitter AH and Garbaye J (1994). Interactions between mycorrhizal fungi and other soil microorganisms. *Plant Soil* 159: 123–132.

Flores A, Chet I and Herrera-Estrella A (1997). Improved biocontrol activity of Trichoderma harzianum by over-expression of the proteinase-encoding gene prb1. *Curr. Genet.* 31: 30–37.

Florida. Ignoffo CM and Conch TL (1981). The nucleopolyhedrasis virus of Heliothis species as a microbial insecticide. In: Microbial control of pests and plant diseases 1970-1980 (H.D.Burges, ed.), Academic Press, London [This article provides data concerning the nucleopolyhedrosis virus as insecticide agent]

Fravel D (2005). Commercialization and implementation of biocontrol. *Annu. Rev. Phytopathol.* 43: 337–359.

Gangopadhyay S and Joshi RK (2000). Efficacy of Trichoderma in controlling root rot of cotton and chickpea. In: Proceedings of International Conference on Integrated Plant Disease Management for Sustainable Agriculture, Indian Phytopathological Society. New Delhi, 325–326.

Garcia-Garrido JM and Ocampo JA (1989). Effect of VA mycorrhizal infection of tomato on damage caused by *Pseudomonas syringae. Soil Biol. Biochem.* 21: 165–167.

Gattuso D (2000). Understanding the benefits of pesticides. *Consum. Res. Mag.* 83 (2), 34.

Gianessi L (1999). Beneficial impacts of pesticide use for consumers. In: Ragsdale, N., Seiber, J. (Eds.), Pesticides: Managing Risks and Optimizing Benefits. American Chemical Society Symposium Series #734. American Chemical Society, Washington, DC, United States of America, 207.

Georgis R, Kaya H and Gaugler R (1991). Effect of stermatid and heterohabditid Nematodes on nontarget arthropods. Environ. Entomol. 20, 815–822 [This efforts provide data concerning the safety of Nematode insecticides on non-target arthropods]

Geremia R, Goldman G, Jacobs D, Ardiles W Vila, S Van Montagu M and Herrera-Estrella A (1993). Molecular charactericzation of proteinase-encoding gene prb1 related to mycoparasitism of *Trichoderma harzianum. Mol. Microbiol.* 8: 603–613.

Gill SS, Cowles EA and Pietrantonio PV (1992). The mode of action of Bacillus thuringiensis endotoxins. Annu. Rev. Entomol. 37, 615–636 [This article presents the mode of action of delta-endotoxin of *B. thuringiensis*]

Glandorf DC, Verheggen P, Jansen T, Jorritsma JW, Smit E, Leefang P, Wernars K, Thomashow LS, Laureijs E, Thomas-Oates JE, Bakker PA, and Van Loon LC (2001). Effect of genetically modified *Pseudomonas putida* WCS358r on the fungal rhizosphere microflora of field-grown wheat. *Appl. Environ. Microbiol.* 67: 3371–3378.

Golberg LJ and Margalit J (1997). A bacterial spore demonstrating rapid larvicidal activity against *Anopheles sergentii, Vronataenia usergenngiuculata, Culex univitattus, Aeded agypti* and *Culex pipiens.* Mosq. News 37, 355–358 [This article presents the detail on larvicidal activity of the bacterial spore of B. thuringiensis var. israelonsisi against different insects]

Goswami BK, Pandey, Rajesh Kumar, Kabindra Singh Rathour and Singh Lokendra (2006). Integrated application of some compatible biocontrol agents along with mustard cake and furadan on Meloidogyne incognita infecting tomato plants. Journal of Zhejiang University Science (B) 7(11): 873–875. (www.springerlink.com)

Goswami BK, Pandey, Rajesh Kumar, Bhattacharya, Chaitali and Lokendra Singh (2005). Evaluation of six isolates of *Trichoderma harzianum* against *Fusarium oxysporum f.* sp. *lycopersici* and *Meloidogyne incognita. International J. of Namatol (UK)* 15 (1): 83–86.

Granados RR and Federici BA (ed) (1986). The biology of Baculoviruses. CRC Press, Inc. Boca Raton.

Gratz NG (1994). What role for insecticides in vector control programmes? *Am. J. Trop. Med. Hyg. US*, 50 (6), 11–20.

Guetsky R, Shtienberg D, Elad, Y and Dinoor, A (2001). Combining biocontrol agents to reduce the variability of biological control. *Phytopathology* 91: 621–627.

Haas D and Defago G (2005). Biological control of soil-borne pathogens by fluorescent pseudomonads. *Nature Rev. Microbiol.* 3: 307–319.

Hader Y, Chet I and Henis Y (1979). Biological control of Rhizoctonia solani damping off with wheat bran culture of *Trichoderma harzianum. Phytopathology* 69: 64–68.

Hainsworth SD, Eden-Green SJ (Eds.) (2000). In: Sustaining Change: Proceedings of a Workshop on the Factors Affecting Uptake and Adoption of Department for International Development (DFID) Crop Protection Programme (CPP) Research Outputs.

Harman GE (2000). Myths and Dogmas of Biocontrol-Changes in perceptions derived from research on *Trichoderma harzianum* T-22. Pl. Dis. 84: 377–393.

Harman GE, Howell CR, Viterbo A, Chet I and Lorito M (2004). Trichoderma species-opportunistic, avirulent plant symbionts. *Nature Rev. Microbiol.* 2: 43–56.

Harman GE, Howell CR, Vitarbo A, Chet I and Lorito M (2004). Trichoderma species - opportunistic, avirulent plant symbionts. *Nature Rev. Microbiol.* 2: 43–56.

Harman GE, Hayer CK, Lorito M, Broadway RM, Di-Pietro A, Peterbauer C and Tronsmo A (1993). Chitinolytic enzymes of *Trichoderma harzianum* purification of chitobiosidase and endochitinase. *Phytopathology* 83: 313–318.

He P, Chintamanani S, Chen Z, Zhu L, Kunkel BN, Alfano JR, Tang X and Zhou JM (2004). Activation of a COI1-dependent pathway in Arabidopsis by *Pseudomonas syringae* type III effectors and coronatine. *Plant J.* 37: 589–602.

Henry MSJ (1994). *J. Environ. Health* 12, 65–70.

Hibbitt C (1990). Putting pesticides in perspective. Agricultural Engineer. Summer (1990), 61.

Hoffman IF, Jere CS, Taylor TE, et al. (1999). The effect of *Plasmodium falciparum* malaria on HIV-1 RNA blood plasma concentration. AIDS 13, 487–494.

Hoitink HAJ and Boehm MJ (1999). Biocontrol within the context of soil microbial communities: a substrate dependent phenomenon. *Annu. Rev. Phytopathol.* 37: 427–446.

Homma Y, Kato Z, Hirayama F, Konno K, Shirahama H and Suzui T (1989). Production of antibiotics by *Pseudomonas cepacia* as an agent for biological control of soilborne plant pathogens. *Soil Biol. Biochem.* 21: 723–728.

Howell CR and Stipanovic RD (1980). Suppression of *Pythium ultimum* induced damping – off of cotton seedlings by Pseudomonas fluorescens and its antibiotic, pyoluterin. *Phytopathology* 70: 712–715.

Howell CR, Beier RC and Stipanovic RD (1988). Production of ammonia by Enterobacter cloacae and its possible role in the biological control of Pythium pre-emergence damping-off by the bacterium. *Phytopathology* 78: 1075–1078.

Iavicoli A, Boutet E, Buchala A and Métraux JP (2003). Induced systemic resistance in Arabidopsis thaliana in response to root inoculation with Pseudomonas fluorescens CHA0. *Mol. Plant-Microbe Interact.* 16: 851-858.

IPPC (1997). International Plant Protection Convention—New Revised Text Approved by the FAO Conference at its 29th Session— November 1997.

Islam TM, Hashidoko Y, Deora A, Ito T and Tahara S (2005). Suppression of damping-off disease in host plants by the rhizoplane bacterium Lysobacter sp. strain SB-K88 is linked to plant colonization and antibiosis against soilborne peronosporomycetes. *Appl. Environ. Microbiol.* 71: 3786-3796.

JM, Slutsker L (2004). Sustainability of reductions in malaria trans-mission and infant mortality in western Kenya with use of insecticide-treated bednets: 4 to 6 years of follow-up. JAMA 291 (21),

Jacobsen BJ, Zidack NK and Larson BJ (2004). The role of Bacillus-based biological control agents in integrated pest management systems: Plant diseases. *Phytopathology* 94: 1272–1275.

Janisiewicz W and Korsten L (2002). Biological control of postharvest diseases of fruits. *Annu. Rev. Phytopathol.* 40: 411–441.

Janisiewicz WJ and Peterson DL (2004). Susceptibility of the stem pull area of mechanically harvested apples to blue mold decay and its control with a biocontrol agent. *Plant Dis.* 88: 662–664.

Jebakumar RS, George G, Baba S, Raghuchander T, Gopalaswamy G, Vidhyasekaran P. Samiyappan R, Raja JAJ and Balasubramanian P

(2000). Purification and N-terminal sequencing of a 42 KDa *Trichoderma viride* Chitinase isoform effective against *R. solani*. *Indian Phytopath.* 53: 157–161.

Jha D and Chand R (1999). National Centre for Agricultural Economics and Policy Research (ICAR), New Delhi, India from Agro-Chemicals News in Brief Special Issue, November.

Johnson KB, Stockwell VO, Sawyer TL and Sugar D (2000). Assessment of environmental factors influencing growth and spread of Pantoea agglomerans on and among blossoms of pear and apple. *Phytopathology* 90: 1285–1294.

Jones RW and Prusky D (2002). Expression of an antifungal peptide in Saccharomyces: A new approach for biological control of the post harvest disease caused by C. coccodes. *Phytopathology* 92: 33–37.

Joshi M (2001). The Future Role of Pesticides in US Agriculture. Committee on the Future Role of Pesticides in US Agriculture, Board of Agriculture and Natural Resources and Board on Environmental.

Joshi R and McSpadden Gardener B (2006). Identification and characterization of novel genetic markers associated with biological control activities of *Bacillus subtilis*. *Phytopathology* 96: 145–154.

Kageyama K and Nelson EB (2003). Differential inactiviation of seed exudates stimulation of *Pythium ultimum* sporangium germination by Enterobacter cloacae influences biological control efficacy on different plant species. *Appl. Environ. Microbiol.* 69: 1114–1120.

Kamuanga M (2001). Farmers' perceptions of the impacts of tsetse and trypanosomosis control on livestock production: evidence from southern Burkina Faso. *Trop. Anim. Health Prod.* 33 (2), 141.

Kaya HK (1990). Entomopathogenic Nematodes. In: Biological control of insects (Baker R.R. & Dunn P.E., eds), 189–198. FAO (1989), CTA/ FAO Symposium, Luxemburg, 26–30 June 1989. [This work presents the biological activities of nematodes as an entemopathogenic agent]

Keel C, Voisard C, Berling CH, Kahir G and Defago G (1989). Iron sufficiency is a prerequisit for suppression of tobacco black root rot by *Pseudomonas fluorescnes* strain CHA0 under genotobiotic contiditions. *Phytopathology* 79: 584–589.

Keiserukhsky MG, Kashirsky OP (1975). Economics of plant protec tection in the USSR. In: VIII International Congress of Plant Protection, Papers at Sessions, Moscow, USSR, *vol.* 2. 12.

Kerr A (1980). Biological control of crown gall through production of agrocin 84. *Plant Dis.* 64: 25–30.

Kerry B (2000). Rhizosphere interactions and the exploitation of microbial agents for the biological control of plant parasitic nematodes. *Annu. Rev. Phytopathol.* 38: 423–441.

Kerwin JL, Dritz DD and Washino RK (1994). Pilot scale production and applicatoin in wildlife ponds of *Lagenidium giganteum* (Oomycetes: Lagenidiales). J. Amer. Mosq. Control Assoc. 10, 451-455 [This article provides the pilot scale production and application of the fungus Lagenidium giganteum in wildlife ponds].

Kessel GJT, Köhl J, Powell JA, Rabbinge R and Van der Werf, W (2005). Modeling spatial characteristics in the biological control of fungi at the leaf scale: Competitive substrate colonization by Botrytis cinerea and the saprophytic antagonist *Ulocladium atrum*. *Phytopathology* 95: 439–448.

Kilic-Ekici O and Yuen GY (2003). Induced resistance as a mechanisms of biological control by Lysobacter enzymogenes strain C3. *Phytopathology* 93: 1103–1110.

Kiss L (2003). A review of fungal antagonists of powdery mildews and their potential as biocontrol agents. Pest Manag. Sci. 59: 475-483.

Kloepper JW, Ryu CM and Zhang S (2004). Induce systemic resistance and promotion of plant growth by Bacillus spp. *Phytopathology* 94: 1259–1266.

Kloepper JW, Leong J, Teintze M and Schroth MN (1980). Pseudomonas siderophores: A mechanism explaining disease suppression in soils. *Current Microbiol.* 4: 317–320.

Kolbe W (1982). Effect of different crop protection programmes on yield and quality of apples II (1967–1981). *Pflanzenschutz-Nachrichten Bayer* 35 (2), 189.

Koumoutsi A, Chen XH, Henne A, Liesegang H, Gabriele H, Franke P, Vater J and Borris R (2004). Structural and functional characterization of gene clusters directing nonribosomal synthesis of bioactive lipopeptides in *Bacillus amyloliquefaciens* strain FZB42. *J. Bact.* 186: 1084–1096.

Kovach J, Petzoldt R and Harman GE (2000). Use of honey and bumble bees to disseminate *Trichoderma harzianum* 1295-22 to strawberries for Botrytis control. *Biol. Control* 18: 235–242.

Kucharik CJ and Ramankutty N (2005). Trends and variability in US corn yields over the 20th century. *Earth Interactions* 9, 1–29.

Kumar, Deepak and Dubey SC (2001). Management of collar rot of pea by the integration of biological and chemical method. *Indian Phytopath.* 54: 62–66.

Lacey LA and Mulla MS (1990). Safety of *Bacillus thuringiensis* and *Bacillus sphaericus* to non-target organisms in the aquatic environment. In: Safety of microbial insecticides (ed. Lacy L.A. and Davidson E.W.) C.R.C. Press, Baton. [This article presents the safety of both organisms when applied in the aquatic environment].

Lafontaine PJ and Benhamon N (1996). Chitosan treatment: an emerging strategy for enhancing resistance of greenhouse tomato to infection by *Fusarium oxysporum* f.sp. *radici-lycopersici. Biocontrol Sci. Technol* 6: 111–124.

Leadbetter ER (2002). Prokaryotic Diversity: Form, Ecophysiology, and Habitat. Pages 19-32 in: Manual of Environmental Microbiology (2nd ed.), ASM Press, Washington DC.

Leclere V, Bechet M, Adam A, Guez JS, Wathelet B, Ongena M, Thonart P, Gancel F, Chollet-Imbert M and Jacques P (2005). Mycosubtilin overproduction by *Bacillus subtilis* BBG100 enhances the organism's antagonistic and biocontrol activities. *Appl. Environ. Microbiol.* 71: 4577–4584.

Leeman M, Van Pelt JA, Den Ouden FM, Heinbroek M and Bakker PAHM (1995). Induction of systemic resistance by *Pseudomonas fluorescens* in radish cultivars differing in susceptibility to Fusarium wilt, using novel bioassay. *Eur. J. Plant Pathol.* 101: 655–664.

Leeman M, Van Pelt JA, Hendrickx MJ, Scheffer RJ and Bakker PAHM (1995). Biocontrol of Fusarium wilt of radish in commercial greenhouse trials by seed treatment with *Pseudomonas fluorescens* WCS374. *Phytopathol* 85: 1301–1305.

Lewis JA, Larkin RP and Rogers DL (1998). A formulation of Trichoderma and Gliocladium reduce damping off caused by *Rhizoctonia solani* and saprophytic growth of the pathogen in the soil less mix. *Plant Dis.* 82: 501–506.

Lewis JA, Lumsden RD, Fravel DR and Shasha BS (1995). Application of pregelatinized starch flour granules formulation of biocontrol fungi to control damping off disease caused by Rhizoctonia solani. *Biol. Control.* 5: 397–404.

Lewis JA and Larkin RP (1997). Extruded granular formulation with biomass of bio-control *Gliocladium virens* and Trichoderma spp. to reduce damping off of eggplant caused by Rhizoctonia solani and saprophytic growth of the pathogen in soil-less mix. *Biocontrol. Sci. Tech.* 7: 49–60.

Lewis JA and Papavizas GC (1987). Application of Trichoderma and Gliocladium in alginate pellets for control of Rhizoctonia damping off. *Plant Pathol.* 36: 397–404.

Lewis N and Ruud J (2004). Apples in the American diet. *Nutr. Clin. Care* 7(2), 82.

Lewis N and Ruud J (2005). Blueberries Am. Diet, Nutr. Today 40 (2), 92 March–April.

Likanen E (2001). The Challenge of Ensuring Safer Cars. European Commission Responsible for Enterprise and the Information Society SPEECH/01/56.

Lindblade KA, Eisele TP, Gimnig JE, Alaii JA, Odhiambo F, ter Kuile FO, Hawley WA, Wannemuehler KA, Phillips-Howard PA, Rosen DH, Nahlen BL, Terlouw DJ, Adazu K, Vulule.

Linderman RG (1994). Role of AM fungi in biocontrol. Pages 1-25 in: Mycorrhizae and Plant Health. F. L. Pfleger and R. G. Linderman, eds. APS Press, St. Paul, MN.

Lindsay WL (1979). Chemical Equilibria in Soils. John Wiley & Sons, Inc., New York.

Lomborg B (2001). The Skeptical Environmentalist. Cambridge University Press, reprint ed. (August 28, 2001). ISBN: 0521010683.

Loper JE and Buyer JS (1991). Siderophores in microbial interactions of plant surfaces. *Mol. Plant-Microbe Interact.* 4: 5–13.

Lumsden RD, Walter JF and Baker CP (1996). Development of Gliocladium virens for damping off disease control. *Can. J. Plant. Pathol.* 18: 463-468.

Maiti D and Sen D (1987). In: Biological control of plant diseases, National Workshop, March, 10-12, 1987, Coimbatore, 7 pp.

Maramorosh K and Sherman KE (eds) (1985). Viral Insecticides for Biological Control, Academic Press, Orlando. Florida. [This book contributes extensive data concerning viral insecticides]

Maurhofer M, Hase C, Meuwly P, Métraux JP and Defago G (1994). Induction of systemic resistance to tobacco necrosis virus by the root-colonizing *Pseudomonas fluorescens* strain CHA0: influence of the gacA gene and of pyoverdine production. *Phytopathology* 84: 139–146.

McCoy CW (1990). Entomogenous fungi as microbial pesticides. In: Baker R.R. and Dunn P.E. (eds), pp 139-160. FAO 1989, CTA/FAO Symposium, Luxemburg, 26-30 June 1989. [This is a comprehensive discussion on the use of entromogenous fungi as microbial pesticide]

McKellar ME and Nelson EB (2003). Compost-induced suppression of Pythium damping-off is mediated by fatty-acid-metabolizing seed-colonizing microbial communities. *Appl. Environ. Microbiol.* 69: 452–460.

McNeely Jeffrey A and Scherr Sara J, (2003). Ecoagriculture: Strategies to Feed the World and Save Wild Biodiversity. *Island Press.* pp. 279

McSpadden Gardener BB and Fravel DR (2002). Biological control of plant pathogens: Research, commercialization, and application in the USA. Online. Plant Health Progress doi:10.1094/PHP-2002-0510-01-RV. (Also online http://www.apsnet.org/online/feature/biocontrol/top.html)

McSpadden Gardener BB, Gutierrez LJ, Joshi R, Edema R and Lutton E (2005). Distribution of phlD+ bacteria in corn and soybean fields. *Phytopathology* 95: 715–724.

McSpadden Gardener B and Fravel D (2002). Biological control of plant pathogens: Research commercialization, and application in the USA. Online. Plant Health Progress doi:10.1094/PHP-2002-0510-01-RV.

McSpadden Gardener B and Weller D (2001). Changes in populations of rhizosphere bacteria associated with take-all disease of wheat. *Appl. Environ. Microbiol.* 67: 4414–4425.

Mercier J and Lindow SE (2001). Field performance of antagonistic bacteria identified in a novel assay for biological control of fireblight. *Biol. Control* 22: 66–71.

Meziane H, Van der Sluis I, Van Loon LC, Hofte M and Bakker PAHM (2005). Determinants of *Pseudomonas putida* WCS358 involved in inducing systemic resistance in plants. *Mol. Plant Pathol.* 6: 177–185.

Milgroom MG and Cortesi P (2004). Biological control of chestnut blight with hypovirulence: a critical analysis. *Annu. Rev. Phytopathol.* 42: 311–338.

Milgroom M and Cortesi P (2004). Biological control of chestnut blight with hypoviulence: A critical review. *Annu. Rev. Phytopathol.* 42: 311–338.

Miller Stanley F (1982). The effects of weed control technology change on rural communities. *Outlook Agric.* 11 (4), 172.

Morris PF and Ward EWR (1992). Chemoattraction of zoospores of the plant soybean pathogen, *Phytophthora sojae,* by isoflavones. *Physiol. Mol. Plant Pathol.* 40: 17–22.

Morton JB and Benny GL (1990). Revised classification of arbuscular mycorrhizal fungi (zygomycetes): a new order glomales, two new suborders, glomineae and gigasporineae and gigasporaceae, with an amendation of glomaceae. *Mycotaxon* 37: 471–491.

Moyne AL, Shelby R, Cleveland TE and Tuzun S (2001). Bacillomycin D: an iturin with antifungal activity against *Aspergillus flavus. J. Appl. Microbiol.* 90: 622–629.

Mukhopadhyay AN (1987). Bio-control of soil borne plant pathogens by Trichoderma spp. *Indian J. Mycol Pl. Pathol.* 17: 1–10.

Mukhopadhyay AN (1994). Bio-control of soil borne fungal plant pathogens-current status, future prospect and potential limitations. *Indian Phytopath.* 47: 119–126.

Mulla MS (1990). Activity, field efficacy, and use of Bacillus thuringiensis var israelensis In: Bacterial control of mosquitoes and blackflies (H deBarjac and D.J. Sutherland, eds). Rutgers University Press, New Brunswick, N.J., 134–160. [This book presents comprehensive data concerning different aspects of the entomopathogenic bacterium, Bacillus thuringiensis israelensis]

Mulla MS and Federici BA (1985). Sporulation and toxin production of *Bacillus thuringiensis* var israelensis in cadavers of mosqiuto larvae (Diptera culicidaes). *J. Inverteb. Pathology* 46, 251-258 [This work provides data concerning sporulation and toxin production of Bacillus thuringiensis]

Murphy RC and Steven ES (1992). Cloning and expression of the Cry/IDV gene of Bacillus thuringiensis subsp. Israelensis in the cyanobacterium Agmenellum quadraplicatum PR-6 and its resulting larvicidal activity. *Appl. Environ. Microbiology* 58, 1650–1655 [This paper provides data concerning genetic technology for new biopesticide production]

Murray MG and Thampson WF (1980). Rapid isolation of high molecular weight DNA. *Nuclic Acids Research* 8: 4321–4325.

Neilands JB (1981). Microbial iron compounds. *Annu. Rev. Biochem.* 50: 715-731.

Neuenschwander P (2001). Biological control of the cassava mealybug in Africa: A review. *Biol. Control* 21, 214–229.

Notz R, Maurhofer M, Schnider-Keel U, Duffy B, Haas D and Defago G (2001). Biotic factors affecting expression of the 2,4-diacetylphloroglucinol biosynthesis gene phlA in Pseudomonas fluorescens biocontrol strain CHA0 in the rhizosphere. *Phytopathology* 91: 873-881.

Numberger T, Brunner F, Kemmerling B and Piater L (2004). Innate immunity in plants and animals: striking similarities and obvious differences. *Immunological Rev.* 198: 249–266.

Odum EP (1953). Fundamentals of Ecology. W. B. Saunders, Philadelphia–London.

Ongena M, Duby F, Rossignol F, Fouconnier ML, Dommes J and Thonart P (2004). Stimulation of the lipoxygenase pathway is associated with systemic resistance induced in bean by a nonpathogenic Pseudomonas strain. *Mol. Plant-Microbe Interact.* 17: 1009–1018.

Ordentlich A, Elad Y and Chet I (1988). The role of chitinase of Serratia marcescens in the biocontrol of Sclerotium rolfsii. *Phytopathology* 78: 84–88.

Palumbo JD, Yuen GY, Jochum CC, Tatum K and Kobayashi DY (2005). Mutagenesis of beta-1,3-glucanase genes in Lysobacter enzymogenes strain C3 results in reduced biological control activity toward Bipolaris leaf spot of tall fescue and Pythium damping-off of sugar beet. *Phytopathology* 95: 701–707.

Pandey, Gopal, Pant, Hemlata; and Pandey, Rajesh Kumar (2003). Screening of chickpea cultivars against *Meloidogyne incognita. Annals of Plant Protection Sciences* 11 (2): 409: 410.

Pandey, Gopal, Pandey, Rajesh Kumar and Pant, Hemlata (2002). Effect of different plant extract of different concentrations on root-knot nematode and its rhizosphere soil in chickpea (Cicer arietinum L.) *Vijnana Parishad Anusandhan.* 45 (4): 317–324.

Pandey, Gopal, Pandey, Rajesh Kumar and Pant, Hemlata (2003). Efficacy of different levels of Trichoderma viride against root-knot nematode in chickpea (Cicer arietinum L.) Annals of Plant Protection Sciences, 11(1): 101–103.

Pandey RC Pandey, Rajesh Kumar, Sant AK and Dwivedi BK (2004). Influence of hydrogen ion concentration on population dynamics of *Hoplolaimus indicus* on citrus tree and management with VAM fungus. *Pakistan J. of Namatol.* 22 (2): 45–49.

Pandey, Rajesh Kumar and Goswami BK (2008). Management of root knot nematode on tomato through application of fungal antagonists, Acremonium strictum and *Trichoderma harzianum.* Journal of Environmental science and Health Part-B, 44(3) 237–240.[Taylor & Francis http://www.tandf.co.uk/journals/titles/03601234.asp]

Pandey, Rajesh Kumar, Bhattacharya, Chaitali, Goswami BK and Singh Lokendra (2006). Management of root knot nematode infecting brinjal through combination of fungal bioagents, *Aspergillus fumigatus* and *Trichoderma harzianum. Indian Phytopathology* 59 (2): 82–86.

Pandey, Rajesh Kumar and Goswami BK (2005). Bacillus subtilis: An Ecofriendly Effective Bacterial Antagonist of Fusarium udum Butler Causing Wilt Disease of Pigeon Pea (Cajanus cajan Millsp). *The Proceedings of the National Academy of Sciences, Section B-Biological Sciences* 75 (IV): 234–237.

Pandey, Rajesh Kumar, Goswami BK and Singh S (2004). Influence of soil pH on Tylenchulus semipenetrans infecting citrus and its Biomanagement using VAM fungi. *International J. of Nematol* 14(2): 174–176. (UK)

Pandey, Rajesh Kumar, Goswami BK and Singh S (2005). Management of root knot nematode and Fusarium wilt disease complex by fungal bioagents, neem oilseed cake and/or VA-Mycorrhiza on chickpea. International Newsletters of Chickpea and Pigeonpea, ICRISAT, 12: 32–34.

Pandey, Rajesh Kumar, Pant H, Pandey G, Dwivedi BK, Devi G and Singh JP (2003). Survey of root-knot disease of chickpea (*Cicer arietinum* L.) in and around of district Allahabad U.P. *Pakistan Journal of Nematol.* 21 (2): 97–102.

Pandey, Rajesh Kumar, Pant H, Yadav S, Varshney S, Pandey G and Dwivedi BK (2004). Application of biocontrol agents and neem cake on *Meloidogyne incognita* on chickpea (Cicer arietinum L.). *Pakistan Journal of Nematol.* 22 (2): 57–60.

Pandey, Rajesh Kumar, SR Singh, PK Gupta, BK Goswami and Yogita Ghade (2011). Effect of different bioformulations of Paecilomyces lilacinus against root-knot nematode (*Meloidogyne incognita*) infecting tomato. *Indian Journal of Agricultural Sciences* 81 (3): 65–71.

Papavizas GC (1985). Trichoderma and Gliocladium: Biology, Ecology and Potential for bio-control. *Annu. Rev. Phytopathol.* 23: 23–54.

Paulitz T and Belanger R (2001). Biological control in greenhouse systems. *Annu. Rev. Phytopathol.* 39: 103–133.

Petrusheva NI (1975). Protection of horticultural crops in the USSR and future perspectives. In: Papers at Sessions, VIII International Congress of Plant Protection, Moscow, USSR, 2, 124.

Phillips DA, Fox TC, King MD, Bhuvaneswari TV and Teuber LR (2004). Microbial products trigger amino acid exudation from plant roots. *Plant Physiol.* 136: 2887–2994.

Pieterse CMJ, Van Wees SCM, Ton J, Van Pelt JA and Van Loon LC (2002). Signalling in rhizobacteria induced systemic resistance in *Arabidopsis thaliana. Plant Biol.* 4: 535–544.

Pimentel D (1997). Techniques for Reducing Pesticide Use. Economic and Environmental Benefits. Wiley, USA.

Pimentel D, Acquy H, Biltonen M, Ric P, Silva M, Nelson J, Lipner V, Giordano S, Horowitz A and D'Amore M (1992).

Prasad RD and Rangeshwaran R (1999). Granular formulation of Trichoderma and Gliocladium spp. in Biocontrol of *Rhizoctonia solani* of chickpea. *J. Mycol. Pl. Pathol.* 29: 222–226.

Prasad RD and Rangeshwaran R (2000), Effect of soil application of granular formulation of Trichoderma on Rhizoctonia solani incited seed rot and damping off of chickpea. *J. Mycol. Pl. Pathol.* 30: 216–220.

Press CM, Loper JE and Kloepper JW (2001). Role of iron in rhizobacteria mediated induced systemic resistance of cucumber. *Phytopathology* 91: 593–598.

Product Safety Commission (2006). At: /http://www.cpsc.gov/LIBRARY/ FOIA/FOIA05/os/Electrocutions2001.pdfS.

Raaijmakers JM, Vlami M and De Souza JT (2002). Antibiotic production by bacterial biocontrol agents. *Anton. van Leeuw.* 81: 537–547.

Raaijmakers JM, Vlami M and De Souza JT (2002). Antibiotic production by bacterial biocontrol agents. *Anton. van Leeuw.* 81: 537–547.

Ramette A, Moënne-Loccoz Y and Défago G (2003). Prevalence of fluorescent pseudomonads producing antifungal phloroglucinols and/or hydrogen cyanide in soils naturally suppressive or conducive to tobacco root rot. FEMS Microb. *Ecol.* 44: 35–43.

Rathour KS, Sharama S, Ganguly S and Pandey, Rajesh Kumar (2007). Community analysis of plant parasitic nematodes associated with some ornamental, medicinal and aromatic plants in Bareilly District of Uttar Pradesh. *International J. Nematology* 17: 9–12. [UK]

Raupach GS and Kloepper JW (1998). Mixtures of PGPR enhance biological control of multiple cucumber pathogens. *Phytopathology* 88: 1158-1164.

Ravlin FW and Robinson WH (1985). Bull. Entomol. *Soc. Am.* 31, 45–50.

Ridsill-Smith J, and Pavri C (2000). Single spring spray protects pastures. Farming Ahead No. 103, July 2000, 60.

Rodrigues LC and Niemeyer HM (2005). Integrated pest management, semiochemicals and microbial pest-control agents in Latin American agriculture. *Crop Protection* 24: 615–623.

Ross EW and Marx DM (1972). Susceptibility sand pine to Phytophthora cinnamomi. *Phytopathology* 62: 1197–1200.

Ross G (2005). Risks and benefits of DDT. Lancet 366 (November (9499)), 1771.

Roy AK (1977). Parasitic activity of *Trichoderma viride* on sheath blight fungus of rice (Corticium sasakii) *J. Pl. Dis. Prot.* 84: 675–683.

Ryu CM, Farag MA, Hu CH, Reddy MS, Kloepper JW and Pare PW (2004). Bacterial volatiles induce systemic resistance in Arabidopsis. *Plant Physiol.* 134: 1017–1026.

Ryu CM, Farag MA, Hu CH, Reddy MS, Wei HX, Paré PW and Kloepper JW (2003). Bacterial volatiles promote growth in Arabidopsis. *Proc. Nat. Acad. Sci.* 100: 4927–4932.

Samson RA and Rombach MC (1985). Biology of the fungi Verticillium and Aschersonia. In: Biological pest control: The glasshouse experience (Hussey N.Y. & Scepes N., eds). Sorset Blanford Press, 34–42 [This article presents the details on the biology of both fungi]

Sandra AI, Wright CH, Zumoff LS and Steven VB (2001). Pantoea agglomerans strain EH318 produces two antibiotics that inhibit Erwinia amylovora in vitro. *Appl. Environ. Microbiol.* 67: 282–292.

Scheuerell SJ, Sullivan DM and Mahaffee WF (2005). Suppression of seedling damping-off caused by *Pythium ultimum, P. irregulare,* and *Rhizoctonia solani* in container media amended with a diverse range of Pacific Northwest compost sources. *Phytopathology* 95: 306–315.

Schouten A, Van den Berg G, Edel-Hermann V, Steinberg C, Gautheron N, Alabouvette C, De Vos CH, Lemanceau P and Raaijmakers JM (2004). Defense responses of *Fusarium oxysporum* to 2,4-DAPG, a broad spectrum antibiotic produced by Pseudomonas fluorescens. *Mol. Plant-Microbe Interact.* 17: 1201–1211.

Sen Bineeta (2000). Biological control: A success story. *Indian Phytopath.* 53: 243–249.

Services, U.S. Department of Agriculture, www.healthierus.gov/dietaryguidelines.

Shakil NA, Pankaj, Kumar J, Pandey, Rajesh Kumar and Saxena DB (2008). Nematicidal prenylated flavanones from *Phyllanthus niruri*. *Phytochemistry*. 69(3): 759-764 [www.elsevier.com/locate/phytochem]

Shanahan P, O'Sullivan DJ, Simpson P, Glennon JD and O'Gara F (1992). Isolation of 2, 4-Diacetylphloroglucinol from a fluorescent pseudomonad and investigation of physio-logical parameters influencing its production. *Appl. Environ. Microbiol.* 58: 353–358.

Shanker P and Jeyarajan R (1996). Seed treatment of Trichoderma and Gliocladium for biological control of Macrophomina phaseolina in sesamum. *Indian Phytopath.* 49: 148–151.

Sheo Raj, Singh RK, Prajapati SSL, Srivastava Pandey, Rajesh Kumar and PK Gupta (2007). Evaluation of different botanicals and non-target pesticides against Sclerotium rolfsii causing collar rot of lentil. *Indian Phytopathology.* 60 (4): 499–501

Shishido M, Miwa C, Usami T, Amemiya Y and Johnson KB (2005). Biological control efficiency of fusarium wilt of tomato by nonpathogenic F. oxysporum Fo-B2 in different environments. *Phytopathology* 95: 1072–1080.

Sikora R (1992). Management of antagonistic potential in agricultural ecosystems for the biological control of plant parasitic nematodes. *Annu. Rev. Phytopathol.* 30: 245–270.

Silva HSA, Romeiro RS, Macagnan D, Halfeld-Vieira BA, Pereira MCB and Mounteer A (2004). Rhizobacterial induction of systemic resistance in tomato plants: non-specific protection and increase in enzyme activities. *Biol. Control.* 29: 288–295.

Singh, Sheo Raj, Prajapati RK, Srivastava SSL and Pandey, Rajesh Kumar (2007). Influence of soil texture, temperature and relative humidity on development of collar rot of lentil. *Indian Phytopathology.* 60 (2): 273–274.

Smith KP, Havey MJ and Handelsman J (1993). Suppression of cottony leak of cucumber with *Bacillus cereus* strain UW85. *Plant Dis.* 77: 139–142.

Sneh B, Dupler M, Elad Y and Baker R (1984). Chlamydospore germination of Fusarium oxysporum f. sp. cucumerinum as affected by fluorescent and lytic bacteria from Fusarium suppressive soils. *Phytopathology* 74: 1115–1124.

Spadaro D and Gullino ML (2005). Improving the efficacy of biocontrol agents against soilborne pathogens. *Crop Prot.* 24: 601–613.

Stevens C, Khan VA, Rodriguez-Kabana R, Ploper LD, Backman PA, Collins DJ, Brown JE, Wilson MA and Igwegbe ECK (2003). Integration of soil solarization with chemical, biological, and cultural control for the management of soilborne disease of vegetables. *Plant Soil* 253: 493-506.

Stockwell VO, Johnson KB, Sugar D and Loper JE (2002). Antibiosis contributes to biological control of fire blight by Pantoea agglomerans strain Eh252 in orchards. *Phytopathology* 92: 1202–1209.

Studies and Toxicology, Commission on Life Sciences. National Academy Press, 31.

Tari PH and Anderson AJ (1988). Fusarium wilt suppression and agglutinability of *Pseudomonas putida. Appl. Environ. Microbiol.* 54: 2037–2041.

Taylor AG and Harman GE (1990). Concept and technologies of selected seed treatment. *Ann. Rev. Phytopathology.* 28: 312–339.

Templeton SR, Zilberman D, Jick Too S (1998). An economic perspective on outdoor residential pesticide use. *Policy Anal.* 32 (17), 416 A–423 A.

Tewari AK and Mukhopadhyay AN (2001). Testing of different formulation of Gliocladium virens against chickpea wilt complex. *Indian Phytopath.* 54: 67–71.

Thomashow LS and Weller DM (1988). Role of a phenazine antibiotic from Pseudomonas fluorescens in biological control of Gaeumannomyces graminis var. tritici. *J. Bact.* 170: 3499–3508.

Thomashow LS, Bonsall RF and Weller DM (2002). Antibiotic production by soil and rhizosphere microbes in situ. Pages 638-647 in: Manual of Environmental Microbiology (2nd ed.), ASM Press, Washington DC.

Thomashow LS, Weller DM, Bonsall RF and Pierson LS III (1990). Production of the antibiotic phenazine-1-carboxylic acid by fluorescent pseudomonas in the rhizosphere of wheat. *Appl. Environ. Microbiol.* 56: 908–912.

Thomson SV and Gouk SC (2003). Influence of age of apple flowers on growth of Erwinia amylovora and biological control agents. *Plant Dis.* 87: 502–509.

Timms-Wilson TM, Kilshaw K and Bailey MJ (2004). Risk assessment for engineered bacteria used in biocontrol of fungal disease in agricultural crops. Plant Soil 266: 57-67.

Townson H, Nathan MB and Zaim M et al. (2005). Exploiting the potential of vector control for disease prevention. *Bull. World Health Organ.* 83 (12), 942–947.

US Environmental Protection Agency (1996). Microbial pesticide test guidelines. OPPTS 885.0001. Overview for microbial pest control agents. EPA 712-C-96-280.

US Environmental Protection Agency (1996). Microbial pesticide test guidelines. OPPTS 885.5000. Background for microbial pesticide testing. EPA 712-C-96-056.

Upadhyay JP and Mukhopadhyay AN (1986). Biological control of Sclerotium rolfsii by *Trichoderma harzianum* in sugarbeet. *Tropical Pest Management.* 32: 215–220.

Upadhyay JP, HC Lal and Ojha KL, (2000). Advances in biological Management of Plant diseases. In:. Advances in plant disease management. (Eds. Narain et al.) Advance Publishing Concept New Delhi. 279–288.

Upton AC (1982). The biological effects of low level ionising radiation. *Sci. Am.* 246, 41–49.

US Congress Office of Technology Assessment. 1995. Biologically-based technologies for pest control. OTA-ENV-636. US Government Printing Office, Washington, DC.

US Dietary Guidelines, 2005. U.S. Department of Health and Human.

Vallad GE and Goodman RM (2004). Systemic acquired resistance and induced systemic resistance in conventional agriculture: review and interpretation. *Crop Sci.* 44: 1920–1934.

Vallad GE and Goodman RM (2004). Systemic acquired resistance and induced systemic resistance in conventional agriculture: review and interpretation. *Crop Sci.* 44: 1920–1934.

Van Dijk K and Nelson EB (2000). Fatty acid competition as a mechanism by which Enterobacter cloacae suppresses Pythium ultimum sporangium germination and damping-off. *Appl. Environ. Microbiol.* 66: 5340–5347.

Van Dijk K and Nelson EB (2000). Fatty acid competition as a mechanism by which Enterobacter cloacae suppresses Pythium ultimum sporangium germination and damping-off. *Appl. Environ. Microbiol.* 66: 5340–5347.

Van Lenteren JC, Babendreier D, Bigler F, Burgio G, Hokkanen HMT, Kuske S, Loomans AJM, Menzler-Hokkanen I, Van Rijn PCJ, Thomas MB, Tommasini MG and Zeng QQ (2003). Environmental risk assessment of exotic natural enemies used in inundative biological control. *BioControl* 48: 3–38.

Van Loon LC, Bakker PAHM and Pieterse CMJ (1998). Systemic resistance induced by rhizosphere bacteria. *Annu. Rev. Phytopathol.* 36: 453–483.

Van Peer R and Schippers B (1992). Lipopolysaccharides of plant-growth promoting *Pseudomonas* sp. strain WCS417r induce resistance in carnation to Fusarium wilt. *Neth. J. Plant Pathol.* 98: 129–139.

Van Wees SCM, Pieterse CMJ, Trijssenaar A, Van't Westende Y and Hartog F (1997). Differential induction of systemic resistance in Arabidopsis by biocontrol bacteria. *Mol. Plant-Microbe Interact.* 10: 716–724.

Varshney S, Pandey RC, Pandey, Rajesh Kumar, Dwivedi BK and Singh V (2004). Interaction between mungbean yellow mosaic virus and root knot nematode on growth of mungbean plants. *Pakistan Journal of Nematol.* 22 (2): 72–78.

Vidhyasekaran P, Sethuraman K, Rajappan K and Vasumathi K (1997). Powder formulations of Pseudomonas fluorescens to control pigeonpea wilt. *BioCont.* 8: 166–171.

Vishwa Dhar and Chaudhary RG (2001). Disease resistance in pulse crops-current status and future approachs. In: Role of Resistance in Intensive Agriculture (Edt. Nagarajan, S. and Singh, D.P.). Kalyani Publishers, New Delhi. 144–157.

Voisard C, Keel C, Haas D and Defago G (1989). Cyanide production by Pseudomonas fluorescens helps suppress black root of tobacco under gnotobiotic conditions. *EMBO J.* 8: 351–358.

Wang H, Hwang SF, Chang KF, Turnbull GD and Howard RJ (2003). Suppression of important pea diseases by bacterial antagonists. *BioControl* 48: 447–460.

Warren GF (1998). Spectacular increases in crop yields in the United States in the twentieth century. *Weed Technol.* 12, 752.

Webster JPG and Bowles RG (1996). Estimating the economic costs and benefits of pesticide use in applies. *In: Proceedings of BCPC Conference, Brighton.* 325–330.

Webster JPG, Bowles RG and Williams NT (1999). Estimating the economic benefits of alternative pesticide usage scenarios: wheat production in the United Kingdom. *Crop Prot.* 18, 83.

Weller DM and Cook RJ (1983). Suppression of take-all of wheat by seed treatments with fluorescent pseudomonads. *Phytopathology* 73: 463–469.

Weller DM, Raaijmakers J, McSpadden Gardener B and Thomashow LM (2002). Microbial populations responsible for specific soil suppressiveness to plant pathogens. *Annu. Rev. Phytopathol.* 40: 309–348.

Weller DM, Raaijmakers J, McSpadden Gardener B and Thomashow LS (2002). Microbial populations responsible for specific soil suppressivenes. *Annu. Rev. Phytopathol.* 40: 309–348.

Wilhite SE, Lumsden RD and Strancy DC (2001). Peptide synthetase gene in Trichoderma virens. *Appl. Environ. Microbiol.* 67: 5055–5062.

World Health Organization (WHO) (1979). Data sheet on the biological control agents Bacillus thuringiensis serotype H-14. WHO/VBC/ 79.750 Rev. 1; VBC/BCDS/ 79.01 [This article provides a data sheet on the biological control agent *B. thuringiensis* serotype H-14]

World Health Organization (WHO) (1982). Guidelines for production of Bacillus thuringiensis H-14. Proceedings of a consultatoin held in Geneva, Switzerland, 25–28 October 1982, 124 [This guideline provides extensive data for the production of B. thuringiensis H-14]

World Health Organization (WHO) (1984). Characteristics of IPS-82 as standard for biological assay of Bacillus thuringiensis H-14 preparation. WHO/VBC/84.892. [This guideline provides data for the standardization of *Bacillus thuringiensis* H-14 products]

Yadav RS, and Sampath RR (2001). Deltamethrin treated bednets for control of malaria transmitted by Anopheles culicifacies (Diptera: Culicidae) in India. *J. Med. Entomol.* 38 (5), 613.

Yancy J and Cecil H (1998). Study Touts Herbicide Benefits, Vol. 32 (no. 11). Southeast Farm Press, 16.

Yin B, Valinsky L, Gao X, Becker JO and Borneman J (2003). Bacterial rRNA genes associated with soil suppressiveness against the plant-parasitic nematode Heterodera schachtii. *Appl. Environ. Microbiol.* 69: 1573–1580.

Yin B, Valinsky L, Gao X, Becker JO and Borneman J (2003). Identification of fungal rDNA associated with soil suppressiveness against Heterodera schachtii using oligonucleotide fingerprinting. *Phytopathology* 93: 1006–1013.

Zakharenko VA (1975). Economics and prospects of pesticide application as related to the intensification of farming. In: Papers at Sessions, VIII International Congress of Plant Protection, Moscow, USSR, 2, 6.

Zettler LJ and Arthur H (2000). Chemical control of stored product insects with fumigants and residual treatments. *Crop Prot.* 19, 577–582.

12

Arbuscular Mycorrhizal Fungi and Its Interactions with Environment

Neeraj* and Kanchan Singh

Botany Department, Feroze Gandhi College, Rae Bareli-229 001
Uttar Pradesh, India
**Email: aj_neer1@rediffmail.com; drneeraj.india@gmail.com*

1 INTRODUCTION

Mycorrhiza is a mutualistic beneficial association between fungi and plant roots. It is more or less a universal phenomenon throughout the plant kingdom (Mosse et al., 1981; Brundrett 2009). More than 80% of all land plant families are thought to have a symbiotic relationship with AM fungi that belong to the Glomeromycota. This interaction of arbuscular-mycorrhizal (AM) symbiosis is the evolutionary precursor of most other mutualistic root-microbe associations (House and Fester 2005). The origin of AM symbiosis is thought to be in the early Devonian period, approximately 400 million years ago hence, AM symbiosis is also called the mother of all plant root endosymbioses (Parniske 2008). This symbiotic relationship benefits both—the individual plant and the fungus (Francis and Read 1995; Smith and Smith 1996). The efficiency of each AM-fungus (AMF) for increasing plant growth, nutrient contents, water-stress tolerance (Vazquez et al., 2001) and providing defense against several pathogenic organisms (Kothamasi et al., 2001) is well documented. In endomycorrhiza, the fungus grows inter- and/or intra-cellularly. Specific fungal structures of endomycorrhiza are produced within the cortical cells by non-septate fungi which are commonly known as vesicles and arbuscules hence, it was called vesicular-arbuscular mycorrhiza (VAM). As many endomycorrhiza forming fungi do not necessarily form internal vesicles, the abbreviated term 'VAM' was suggested to be replaced by 'AM' by Daft and Nicolson (1974). The establishment of the mycorrhizal network offers a number of basic advantages for the acquisition of mineral nutrients: (i) fungal hyphae extend beyond the area of nutrient depletion surrounding the root; (ii) fungal hyphae greatly increase the surface area for the absorption of nutrients relative to non-mycorrhizal roots; (iii) hyphae are able to extend into soil pores that are too small for roots to enter; and (iv) some mycorrhizal fungi can access forms of N and P that are unavailable to non-mycorrhizal plants, particularly organic forms of these nutrients (Morgan et al., 2005).

2 TYPES OF MYCORRHIZAE

Historically, mycorrhizae have been categorized into eight groups (Arbuscular mycorrhiza, Endomycorrhiza, Ectomycorrhiza, Ectendomycorrhiza, Ericoid mycorrhiza, Arbutoid mycorrhiza, Monotropoid mycorrhiza and Orchid mycorrhiza) based primarily on the structural characteristics.

Roots of most extant plants are able to interact with a small group of fungi. Armstrong and Peterson (2002) described two major structural classes of AM symbioses – Arum-types and Paris-types, which have been distinguished from fungal morphological differences observed in the roots of host plant. Although Paris-type is the most common, it is the Arum-type that has been most intensively studied in terms of structure and function because of its prevalence in agronomical important plant species. The interface between the host cell cytoplasm and intracellular hyphae, i.e., extensive hyphal coils and arbusculate coils, typify the Paris-type mycorrhiza. Dramatic changes in the cytoskeleton in colonized cells were also observed between the fungus and host cell cytoplasm. These were less adjacent to the fine branches of arbusculate coils compared to the hyphal coils. These observations show some similarities to those seen with Arum-type mycorrhizae. It has also been observed that the time required for the development of Paris-type AM is less than that for the Arum-type (Cavagnaro et al., 2001). Colonizing strategy of AM fungi differs considerably and this variation is used for classifying families (Hart and Richard 2002).

2.1 A Family has Now become a Phylum

Earlier, endomycorrhiza forming fungi were described as genus *Endogone* and placed in family endogonaceae of class Zygomycotina. Later, it was raised to order Glomales and then several described genera like *Glaziella, Modicella, Complexipes* and *Endogone* were removed (INVAM website; www.http//invam.caf.wvu.edu). The order Glomales was divided into four families – Glomaceae (genera *Glomus, Sclerocystis*), Acaulosporaceae (genus *Acaulospora*), Gigasporaceae (genara *Gigaspora, Scutellospora*) and Entrophosporaceae (genus *Entrophospora*) (INVAM website; www.http//invam.caf.wvu.edu). Later on, Morton and Redecker (2001) raised its status to phylum Glomeromycota which was divided into two groups Glomineae and Gigasporineae. Glomineae had four families Glomaceae (genus *Glomus*); Acaulosporaceae (genera Acaulospora, *Entrophospora*); Archaeosporaceae (genus *Archaeospora*); Paraglomaceae (genus *Paraglomus*) whereas Gigasporaneae had only one family Gigasporaceae (genera *Gigaspora, Scutellospora*). Later, Schussler et al. (2001) corrected the previously used, grammatically incorrect ordinal name 'Glomales' to 'Glomerales' and divided the Glomeromycota into four orders:

I. Glomerales consisting of two families: (i) Glomeraceae (Glomus - group A); (ii) Glomeraceae (Glomus - group B);

II. Diversisporales consisting of three families: (i) Gigasporaceae, (ii) Diversisporaceae and (iii) Acaulosporaceae;

III. Archaeosporales consisting of two families: (i) Archaeosporaceae and (ii) Geosiphonaceae, and

IV. Paraglomerales consisting of only one family– Paraglomeraceae.

Again, Glomeraceae was divided into *Glomus*–group Aa and *Glomus* sp W3347, *Glomus*-group Ab and *G. intraradices*, *Glomus*-group B and *G. etunicatum* (INVAM website). A new family Pacisporaceae was added to Diversisporales and another family Ambisporaceae was added to Archaeosporales.

AM fungi have limited morphological characters for species definition and recognition, but through molecular analyses the phylogenetic relationships within these organisms are being elucidated. Combined morphological, cytological and molecular evidences reveal a new clade within the Glomeromycota. Reinforcing the value of molecular data-sets in providing a more clear understanding of phylogenetic relationships Morton and Redecker (2001) believed that it can lead to a more robust taxonomy.

Wubet et al. (2003) studied 5.8 S/ITS 2 and DNA sequence analyses which indicated high AM fungal diversity and dominance of *Glomus* type. There were 20 sequence types which belonged to family Glomeraceae and one each to the Diversisporaceae and Archaeosporaceae. The combination of DNA sequences, fatty acid profiles, immunological reactions against specific monoclonal antibodies and mycorrhizal morphology, provide the base for recognizing two distinct genera, viz., *Archaeospora* and *Paraglomus* (these two ancestral clades of AM fungal species were separated due to widely divergent ribosomal DNA sequences).

Gerdemanniaceae has only one genus *Gerdemannia* (Walker et al., 2004). This family differs from other families in the Glomeromycota by possession of glomoid spores with a combination of amyloid inner wall components and germination by means of a germination shield, coupled with the formation of VAM and by the possession of the SSU and RNA gene sequence signature TTATCGGTTRAATIC.

A new genus described in the family Glomeraceae, order Glomerales of class Glomeromycetes was named *Pacispora* with the type species of *P. scinlilians* which was formerly placed in the genus *Glomus* of the Glomeraceae as *P. dominikii* and *P. chimnobambusae* (Fritz and Sieverding 2004). Four other new species of the new genus are: *P. franciscana,*

P. robigina, P. coralloidea and *P. boliviana.* Spores of the new genus are formed terminally on hyphae, a feature they share only with *Glomus* and *Paraglomus.* An inner, usually three-layered wall is present inside the spores, from which the spores germinate directly through the outer spore wall, which usually is also three layered. This germination characteristic is shared with *Scutellospora, Acaulospora* and *Entrophospora* but not with *Glomus* and *Paraglomus.* Thus, *Pacispora* is currently placed in a separate family Pacisporaceae.

Five mycorrhizal fungal species of the genus, *Entrophospora* have also been revised. They all form their spores within the hyphal stalk directly beneath or at some distance of a soporiferous saccule formed intercalary or terminally in the mycelium. Based on differences, respective similarities in spore morphologies and root infection characteristics only *Entrophospora infrequens* and *Entrophospora baltica* are retained in this genus while *Kuklospora* gen. nov. with *Kuklospora columbiana* and *Kuklospora kentinensis* (formerly *Entrophospora columbiana* and *Entrophospora kentinensis,* respectively) are placed into the family Acaulosporaceae. *Intraspora* gen. nov. so far contains only *Intraspora schenckii* (formerly, *Entrophospora schenckii*) and is included into the family Archaeosporaceae (Sieverding and Fritz 2006).

Another new AM fungal genus *Appendicispora* (Archaeosporaceae) was described by Spain et al. (2006). The names of new species are (i) *Ap. appendicula,* the type species derived from the dimorphic *Acaulospora appendicula,* a synonym of *Ac. gerdemannii* erroneously thought to represent the acaulosporoid form of *Archaeospora leptoticha* (ii) *Ap. gerdemannii,* based on *Archaeospora* (*Glomus*) *gerdemannii;* and (iii) *Ap. jimgerdemannii,* a nom. nov. for the former *Ac. gerdemannii. Archaeospora trappei,* which remains alone in the now monotypic genus *Archaeospora* differs from the *Appendicispora* species in spore development, wall morphology, germination, and root colonization structures. Its separation is further supported by molecular data (Spain et al., 2006).

The family Archaeosporaceae and the genus *Archaeospora* are rendered paraphyletic by the relationship with the Geosiphonaceae. Members of the Archaeosporaceae were described as forming both glomoid and acaulosporoid spores, or solely acaulosporoid spores. However, it was found that *Glomus callosum* fell into the same phylogenetic clade as *A. leptoticha* and *A. gerdemannii,* but exclusively formed glomoid spores (Walker et al., 2007). To resolve these inconsistencies, a genus, *Ambispora* gen. nov., typified by *Ambispora fennica* sp. nov. was erected based on morphological evidence and SSU and ITS region rDNA data. *Ambispora* contains three species known to produce both acaulosporoid and glomoid spores: *A. fennica, A. leptoticha* comb. nov. (basionym *G. leptotichum*) and *A. gerdemanii* comb. nov. (basionym *G. gerdemannii*). Another species *A. callosa* comb. nov. (basionym *G. callosum*) known to form only glomoid spores *Ambispora* is placed in a new family, the Ambisporaceae. fam. nov. (Walker et al., 2007).

Palenzuela et al. (2008) created a new genus *Otospora bareai* characteristically brown spores (140–210 mm diam.) formed laterally on a persistent, brown stalk (-neck) of a soporiferous saccule. They have two walls without ornamentation— a brown, 3 to 4 layered outer wall and another hyaline, 2 to 3 layered inner wall. Spore formation is similar to that of *Acaulospora* spp. and *Archaeospora trappei*, but *Acaulospora* spp. has 3 spore walls with a characteristic 'beaded' wall and the outer wall of *Archaeospora trappei* is simple, thin, hyaline and only bilayered. Spore wall structure of this new fungus is similar to that of *Entrophospora infrequens*. However, this fungus forms its spores internally, inside the hyphal stalk of the new fungus next to *Diversispora spurca* which forms one-walled glomoid spores (i.e., terminally on hyphae). Based on these analyses Palenzuela et al. (2008) placed this fungus into a new genus in the family Diversisporaceae under the epithet *Otospora bareae*.

Further, a new genus *Indiana* was added to the family Pacisporaceae, with spores having characteristics of two genera — *Glomus* and *Gigaspora*. The genus is named *Indiana* since the type species was reported from India (Khade 2010). Fungi are hypogeous or partly epigeous, forming arbuscular mycorrhizae with or without vesicles and/or auxillary cells. In nature, the spores of this genus might have been formed due to the fusion of sporogenous hypha of the two genera—*Glomus* and *Gigaspora*. The spores, 60-150 mm in diameter, are light yellow to yellow with reddish-brown spore walls; consisting of two recurved hyphal attachments at opposite ends — one *Glomus* like and the other one *Gigaspora* like (Khade 2010).

As on today, phylum Glomeromycota has a single class Glomeromycetes which has been divided into four orders: (a) Glomerales with only one family Glomeaceae having only one genus *Glomus*; (b) Diversisporales which has six families, namely, Acaulosporaceae with two genera *Acaulospora, Kuklospora*; Diversisporaceae with two genera *Diversispora, Otospora*; Gigasporaceae with two genera *Gigaspora, Scutellospora*; Entrophosporaceae with genus *Entrophospora*; Gerdemanniaceae with genus *Gerdemannia*; and Pacisporaceae with two genera *Pacispora* and *Indiana*; (c) Paraglomerales which has only one family Paraglomeraceae with genus *Paraglomus* and (d) Archaeosporales which has three families, namely, Ambisporaceae with one genus *Ambispora*; Archaeosporaceae with two genera *Archaeospora, Intraspora* and Geosiphonaceae with genus *Geosiphon*.

Till 2009 fourteen new species were reported from India (Muthukumar 2010). Recently, one new genus named *Indiana,* having the combined characteristics of *Glomus* and *Gigaspora* has been described from Goa. These have been summarized in Table 1.

Table 1 AM Fungi reported new from India (after Muthukumar 2010)

1976	*Glomus multicaulis* Gerd. and Bakshi	Dehradun
	Sclerocystis sinuosa Gerd. and Bakshi	Dehradun, Nainital
1980	*Glomus reticulatum* Bhattacharjee and Mukerji	Bangalore
	Sclerocystis indicus Bhattacharji and Mukerji	Nainital
1982	*Gigaspora candida* Bhattacharji, Mukerji and Tewari	Delhi
1983	*Glomus delhiense* Mukerji, Bhattacharji and Tewari	Delhi
	Glomus multisubstensum Mukerji, Bhattacharji and Tewari	Delhi
1992	*Glomus sterilum* Mehrotra and Bajaj	Allahabad
1993	*Gigaspora tuberculata* Neeraj, Mukerji, Sharma and Varma	Alwar
1997	*Glomus bagyarajii* Mehrotr	Unknown
2002	*Glomus indica* Manoharachary, Sharathbabu and Adholeya Hyderabad	
	Glomus monovesiculatum, Khan and Uniyal	Haryana
2004	*Glomus hyderabadensis* Swarupa, Kunwar, Prasad and Manoharachary	Hyderabad
2009	*Glomus goaensis* Khade	Goa
2010	*Indiana bulbigoa Khade gen. et* sp. *nova* (Pacisporaceae)	Goa

A new phylogenetic classification of fungi given by 66 authors is based on molecular data. The first table of this classification deals with the older and simpler groups, including the Chytrids, the Zygomycetes and the Glomeromycota (Hibbett et al., 2007; Fig. 1). In December 2010 the classification of Glomeromycota has again been revised by Schussler and Walker.

Recent Classification of AM fungi

Phylum Glomeromycota

Class Glomeromycetes

Order Glomerales

Family Glomeraceae

Genus *Glomus*

Order Diversisporales

Family Gigasporaceae

Genus *Gigaspora*

Genus *Scutellospora*

Family Acaulosporaceae

Genus *Acaulospora*

Genus *Kuklospora*

Family Entrophosporaceae

Genus *Entrophospora*

Family Pacisporaceae

Genus *Pacispora*

Genus *Indiana*

Family Diversisporaceae

Genus *Diversispora*

Genus *Otospora*

Family Gerdemanniaceae

Genus *Gerdemannia*

Order Paraglomerales

Family Paraglomeraceae

Genus *Paraglomus*

Order Archaeosporales

Family Geosiphonaceae

Genus *Geosiphon*

Family Ambisporaceae

Genus *Ambispora*

Family Archaeosporaceae

Genus *Archaeospora*

Genus *Intraspora*

3 NUTRIENT TRANSPORTERS HELP AM FUNGI

Improvement in nutrient uptake by mycorrhizal plants is attributed to transporter macromolecules (Yadav et al., 2005). Various nutrient transporters in the mycorrhizal tissues have been identified (Chalot et al., 2002). Six transporters from mycorrhizal fungi have been fully characterized (Fig. 2) and many more are being characterized.

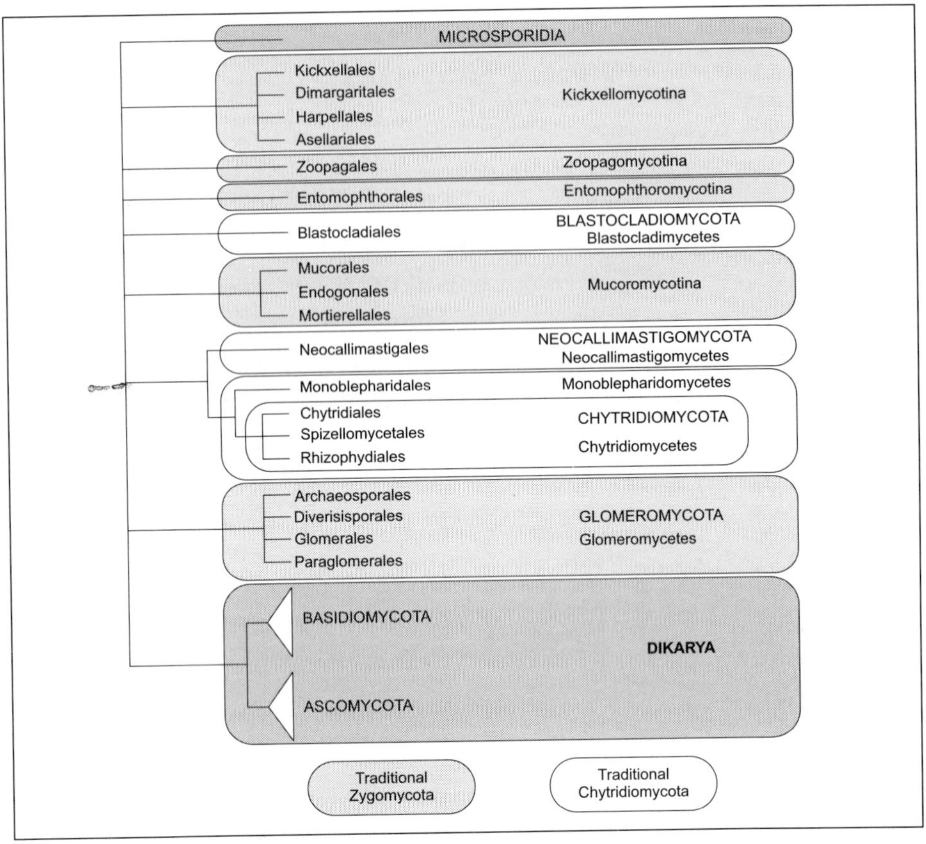

Fig. 1 Position of Glomeromycota in the classification given by Hibbett et al. (2007)

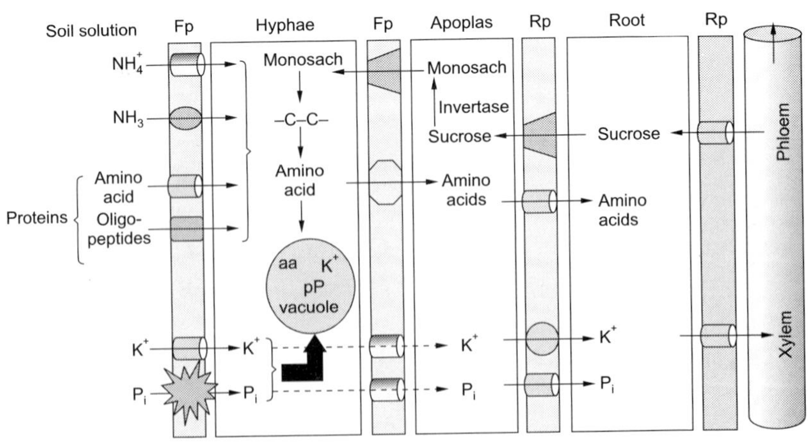

Fig. 2 Summary of current understanding of transporters in mycorrhizal tissues. Fp: fungal plasma membrane, Rp: root plasma membrane, pP: polyphosphate. Black cylinder: one member of the transporter family has been fully characterized, i.e., functional complementation of a yeast deficient strain; grey cylinder: putative transporters (candidate genes available in i.e., EST resources); white cylinder: hypothetical transporters (c.f. Chalot et al., 2002).

Nutrient fluxes from fungus to host largely depend upon coordination with nutrient release by the fungus into, and nutrient re-uptake by the root cells from the apoplastic interface (Fig. 3). The fungal transporters include: ammonium transporters, amino acid transporters, potassium transporters, phosphorus transporters, etc.

Besides these, two more plant transporters have been confirmed but sugar transporters are more important because:

1. Fungus gets reduced carbon (sugar) from host.
2. Sucrose is delivered into apoplast of the plant fungal interface and hydrolyzed via a plant derived invertase.

Laser micro-dissection (LMD) reveals that transcripts for five plant and one fungal phosphate transporter (PT) genes are contemporaneously present in arbusculated cells (Balestrini et al., 2007). In *Lycopersicon esculentum* transcripts for plant phosphate transporters (LePTs), selected as molecular markers for a functional symbiosis, have been detected by reverse-transcriptase polymerase chain reaction assays and associated to distinct cell types, leading to novel insights into the distribution of LePT mRNAs. In fact, the transcripts of the 5 phosphate transporters (PTs) have been detected contemporaneously in the same arbusculated cells, unlike from the neighboring non-colonized cells. In addition, fungal H^+ATPase (Gm HA5) and phosphate transporter (Gmos PT) mRNAs were found exclusively in arbusculated cells. The discovery that five plant and one fungal PT genes are consistently expressed inside the arbusculated cells provides a new scenario for plant fungus nutrient exchanges (Balestrini et al., 2007).

The AM fungal symbiont colonizes root cortical cells, where it establishes differentiated hyphae called arbuscules. As each arbuscule develops, the cortical cell undergoes a transient reorganization and envelops the arbuscule in a novel symbiosis-specific membrane, called the periarbuscular membrane (Pumplin and Harrison 2009). This membrane is continuous with plant plasma membrane of the cortical cell, and forms a key interface in the symbiosis. The periarbuscular membrane is composed of at least two distinct domains, an 'arbuscule branch domain' that contains the symbiosis-specific phosphate transporter, Mt PT4, and an 'arbuscule trunk domain' that contains Mt Bcp1. This suggests a developmental transition from plasma membrane to periarbuscular membrane, with biogenesis of a novel membrane domain associated with the repeated dichotomous branching of the hyphae (Pumplin and Harrison 2009).

Two *Medicago truncatula* half-ABC transporters are reported to be essential for arbuscule development in arbuscular mycorrhizal symbiosis. There is a report that a *Medicago truncatula* mutant, stunted arbuscule (Str), exists in which arbuscule development is impaired and AM symbiosis

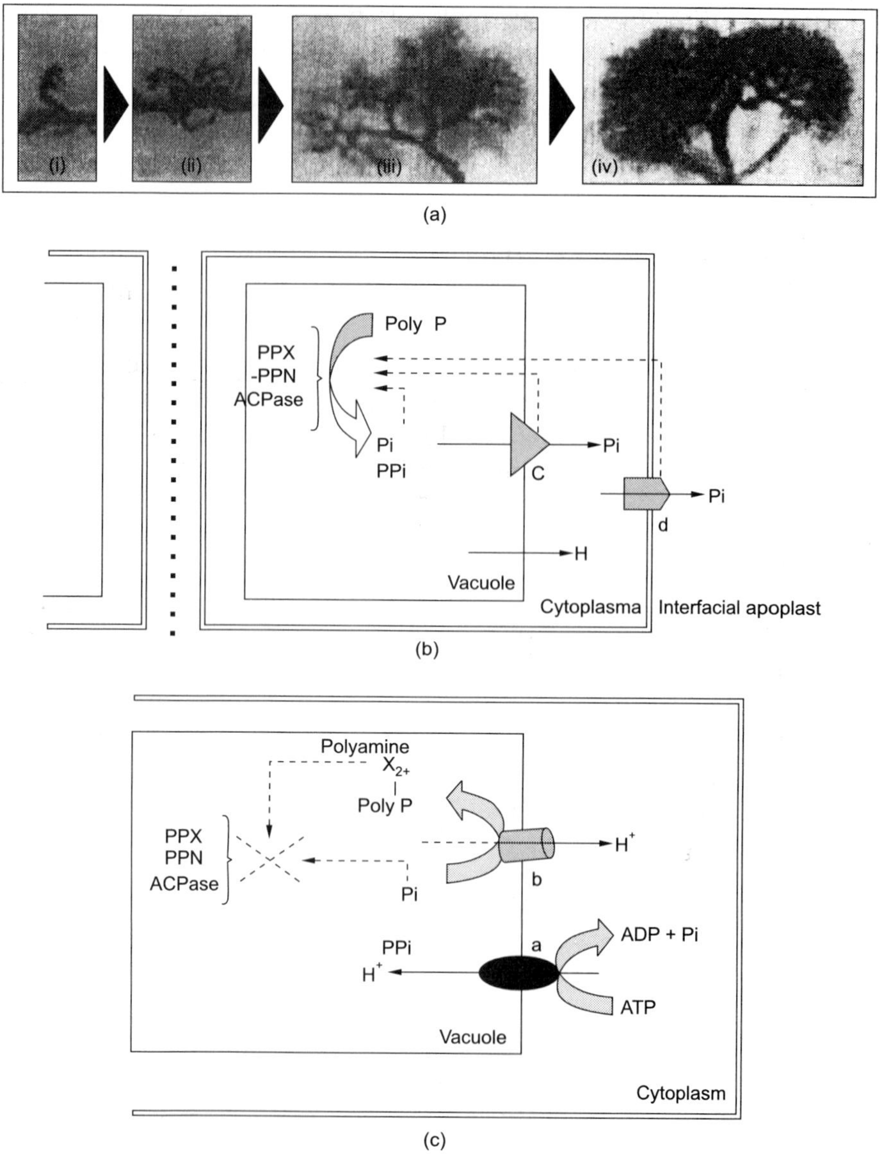

Fig. 3 Hypothetical scheme of Pi efflux from arbuscules. Efflux may be regulated at the following three steps; arbuscule formation, Polyphosphate hydrolysis and Pi efflux through the membrane (a) Development of an arbuscule from initiation of maturity (i) the first trunk of arbuscule is initiated (ii) two major branches are formed (iii) fine branches are being formed (iv) fully developed arbuscule (mature) showing high activity of alkaline phosphates (b) Polyphosphate (poly-P) metabolism in the initial phase of arbuscule formation. V-ATPase (a) may still be active and able to energize poly-P synthesis (b). (c) Poly P hydrolysis during maturation of an arbuscule (c.f. Ezawa et al., 2001).

fails. In contrast with legume symbiosis mutants reported previously, STR shows a wild-type nodulation phenotype. The STR was identified by positional cloning and encodes a half-size ATP binding cassette (ABC) transporter. STR is a representative of a novel clade and its orthologs are highly conserved throughout the vascular plants but absent from *Arabidopsis thaliana* (Zhang et. al., 2010). The STR clade is unusual in that it lacks the taxon-specific diversification that is typical of the ABCG gene family. This distinct phylogenetic profile enabled the identification of a second AM symbiosis-induced half-transporter, STR2. Silencing of STR2 by RNA interference results in a stunted arbuscule phenotype identical to that of STR. Both, STR and STR2 are co-expressed constitutively in the vascular tissue, and expression is induced in cortical cells containing arbuscules. STR heterodimerizes with STR2 and the resulting transporter (a dimer) is located in the periarbuscular membrane where its activity is required for arbuscule development and consequently a functional AM symbiosis.

4 AM FUNGI ARE ASSISTED BY SIGNALLING MOLECULES

Bucher et al. (2009) reported that successful colonization and a functional interaction between host plant and mycobiont are based upon exchange of signalling molecules at different stages of symbiosis development. Strigolactones, a novel class of plant hormones, are secreted by plant roots stimulating pre-symbiotic growth of AM fungi (Akiyama et al., 2005, 2010). The fungi release soluble signalling molecules, the enigmatic 'Myc factors' that activate early symbiotic root responses. Lysophosphatidylcholine is a lipophilic intraradical mycorrhizal signal triggering plant phosphate transporter gene expression regulating AM development through a P-controlled transcriptional mechanism. This enables uptake of orthophosphate released from the AM fungus. Although strigolactones play a critical role as rhizospheric signalling molecules for the establishment of AM symbiosis, the susceptibility of plants to AM fungi and the possible involvement of strigolactones in this AM susceptibility is still a matter of research. The role of strigolactones for the formation and the regulation of the AM symbiosis as well as the possible implication of these compounds as plant signals in other soil-borne plant-microbe interactions are now being worked out. Nowadays, the strigolactones are being considered as the signals not only for arbuscular mycorrhizal fungi but also for parasitic plants (Garcia-Garrido et al., 2009). Two strigolactones (Orobanchol and Orobanchyl acetate) are found in *Arabidopsis thaliana* which is not a host for AMF. Kohlen et al., (2010) showed that at least one of them (orobanchol) was detected in xylem sap and it is up-regulated under phosphate deficiency, which is consistent with the idea that root-derived strigolactones are transported to the shoot, where they regulate branching.

Moreover, two additional putative strigolactone-like compounds were detected in xylem sap, one of which was not detected in root exudates. Thus, xylem transported strigolactones contribute to the regulation of shoot architectural response to phosphate limiting conditions. Lapez-Raez et al. (2011) found that strigolactone production is significantly reduced upon AM symbiosis, considering the dual role of the strigolactones in the rhizosphere as signals for AM fungi and parasitic plants. Application of the synthetic strigolactone analog GR24 has shown that strigolactones are able to modulate local auxin levels and that the net result of strigolactone action is dependent on the auxin status of the plant. On this basis Ruyter-Spira et al. (2011) postulated that the tightly balanced auxin-strigolactone interaction is the basis for the mechanism of the regulation of the plants' root-to-shoot ratio.

Balzergue et al. (2011) found that root exudates of plants grown under high phosphate lost their ability to stimulate AM fungi and lacked strigolactones. In addition, a systemic down-regulation of strigolactone release by high phosphate supply was demonstrated using split-root systems. Nevertheless, supplementation with exogenous strigolactones failed to restore root colonization under high phosphate. They suggested that this observation does not exclude a contribution of strigolactones to the regulation of AM symbiosis by phosphate, but indicates that they are not the only factors involved.

For arbuscule formation, the expression of a *Medicago truncatula* gene named 'Vapyrin' has been reported essential (Pumplin et al., 2010). It is essential also for efficient epidermal penetration by AM fungi. Vapyrin is a cytoplasmic protein, and in cells containing AM fungal hyphae, it accumulates in small amounts that move through the cytoplasm. This novel protein is composed of two domains that mediate protein-protein interactions: (i) N-terminal VAMP – associated protein (VAP)/major sperm protein (MSP) domain and (ii) C-terminal ankyrin-repeat domain. Putative vapyrin orthologs exist widely in the plant kingdom but not in *Arabidopsis*, or in non-plant species. Vapyrin plays a vital role in cellular remodeling to support the intracellular development of fungal hyphae during AM symbiosis (Pumplin et al., 2010).

5 ENDOSYMBIOTIC BACTERIA INHABIT AMF HYPHAE AND SPORES

Arbuscular mycorrhizal fungi have been symbionts of land plants for at least 450 million years. It is now known that some AM fungi in their cytoplasm host rod-shaped, Gram-positive endobacteria called bacterium like organisms (BLOs) are of unknown phylogenetic origin (Bianciotto et al., 2003). The obligate BLOs have been demonstrated inside spores of AM fungus *Gigaspora margarita*. The bacterium lives and multiplies in the living mycelium, moving along the hyphae even when the fungus colonizes the plant roots. The AM fungi which are themselves obligate plant symbionts, represent a specialized niche for rod-shaped bacteria consistently found in many of the Gigasporaceae through all the steps of fungal life cycle. On the basis of their ribosomal sequences, the endobacteria have been identified as a new taxon and named as *Candidatus Glomeribacter gigasporarum* (Bianciotto et al., 2003). These endobacteria affect the metabolic profile of their host *Gigaspora margarita* and have an impact on the host fungal activity, influencing both protein expression and lipid profile. It suggests that bacterial absence is perceived by *G. margarita* as a stimulus which activates stress responsive proteins (Salvioli et al., 2010). There is no doubt that fungal endobacterial interactions cover multiple aspects from evolutionary biology to applied microbiology, and may offer unexpected tools to investigate key steps in the evolution of modern fungi (Lumini et al., 2006). In the context of nutritional strategies, the mechanisms that operate among members of the consortium and that often promote plant growth. Release of active molecules, including volatiles, and physical contact among partners seem important for the establishment of the bacteria/mycorrhizal fungus/plant network (Bonfante and Julia-Andra 2009). Its existence is supported by analysis of 16S rDNA and DNA amplification. The BLO-AMF symbiosis can, therefore, be dated back at least to the times when AMF formed the ancestral symbiosis with emergent land plants (Naumann et al., 2010).

6 MYCORRHIZAL ASSOCIATIONS OF VASCULAR PLANTS: THE GLOBAL DIVERSITY OF HOST PLANTS

It is well established now that ecologically significant functional diversity exists among AM fungi and there are evidences that suggest that such diversity does have significant ecological consequences. Being the most important microbial symbionts for majority of plants, AM fungi directly affect the composition, development and productivity of plant community. Nutrient uptake, water relations and above-ground productivity are also influenced by these, particularly under the phosphate-limiting conditions.

An estimation of the relative diversity of plants with different types of mycorrhizae provided by assigning mycorrhizal associations to angiosperm families is presented by Brundrett (2009) in Fig. 4. In his analysis, the majority of flowering plants (>99%) belonged to families that could be reliably assigned to mycorrhizal categories with the existing data and there are < 200 families yet to be sampled. Most of these unallocated families are likely to contain AM plants as they are sister to, or nested within clades known to predominantly contain AM plants. Even if families reported to contain both NM and AM species (NM-AM) is added to the NM total, 86% of flowering plants are mycorrhizal (Brundrett 2009).

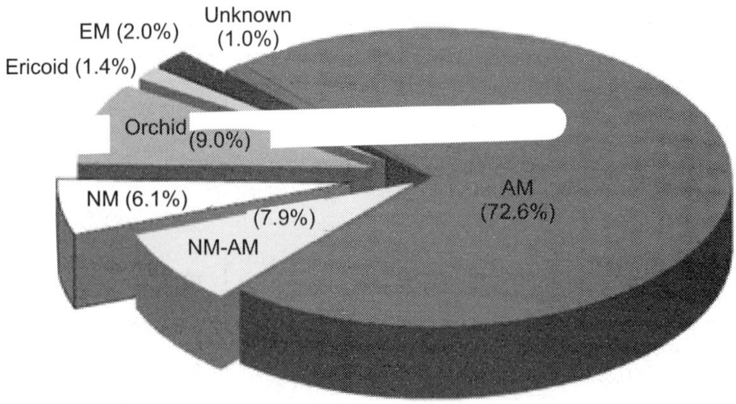

Fig. 4 The relative diversity of angiosperms with different types of mycorrhizal and non-mycorrhizal roots. Less than 1% of angiosperms could not be allocated to any category (unknown). AM- arbuscular mycorrhizae; NM- non-mycorrhizal; NM-AM- variable NM or AM; EM- ectomycorrhizal. (Brundrett 2009).

Kuszala et al. (2001) demonstrated that 20 AMF isolates representing 16 genera were able to sporulate after storage in moist soil for up to 8 months. The ability of mycorrhizal propagules to maintain their viability for relatively longer-periods suggests a degree of resilience that can be beneficial for restoring biodiversity and function of an ecosystem. Despite the longevity of inocula, it is not yet known if mycorrhizal populations are able to reform on plant species re-introduced into a particular habitat after their loss (Johnson et al., 2005).

7 FUNCTION, COMMUNITY STRUCTURE, AND DIVERSITY OF AM FUNGI ARE REGULATED BY PLANTS

This may have arisen through the dominance of a restricted group of fungi that were able to form an extensive and vigorous mycelial network using carbon from the established community of mature plants. In the bare soil, such a supply of carbon would not exist, instead, most colonization of the seedling roots would have occurred from an assemblage of propagules. Because propagules in the bare soil treatment would not have been exposed to these selection pressures, the loss of diversity would be avoided (Johnson et al., 2005). Changes in rhizosphere microbial community may be due to either direct interactions of AM fungi or its effect on host physiology (Solís-Domínguez et al., 2011).

While some plants are likely to be colonized by only a few AMF species, there is now a growing body of evidence suggesting extensive diversity in some grassland species (Johnson et al., 2005). On the other hand, loss of a particular plant species may have disproportionate impacts on the diversity of the AM fungal community (Fig. 5).

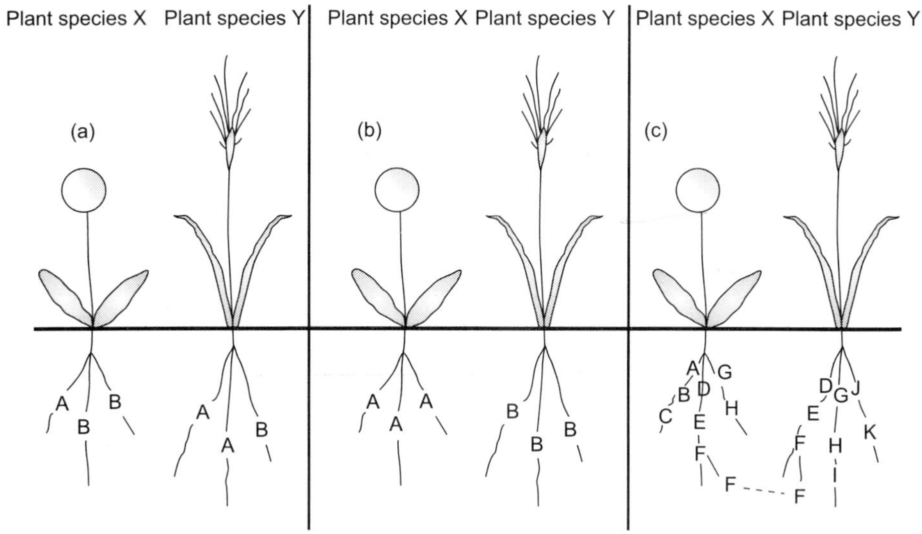

Fig. 5 Models of host specificity in AM fungal symbioses. (a) No host specificity; AM fungal species A and B shared between each host plant. (b) Absolute host specificity; AM fungal species A and B colonize specific plant species. (c) Limited specificity as demonstrated in natural grassland species. Only AM fungal species A-C and I-K are host specific. In this case the length of external mycelium of fungus F differs between plant species, and may or may not form a common network (Johnson et al., 2005).

Chen et al. (2004) constructed a range of plots comprising 0, 1, 2, 4, 8, and 12 species of weeds from a range of plant families including Leguminosae, Ranunculaceae, and Gramineae. The number of AM fungal spores extracted increased progressively with each increment of plant diversity. A similar situation was also observed at the Cedar Creek biodiversity experiment utilizing a similar plant diversity gradient (1, 2, 8, and 16 species; Burrows and Pfleger 2002). In this case, AM fungi in the 16-species plots produced from 30–150% more spores and from 40–70% greater spore volumes than AM fungi in 1-species plots. Although in later experiment soil nitrate concentration was also a significant predictor of AM sporulation in parallel with plant species richness.

8 ELEVATED ATMOSPHERIC CO_2 AFFECTS THE MICROBIAL COMMUNITY STRUCTURE AND ACTIVITIES IN THE RHIZOSPHERE

As the majority of life in soil is heterotrophic and dependent on the input of plant-derived organic C, the activity and functioning of soil organisms will be greatly influenced by changes in the atmospheric CO_2 concentration. The main effects of elevated atmospheric CO_2 on soil micro-biota occur via plant metabolism and root secretion, thereby directly affecting the mycorrhizal, bacterial, and fungal communities in the close vicinity of the root while there is little or no direct effect on the microbial community of the bulk soil (Drigo et al., 2008). Available data indicate that increased

CO_2 generally stimulates mycorrhizal fungi. (Denef et al., 2007; Olsrud et al. 2004). In another experiment, elevated CO_2 increased the species richness of the plant community when AMF were present, but not when they were absent (Johnson et al., 2003; Johnson and Gehring 2007).

The response of soil microorganisms to changes in plant production under elevated CO_2 is highly variable due to very different patterns of plant C allocation in different plant-soil systems. Microbial biomass, gross N mineralization, microbial immobilization, and net N mineralization under elevated CO_2 show a high degree of variability within and between plant life forms (Zak et al., 2000). However, rates of soil and microbial respiration are generally more rapid under elevated CO_2, indicating that enhanced plant growth under elevated CO_2 increases the amount of C entering the soil, thereby stimulating soil microbial activity.

AMF communities have been shown to change in response to CO_2 enrichment (Wolf et al., 2003). Klironomos et al. (2005) demonstrated that when the atmospheric CO_2 levels were abruptly increased from 350 to 550 ppm in one single step, the species composition and functioning of AMF communities changed sharply. However, they also observed that responses did not occur when the CO_2 level was increased gradually over 21 generations. This is an interesting observation, as most studies look at very sudden changes in CO_2 concentration, and this may have a major impact on how results are extrapolated to the real world. Increasingly, studies of relationships between mycorrhizal fungi and global change focus on interactions among multiple factors rather than single factors (Rillig et al. 2002; Johnson and Gehring 2007). This seems to be a realistic approach as few, if any, natural environments will experience only one change, and interactions among multiple factors may generate complex outcomes (Johnson and Gehring 2007). For example, increased availability of photosynthetic C may reduce the consequences of drought and also change the balance between costs and benefits in mycorrhizal symbioses. This was shown in a mesocosm experiment by Johnson et al. (2003). Elevated CO_2 increased the species richness of the plant community when AMF were present, but not when they were absent. The survival of slow-growing mycotrophic forbs was increased at elevated CO_2 levels, suggesting that the treatment ameliorated the C cost of the symbiosis (Johnson et al. 2003; Johnson and Gehring 2007).

9 FUNCTIONAL RESILIENCE OF SOIL MICROBIAL COMMUNITIES: DEPENDS ON BOTH SOIL STRUCTURE AND MICROBIAL COMMUNITY COMPOSITION

The resistance and resilience of microbial community might be related to its composition and properties of the resident microorganisms. Soil type and other environmental factors largely determine microbial community composition or that it is affected by a complex interaction between soil types. The soil-microbe complex can be viewed as a self-organizing system capable of adapting to prevailing conditions. With the single species, there was no recovery and a gradual decline in function compared to the unstressed control, whilst with a community, there was recovery over time (resilience). This may relate to the perceived benefits of biodiversity. (Griffiths et al., 2008).

Studies of Helgason et al. (2007) indicated that some generalist and easily cultured AM fungi are also more resistant to soil perturbation, while other AM fungi (several of them still uncultured) behave differently. An intriguing question, and challenge for the future, is to test in which way such uncultured and specialist AM fungi contribute to plant diversity and productivity in natural communities. Hence, it will be extremely important to cultivate the non-culturable and assess their ecological role. Until now most of the studies have focused on the diversity of ribosomal gene sequences. However, microbes with the same ribosomal gene sequences are often functionally diverse. Thus, for linking microbial diversity to ecosystem function it is necessary to focus on functional traits and functional genes that are important for biogeochemical processes. Ecological genomics (Straalen and Roelofs 2006) and the use of microarrays to detect key genes responsible for important ecosystem processes (e.g., nitrogen fixation, denitrification, decomposition, phosphorus acquisition) are essential in this respect (He et al., 2007) for the description of such a micro-array. These tools will allow us to identify the factors that regulate microbial functional gene diversity in soil and provide insights into its importance for plant community dynamics in terrestrial ecosystems (van der Heijden et al., 2008).

10 CONCLUSION

AM fungal symbioses that originated in early Devonian period are considered as mother of all plant-root symbioses. Besides intra-symbiotic interactions these also interact with other macro- and microbes inhabiting below-ground rhizosphere as well as above-ground atmosphere that include its fauna and flora of free-living, symbiotic or even pathogenic nature with which it may or may not have key connections. These AM fungi have been variously classified from time to time using the traits of morphology, immunology, analyses of 5.8S/ITS-2 RNA and DNA sequences, and RFLP etc. revealing several new clads which are helpful in tracing the phylogeny. Equipped with several intracellular obligate bacterium like organisms, nutrient transporter genes and biogenetically synthesized peri-arbuscular membrane these fungi release soluble signaling molecules in roots to activate host plants, regulating the above-ground components of ecosystem and altering the productivity and biodiversity through reciprocal relationships of competition, facilitation, mutualism, diseases, and predation in complex communities compelling us to consider multi-partite interactions through which AM fungal diversity itself gets affected. Even the abrupt changes in atmospheric CO_2 concentration, climate, land use, agricultural practices, resource availability and various disturbance regimes are stimulatory to AM fungi probably through physiology of host plant communities which may be beneficial to biodiversity, its activities and productivity suggesting ecosystem multi-functionality both above- and below-ground. However, the vice versa selective positive and negative impacts of different ecosystem functional attributes need to be deeply investigated to enrich the inadequate understanding of the dynamics of diverse interactive populations, cycling of nutrients and energy flows where role of inhabitant inhibitory soil biota and apparent insensitivity or inertness of biotic components cannot be simply ignored or underestimated because the magnitude of contribution of different species and abiotic environmental components to the ecological niche may not always be equal for establishment of sustainable equilibrium amongst vegetation representatives for long-term natural management of a particular ecosystem.

References

Akiyama K, Matsuzaki K and Hayashi H (2005). Plant sesquiterpenes induce hyphal branching in arbuscular mycorrhizal fungi. *Nature* 435: 824–827.

Akiyama K, Ogasawara S, Ito S and Hayashi H (2010). Structural requirements of strigolactones for hyphal branching in AM fungi. *Plant Cell Physiol* 51: 1104–1117.

Armstrong L and Peterson RL (2002). The interface between the arbuscular mycorrhizal fungus *Glomus intraradices* and root cells of *Panax quinquefolius* a Paris-type mycorrhizal association. *Mycologia* 94: 587–595.

Balestrini R, Gomez-Ariza J, Lanfranco L and Bonfante P (2007). Laser micro-dissection reveals that transcripts for five plant and one fungal phosphate transporter genes are contemporaneously present in arbusculated cells. *Molecular Plant-Microbe Interactions* 20: 1055–1062.

Balzergue C, Puech-Page`s V, Be´card G and Rochange SF (2011). The regulation of arbuscular mycorrhizal symbiosis by phosphate in pea involves early and systemic signalling events. *J Exptl Bot* 62: 1049–1060

Bianciotto V, Lumini E, Bonfante P and Vandamme P (2003). *Candidatus Glomeribacter gigasporarum* gen. nov. sp. nov, an endosymbiont of arbuscular mycorrhizal fungi. *Int J Syst Evol Microbiol* 53: 121–124.

Bonfante P and Julia-Andra A (2009). Plants, mycorrhizal fungi, and bacteria: A network of interactions. *Ann Rev Microbiol* 63: 363-383

Brundrett MC (2009). Mycorrhizal associations and other means of nutrition of vescular plants: understanding the global diversity of host plants by resolving conflicting information and developing reliable means of diagnosis. *Plant Soil* 320: 37–77.

Bucher M, Wegmaller S and Drissher D (2009). Chasing the structures of small molecules in arbuscular mycorrhizal signalling. *Curr Opinion Plant Biol.* 12: 500–507.

Burrows RL and Pfleger FL (2002). Arbuscular mycorrhizal fungi respond to increasing plant diversity. *Can J Bot* 80: 120–130.

Cavagnaro TR, Smith FA, Lorimer MF, Haskard KA, Ayling SM and Smith SE (2001). Quantitative development of Paris-type arbuscular mycorrhizas formed between *Asphodelus fistulosus* and *Glomus coronatum*. *New Phytol* 149: 105–113.

Chalot M, Javelle E, Blaudez D, Lamboite R, Cooke R, Sentenac H, Wipf D and Botton B (2002). An update on nutrient transport processes in ectomycorrhizas. *Plant Soil* 244: 165–175.

Chen X, Tang JJ, Fang ZG and Shimizu K (2004). Effects of weed communities with various species numbers on soil features in a subtropical orchard ecosystem. *Agric Ecosystems Environ* 102: 377–388.

Daft MJ and Nicolson TH (1974). Arbuscular mycorrhizas in plants colonizing coal wastes in Scotland. *New Phytol* 73: 1129–1138.

Denef K, Bubenheim H, Lenhart K, Vermeulen van Cleemput O, Boeckx P and Muller C (2007). Community shifts and carbon translocation within metabolically-active rhizosphere microorganisms in grasslands under elevated CO_2. *Biogeosciences* 4: 769–779.

Drigo B, Kowalchuk GA and Johannes A van Veen (2008). Climate change goes underground: effects of elevated atmospheric CO_2 on microbial community structure and activities in the rhizosphere. *Biol Fertil Soils* 44: 667–679.

Ezawa T, Smith SE and Smith FA (2001). Differentiation in polyphosphate metabolism between the extra and intraradical hyphae of arbuscular mycorrhizal fungi. *New Phytol* 149: 555–563.

Franics R and Read DJ (1995). Mutualism and antagonism in the mycorrhizal symbiosis, with special reference to impacts on plant community structure. *Can J Bot* 73: 1301–1309.

Fritz O and Sieverding E (2004). *Pacispora,* a vesicular arbuscular mycorrhizal fungal genus in Glomeromycetes; *J Appl Bot-Angew. Botanik* 78: 72–82.

García-Garrido JM, Lendzemo V, Castellanos-Morales V, Steinkellner S and Horst V (2009). Strigolactones, signals for parasitic plants and arbuscular mycorrhizal fungi. *Mycorrhiza* 19: 449–459.

Griffiths BS, Hallett PD, Kuan HL, Gregory AS, Watts CW and Whitmore AP (2008). Functional resilience of soil microbial communities depends on both soil structure and microbial community composition. *Biol Fertil Soils* 44: 745–754.

Hart MM and Richard J (2002). Taxonomic basis for variation in the colonization strategy of arbuscular mycorrhizal fungi. *New Phytol* 153: 335–344.

He Z, Gentry TJ, Schadt CW, Wu L, Liebich J, Chong SC, Huang Z, Wu W, Gu B, Jardine P, Criddle C and Zhou J (2007). GeoChip: a comprehensive microarray for investigating biogeochemical, ecological and environmental processes. *ISME J* 1: 67–77.

Helgason T, Merryweather JW, Young JPW and Fitter AH (2007). Specificity and resilience in the arbuscular mycorrhizal fungi of a natural woodland community. *J Ecol* 95: 623–630.

Hibbett et al. (2007). A higher level phylogenetic classification of the fungi. *Mycological Research* 111: 509–549.

House B and Fester T (2005). Molecular and cell biology of arbuscular mycorrhizal symbiosis. *Planta* 221: 184–196.

INVAM website: www.invam.caf.wvu.edu/fungi/fungindex.htm

Johnson D, Ijdo M, Genney DR, Anderson IC and Alexander IJ (2005). How do plants regulate the function, community structure and diversity of mycorrhizal fungi. *J Expt Bot* 56: 1751–1760.

Johnson NC, Wolf J and Koch GW (2003). Interactions among mycorrhizae, atmospheric CO_2 and soil N impact plant community composition. *Ecol Lett* 6: 532–540.

Johnson NC and Gehring CA (2007). Mycorrhizas: Symbiotic mediators of rhizosphere and ecosystem processes. In: Cardon ZG and Whitbeck JL (eds) The rhizosphere: an ecological perspective. Elsevier Academic Press, London, UK, 73-100.

Khade SW (2010). Arbuscular mycorrhizal fungi in wild banana III: a rare combination of *Glomus* and *Gigaspora*. *Mycorrhiza News* 22: 18–20.

Klironomos JN, Allen M, Rillig MC, Piotrowski J, Makvandi-Nejad S, Wolfe BE and Powell JR (2005). Abrupt rise in atmospheric CO_2 overestimates community response in a model plant–soil system. *Nature* 433: 621–624.

Kohlen W, Charnikhova T, Liu Q, Bours R, Domagalska MA, Beguerie S, Verstappen F, Leyser O, Bouwmeester H and Ruyter-Spira C (2011). Strigolactones are transported through the xylem and play a key role in shoot architectural response to phosphate deficiency in non-AM host *Arabidopsis thaliana*. *Plant Physiol* 155: 974–987.

Kothamasi D, Kuhad RC and Babu CR (2001). Arbuscular mycorrhizal in plant survival strategies. *J Intern Soc Tropical Ecol* 42: 1–13.

Kuszala C, Gianinazzi S and Gianinazzi-Pearson V (2001). Storage conditions for the long-term survival of AM fungal propagules in wet sieved soil fractions. *Symbiosis* 30: 287–299.

Lopez-Raez JA, Charnikhova T, Fernandez L, Bouwmeester H and Pozo MJ (2011). Arbuscular mycorrhizal symbiosis decreases strigolactone production in tomato. *J Plant Physiol* 168: 294–297.

Lumini E, Ghignone S, Bianciotto V and Bonfonte P (2006). *Endobacteria* or bacterial endosymbionts: To be or not to be. New *Phytol* 170 (2): 205–208.

Morgan JAW, Bending GD and White PJ (2005). Biological costs and benefits to plant-microbe interactions in the rhizosphere. *J Expt Bot* 56: 1729–1739.

Morton JB and Redecker D (2001). Two new families of Glomales, Archaeosporaceae and Paraglomaceae, with two new genera *Archaeospora* and *Paraglomus* based on concordant molecular and morphological characters. *Mycologia* 93: 181–195.

Mosse B, Stribley DP and Le Tacon F (1981). Ecology of mycorrhizae and mycorrhizal fungi. In: Alexander M (ed) Advances in Microbial Ecology. Plenum Press, New York, 137–209.

Muthukumar T (2009). Morphology and taxonomy of arbuscular mycorrhizal fungi. Proc. National Workshop on Techniques in Arbuscular Mycorrhizal Fungal Research, Botany Deptt. Goa University, Goa Feb. 25–27.

Naumann M, Schussler A and Bonfante P (2010). The obligate endobacteria of arbuscular mycorrhizal fungi are ancient heritable components related to the molecules. *The ISME Journal* 4: 862–871.

Olsrud M, Melillo JM, Christensen TR, Michelsen A, Wallander H and Olsson PA (2004). Response of ericoid mycorrhizal colonization and functioning to global change factors. *New Phytol* 162: 459–469.

Palenzuela J, Ferrol N, Boller T, Azcon-Aguilar C and Fritz O (2008). *Otospora bareai*, a new fungal species in the Glomeromycetes from a dolomitic shrub land in Sierra de Baza National Park (Granada, Spain) *Mycologia* 100: 296–305.

Parniske M (2008). Arbuscular mycorrhiza: the mother of plant root endosymbioses. *Nature Reviews Microbiology* 6: 763–775.

Pumplin N and Harrison MJ (2009). Live-cell imaging reveals peri-arbuscular membrane domains and organelle location in *Medicago truncatula* roots during arbuscular mycorrhizal symbiosis. *Plant Physiol* 151: 809–819.

Pumplin N, Mondo SJ, Topp S, Starker CG, Gantt JS and Harrison MJ (2010). *Medicago truncatula* Vapyrin is a novel protein required for arbuscular mycorrhizal symbioses. *Plant J* 61: 482–494.

Rillig MC, Wright SF, Shaw MR and Field CB (2002). Artificial climate warming positively affects arbuscular mycorrhizae but decreases soil aggregate water stability in an annual grassland. *Oikos* 97: 52–58.

Ruyter-Spira C, Kohlen W, Charnikhova T, van Zeijl A, van Bezouwen L, de Ruijter N, Cardoso C, Lopez-Raez JA, Matusova R, Bours R, Verstappen F and Bouwmeester H (2011). Physiological effects of the synthetic strigolactone analog GR24 on root system architecture in *Arabidopsis*: another belowground role for strigolactones? *Plant Physiol* 155: 721–734.

Salvioli A, Chiapello M, Fontaine J, Hadiji-Sahraoui AL, Grandmougin-Ferjani A, Lanfranco L and Bonfante P (2010). *Endobacteria* affect the metabolic profile of their host *Gigaspora margarita*, an arbuscular mycorrhizal fungus. *Symbiosis* 12: 2083–2095.

Schussler A, Schwarzott D and Walker C (2001). A new fungal phylum, the Glomeromycota: phylogeny and evolution. *Mycological Research* 105: 1413-1421

Schussler A and Walker C (2010). A species list with new genera and family of Glomeromycota, Published in Index fungorum and to EMBL version of NCBI taxonomic data.

Sieverding E and Fritz O (2006). Revision of *Entrophospora* and description of *Kuklospora* and *Intraspora*, two genera in the arbuscular mycorrhizal glomeromycetes. *J Appl Bot Food Quality* 80: 69–81.

Smith FA and Smith SE (1996). Mutualism and parasitism: diversity in function and structure in the arbuscular mycorrhizal symbiosis. *Adv Bot Res* 22: 1–43.

Spain JL, Sieverding E and Fritz O (2006). *Appendicispora*: a new genus in the arbuscular mycorrhiza-forming Glomeromycetes, with a discussion of the genus *Archaeospora*. *Mycotaxon* 97: 163–182.

Solís-Domínguez FA, Valentín-Vargas A, Chorover J and Maier RM (2011). Effect of arbuscular mycorrhizal fungi on plant biomass and the rhizosphere microbial community structure of mesquite grown in acidic lead/zinc mine tailings. *Sci Total Environ* 409: 1009–1016.

Straalen NM van and Roelofs D (2006). An introduction to ecological genomics. Oxford University Press, Oxford, UK 320.

van der Heijden MCA, Bardgett RD and van Straalen NM (2008). The unseen majority: soil microbes as drivers of plant diversity and productivity in terrestrial ecosystems. *Ecol Lett* 11: 296–310.

Vazquez MM, Azcon R and Barea JM (2001). Compatibility of a wild type and its genetically modified *Sino Rhizobium* strain with two mycorrhizal fungi on *Medicago* species as affected by drought stress. *Plant Sci* 161: 347–358.

Walker C, Błaszkowski J, Schwarzott D and Schüssler A (2004). *Gerdemannia* gen. nov., a genus separated from *Glomus*, and Gerdemanniaceae fam. nov., a new family in the Glomeromycota. *Mycol Res* 108: 707–718.

Walker C, Vestberg M, Demircik F, Stockinger H, Saito M, Sawaki H, Nishmura I and Schussler A (2007). Molecular phylogeny and new taxa in the Archaeosporales (Glomeromycota): *Ambispora fennica* gen. sp. nov. Ambisporaceae fam. nov. and emendation of *Archaeospora* and Archaeosporaceae. *Mycological Research III:* 137–153.

Wolf J, Johnson NC, Rowland DL and Reich PB (2003). Elevated CO_2 and plant species richness impact arbuscular mycorrhizal fungal spore communities. *New Phytol* 157: 579–588.

Wubet T, Wei BM, Kattke I, Teketay D and Oberwiitzler F (2003). Molecular diversity of AM fungi in *Prunus africana* an endangered medicinal tree species in dry Afromontane Forest of Ethiopia. *New Phytol* 161: 517–528.

Yadav V, Sharma S, Verma PK and Varma A (2005). Phosphorus metabolism and regulation in symbiotic fungi. In: Podila GK and Varma A (eds) Basic Research and Applications of Mycorrhizal Fungi. I.K International Pvt. Ltd., New Delhi, India 111–139.

Zak DR, Pregitzer KS, Curtis PS and Holmes WE (2000). Atmospheric CO_2 and the composition and function of soil microbial communities. *Ecological Applications* 10: 47–59.

Zhang Q, Blaylock LA and Harrison MJ (2010). Two *Medicago truncatula* half-ABC transporters are essential for arbuscule development in arbuscular mycorrhizal symbiosis. *Plant Cell* 22: 1483–1497.

13

Arbuscular Mycorrhiza as Bioprotectant: A Review of the Mechanisms Involved

Munawar Fazal*[1], M. Yaqub Bhat[2] and Mohd Ashaq[3]

[1]*Department of Botany, S. Sinha College, Aurangabad, Bihar, India*

[2]*University Department of Botany, University of Kashmir, Srinagar, J & K, India*

[3]*Department of Botany, Govt. College for Women, Gandhi Nagar-180001, Jammu J & K India*

**Email: munawar_fazal@rediffmail.com*

1 INTRODUCTION

The symbiotic association between mycorrhizal fungi and the roots of plants is widespread in natural environment. There are number of different types of fungi that form these associations, but for agriculture production system, it is the arbuscular mycorrhizal fungi (AMF) of Phylum Glomeromycota (Schussler et al., 2001; Sieverding and Oehl 2006) that are most important. AMF appeared on the earth at least 4000 years ago in the early Devonian (Remy et al., 1994). Since then it has developed obligate symbiotic association with about 80% of plant species (Vander Heijden and Sanders 2002) that provide them with organic carbon (C) (Bago et al., 2000). Additionally, AMF interacts with different classes of microorganism in the rhizosphere influencing this ecosystem to such an extent that a new term mycorrhizosphere has been coined (Martin et al., 2000). The AMF consists of an internal phase inside the root and an external phase or extraradical mycelium (ERM) phase, which forms an extensive network within the soil.

Importance of the AMF in agricultural production system received considerable attention in recent years owing to their beneficial response. AMF provides a range of benefits to the host plants. Symbiotic relationship between AMF and the roots of host plants contributes significantly to plant nutrition and growth (Auge 2001), and has been shown to increase the productivity of a variety of agronomic crops (Sylvia et al., 1993). These positive responses in productivity to AM colonization have mainly been attributed to the enhanced uptake by AM of relatively immobile soil ions such as phosphorus (P), potassium (K), calcium (Ca), magnesium (Mg), sulfur (S), iron (Fe), zinc (Zn), copper (Cu), and manganese (Mn) (Marschner and Dell 1994; Azaizeh et al., 1995; Marschner 1995; Liu et al., 2000 a, b; Liu et al., 2007), but also involve the enhanced uptake and transport of far more mobile nitrogen (N) ions (Tobar et al., 1994a,b; Azc´on et al., 1996; Liu et al. 2007). The most widely recognized contribution of AM fungi to host-plant nutrition involves their ability of fungal ERM to extract P from outside the P depletion zone around plant roots (Sanders and Tinker 1971; Koide 1991; George et al., 1995; Marschner 1995; Smith and Read 1997; Miller 2000; Liu et al., 2003; Smith et al., 2003; Liu et al., 2007). In return the fungi receive carbon (C) from the host plant. Other benefits to the host plant includes: increased resistance to foliar feeding insects (Gange and West 1994); protection from soil-borne pathogen (e.g., nematodes, pathogenic fungi, bacteria) (Newsham et al., 1995; Lingua et al., 2002; Pozo et al., 2002; Cardoso and Kuyper 2006; St-Arnaud and Vujanovic 2007), improved soil structural development, aggregate stabilization (Bethlenfalvay and Schüepp, 1994; Miller and Jastrow 2000; Picone 2003; Rillig 2004; Wright 2005); improved drought resistance (Auge et al., 1994; Boomsma et al., 2008); increased tolerance of salinity and heavy metals (Shetty et al., 1995; Diaz et al., 1996; Al-Karaki et al., 2001; Feng et al., 2002; Mohammad et al., 2003) and non-AM-host weed suppression (Picone 2003).

Though AM association offer multiple benefits to the host plant, it may not be obviously mutualistic at all points in time, and it is possible under some condition that AMF may cheat their host plant into supplying C with no apparent benefit to the plant (Gosling et al., 2006). In some cases this can cause a decline in growth (Lerat et al., 2003). However, it is difficult to prove that AMF are actually cheating (Fitter 2001) because of wide range of benefits to the host, which may only become obvious at specific time or under certain environmental conditions or stress (Gosling et al., 2006).

Yet the positive effects of AM fungi on host-plant growth and development are not only noticeable in low soil fertility conditions (Jeffries 1987) but also in drought environments (Sylvia et al., 1993; Picone 2003; Liu et al., 2007). Since nutrient mobility is limited under drought conditions, AM may have a larger impact on overall plant growth and development in dry relative to well-watered conditions (Sánchez-Díaz and Honrubia 1994)

In recent years the role of AMF interacting with plant roots has gained increased attention and its protective role against root associated pathogens have been identified (Dehne 1982; Smith 1988; Schippers et al., 1987; Fitter and Garbane 1994; Harrier and Watson 2004; St-Arnand and Veijanovic 2007; Avis et al., 2008; Vierheilig et al., 2008; Akhtar and Siddiqui 2009). Since chemical pesticide use is more and more restricted due to its risk to human health and the environment, implementation of sustainable agriculture has become imperative in crop production system. The understanding of the mechanism involved in the AM-mediated bioprotection will permit the performance of an adequate management of such agroecosystem and then permit the maximization of AM benefits (Gosling et al. 2006). The present article aims to review some of the mechanisms involved in the bioprotection of diseases induced by root/soil-borne pathogens after root colonization with AMF.

2 FACTOR AFFECTING BIOPROTECTION EFFICACY

Apart from abiotic factors such as temperature, soil moisture, and soil P-content the bioprotective efficacy of AMF also depends upon several biotic factors like host genotype, AMF isolate and degree of mycorrhization.

2.1 Host Genotype

The host genome seems to affect the protective efficacy of AMF as the host genotype seems to result in a differing bio-protective response by the mycorrhizal association. Depending upon the genotype, Mycorrhizal strawberry showed a different susceptibility to *P. fragariae* (Mark and Cassels 1996).

2.2 The AMF

There are few experimental evidences that bio-protective efficacy of AMF is species dependent. Different AMF are known to confer variable tolerance to *Phytophthora parasitica.* Davis and Menge (1980, 1981) used five different AMF species on three different plants. They observed that maximum resistance to the pathogens was due to two AMF, *G. fasiculatum* and *G. constrictum* in Citrus. *Glomus intrardices* has been reported not to protect clover against nematode (Habte et al., 1999) and was not effective in reducing disease symptoms produced by *P. parasitica* infection (Pozo et al., 2002). In contrast, root colonization by *G. mosseae* resulted in clear protective effect against nematodes and *P. parasitica* infection (Habte et al., 1999; Pozo et al., 2002). *Glomus mosseae* and *G. versiforme* were more effective in suppressing *Meloidogyne incognita* than *G. intraradices* (Zhang et al., 2008). Maherali and Klironomos (2007) found taxonomic variation within the phylum Glomeromycota, both in terms of pathogenic protection. Members of the family Glomeraceae reduced pathogen abundance in the roots to a greater extent than the members of Gigasporaceae. This shows that some AMF species exhibit clear bio-protective effect, whereas, other species do not affect the plant-pathogen interaction in terms of bio-protection. Therefore, more detailed comparative studies are needed involving a greater variety of AMF species.

2.3 Time of Mycorrhization

Time of inoculation or mycorrhization has been suggested as an important factor in AM bio-protection. In general, it has been suggested that inoculation of AMF has to be done prior to the inoculation with the soil-borne root pathogens (Sikora and Sitaramaiah 1980; Umesh et al., 1988; Singh et al., 2000; Azcon-Aguilar et al., 2002; Xavier and Boyetchko 2004; Anjos et al., 2010) as it gives enough time for establishment of the association before the contact with the pathogen (Calvet et al., 2001). Caron et al. (1986 a, b) observed that reduction in root necrosis due to F*usarium oxysporium f.* sp. *radicis lycopersici* was more pronounced when *G. intraradices* was inoculated 4 week prior to fungal inoculation. Tomato plants infested by *M. incognita* at the time of transplanting were not protected by *Glomus mosseae* (Talavera et al., 2001). The growth of mycorrhizal coffee plants was also reduced when *Pratylenchus coffeae* was inoculated before the establishment of the symbiosis (Vaast et al., 1998). Early mycorrhizal infection increases tolerance to *P. vulnus* (Camprubi et al., 1993; Calvet et al., 1993; Pinochet et al., 1993, 1995). Pre-inoculation of tomato plants with *Glomus coronatum* stimulated plant growth and reduced *M. incognita* infection (Diedhiou et al., 2003). Inoculation of black gram (*Vigna mungo*) with *Glomus fasciculatum* 15 and 20 days earlier than inoculation with *Meloidogyne incognita* controlled the nematode population and also increased biomass production. But, when *G. fasciculatum* was applied 5 days earlier than nematode mycorrhizal colonization and spore production was reduced (Sankarayanan and Sundarababu 1994). The establishment AMF *Scutellospora heterogama* on the roots of sweet passion fruit plant prior to nematode, *Meliodogyne incognita* infection contributed for the reduction of symptoms severity and reproduction of nematode in infested soil (Anjos et al., 2010). Pre-inoculation of cucumber plants with *Glomus intraradices, G. mosseae* and *G. versiforme* significantly reduced the galling index, number of eggs, number of females and number of eggs per egg mass (Zhang et al., 2008).

3 DEGREE OF MYCORRHIZATION

The first sign of root colonization are visible after a few days of inoculation with AMF. Thereafter mycorrhization can increase manifold until it reaches a final plateau (Wyss et al., 1991; Vierheilig et al., 1994). In several studies it has been reported that a local bio-protectrive effect depends upon the degree of mycorrhization. High degree of AM colonization showed bioprotective effect whereas, intermediate and low levels of colonization showed no bio-protective effect. In mycorrhizal wheat plants a bio-protective effect against *Gaeumannomyces graminis* (Graham and Menge 1982) and in tomato plants against *Fusarium oxysporium* (Caron et al., 1986 a, b) and *P. parasitica* (Cordier et al., 1998b) could only be observed when roots were heavily colonized by AMF, low mycorrhization levels resulted in no bio-protection. Chlamydospore production by *Thielaviopsis basicola* was negatively correlated to per cent mycorrhizal colonization of tomato and alfalfa (Baltruschat and Schoenbeck 1972). Saleh and Sikora

(1984) reported that 38% mycorrhization by *Glomus fasiculatum* was required for control of *Meloidogyne incognita*. In a study by Diedhiou et al., (2003), mycorrhization by *G. coronatum* was 11% after 9 weeks and around 30% after 12 and 15 weeks. While no reductions in gall index was observed after 9 weeks, reductions were significant after 12 and 15 weeks, indicating 30% mycorrhization was threshold level for nematode control. A further increase in mycorrhiozation above that level did not necessarily lead to higher nematode control. Recently, it has been demonstrated that not only the local, but also the systemic, bio-protective effect of mycorrhization depends on the degree of AM root colonization (Khaosaad et al., 2007).

4 MECHANISM

Various mechanisms have been put forward in an attempt to explain the AM-mediated bio-protection against soil-borne root pathogens. These include biochemical and molecular changes in mycorrhizal plant tissues, microbial changes in mycorrhizosphere, altered nutrient status of host, changes in root morphology, direct competition. Some of these suggested mechanisms might play no role, it is generally agreed that protection through mycorrhization is the result of combination of several mechanism (Hooker et al., 1994; Vierheilig et al., 2008).

4.1 Improved Nutrient Status of the Host

It is well documented that AMF can improve nutrient status of the host plants (Smith and Read 2008). Several mineral nutrients, especially P are allocated through the symbiosis to the plant in exchange of carbon (Pearson and Jakobsen 1993; Gosling et al., 2006). There is evidence that plants that took up amounts of nutrients through their AM fungal symbiont have an increased tolerance for pathogenic infection (Bodker et al., 1998; Karagiannidis et al., 2002; Akhtar and Siddiqui 2008; Wehner et al., 2009; Tahat et al., 2010). Reports indicate that P-induced changes in root exudation could reduce the germination of pathogen spores and infection (Graham 1982; Graham et al., 1982; Sharma et al., 2007). It has been hypothesized that direct competition between root pathogens require host nutrition for development and reproduction and this competition may be the cause of their inhibition (Dehne 1982; Smith 1988). Greater tolerance to AM plant is also attributed to increased root growth and phosphate status of plants (Cameron 1986). In addition to P, AMF may increase host tolerance to pathogens by increasing the uptake of essential nutrients like Ca, Cu, Mn, Zn, and S (Pacovsky et al., 1986; Smith and Gianinazzi-Pearson 1988; Gosling et al., 2006) which are otherwise deficient in non-mycorrhizal plants.

Increased tolerance to pathogen infection can also be influenced by nutritional status of the host and fertility status of the soil (Cook and

Baker 1982; Jaizmevega and Pinochet 1997). However, it is not clear whether increased tolerance was a direct consequence of the improved nutrition. In other studies increased nutrient uptake had no effect on pathogen tolerance (Graham and Egel 1988; Shaul et al., 1999; Fritz et al., 2006). Maherali and Klironomos (2007) found taxonomic variation within the phylum Glomeromycota, both in terms of nutrient responses and pathogenic protection. While members of the family Gigasporaceae increased nutrient uptake to a larger extent than those of the family Glomeraceae, the latter reduced pathogen abundance in the roots to a greater extent. This suggests that the enhanced nutrition by AM symbiosis may increase plant development does not account for all protection conferred by AM to the host (Caren et al., 1988; Trolta et al., 1996).

4.2 Anatomical or Architectural Changes in the Root System

AMF colonization influences root morphology of the host plants in most studies by causing a more profusely branched root system, i.e., root system contains shorter, more branched, larger adventitious roost diameter and lower specific root length (Price et al., 1989; Schellenbaun et al., 1991; Berta et al., 1993; Yanoet et al., 1996; Paszkawski et al., 2002; Olah et al., 2005; Tahat et al., 2008; Gietlahr et al., 2009). Such changes in the root morphology increases surface area which can provide more infection sites for the pathogens (Hooker et al., 1994).

Interactions between changes in the root architecture and protection of plant roots from soil pathogen attack have been demonstrated. Dehne et al. (1978) have shown increased lignifications in the endodermal cells of cucumber and tomato plants and speculated that such responses may account for reduced incidence of *Fusarium* wilt. Becker et al. (1976) reported similar results on pink root of onion (*Pyrenochaeta terrestris*). Matsubara et al. (1995) have shown that eggplants colonized by *Glomus etunicatum* or *Gigospora margarita* contained higher lignin concentration in the first and second order roots in presence of *Verticillium dahliae*. In addition, these AMF caused the plant to produce thicker third order roots. Increased lignifications may protect roots from penetration by pathogens while elevating phenolic metabolism within the host plant (Morandi 1996). In contrast to the induced higher branching of the root system, *Glomus mosseae* decreased root branching of tomato under the influence of higher P if plants were attached with *Phytophthora nicotiana* (Trotta et al., 1996). Alternatively branching of the tomato root system was not affected by inoculation with *G. mosseae*, but decreased the number of infection loci of *Phytophthora parasitica* (Vigo et al., 2000).

Abundance of lateral root tips and developing meristems make highly branched root systems more susceptible for pathogen attacks, resulting in an increased demand for AMF to protect those (Newsham et al., 1995b).

Norman et al. (1996) supported this hypothesis as they compared plants with inherently highly branched root system and found that mycorrhizal plants had fewer necrosis compared to non-mycorrhizal ones. If AMF frequently cause profusely branched root system but increased branching in itself leads to higher susceptibility to root pathogen attack, AMF must confer pathogen protection through additional mechanisms. Mycorrhizal plant produced wound barriers at a faster rate than non-mycorrhizal plants wound barrier formation inhibited Thielaviopsis black rust of mycorrhizal holly plants (Wicknand Moore 1984). Decreased branching caused by the nematodes, *Rodopholus similis* and *Pratylenchus coffeae* was counterbalanced by the increased branching caused by the AMF (Elsen et al., 2003).

4.3 Direct Competition

4.3.1 Competition for infection site

A competition effect at infection site has also been suggested. AMF might occupy infection sites on the root surface needed by the root pathogen to penetrate the root or cells in the root already occupied AMF cannot be colonized any further by the pathogens (Cordier 1988a, b). Since AMF and root pathogen occupy similar root tissue there may be competition for space if colonization takes place at the same time (Smith 1988). Dehne (1982) and Azcon-Aguilar and Barea (1996) documented how AMF and root pathogens colonize in the same host tissues and how they develop in different root cortical cells, indicating some sort of competition between AMF and *Phytophthora*. They observed reduced development of *Phytophthora* in AM colonized and adjacent uncolonized root systems, and the pathogen does not penetrate arbuscular containing cells (Cordier et al., 1998). Similarly Aphanomyces was subdued on pea roots by AMF only when the two fungus was present on the same root (Rosendahl 1985). Larsen and Bodker (2001), using signature fatty acids profiles, demonstrated the decrease in biomass and energy reserves of both *G. mosseae* and *A. euteiches* occupying pea roots. A reduction in the extent of mycorrhizal colonization by different plant pathogens has been reported (Davis and Menge 1980; Baath and Hayman 1983; Krishna and Bagyraj 1983) indicating the possible occurrence of competitive interactions. Mycorrhization reduced the number of infection sites and colonization by AMF had no effect on the spread of necrosis (Vigo et al., 2000).

4.3.2 Competition for carbon

Growth of AMF and root pathogens depends on carbon from photosynthesis and they compete for carbon compounds received by roots (Smith and Read 1997). It has been suggested that higher carbon availability in mycorrhizal plants could explain the bio-protective effect of AMF. AMF is dependent on the host plant for carbon source and about 4–20% of the net photosynthate is transferred to the AMF (Smith and Read 1998). Once the plant is colonized by AMF, due to carbohydrates used by the symbiotic AMF, limited carbon could be available for root pathogens (Azcon-Aguilar and Barea 1996; Singh et al., 2000; Azcon-Aguilar et al., 2002; Xavier and Boyetchko et al., 2004; Vierhuilig et al., 2008). Interference competition may also arise if C-availability within the intercellular spaces and rhizosphere is reduced as result of AMF colonization (Graham 2001).

Different AMF species have been reported to exhibit a different C-sink strength in mycorrhizal roots (Lerat et al., 2003a, b) and thus, should exhibit a different bio-protective effect. Interestingly, in a split root system, one side colonized by AMF, disease symptom produced by pathogen on the other side of the split root system were reduced (Cordlier et al., 1998a; Pazo et al., 2002; Zhu and Zao 2004; Khaosaad et al., 2007; Elsen et al., 2008). This would point towards certain C-sink strength of AMF. Moreover, despite its high C-sink strength the AMF *Glomus intraradices* did not provide bio-protection against *P. parasitica* (Pozo et al., 2002; Lerat et al., 2003a).

4.3.3 Inhibition in soil

There are some reports about possible mechanisms of direct action of AM fungi against pathogens in soil (Garcia-Garrido and Ocampo 1989; St-Arnaud et al., 1995; Filion et al., 1999). However, at present the production of antibiotics or inhibitory compounds by AMF has not been reported (Vierheilig et al., 2008).

5 MYCORRHIZOSPHERE

The narrow zone of soil surrounding the living roots is called rhizosphere (Hiltner 1904) characterized by increased microbial activity and by a specific microbial community structure (Duineveld et al., 2001; Kim et al., 2006). Since plant roots in natural and semi-natural ecosystems are commonly mycorrhizal, the rhizosphere concept has been widened to include the fungal component of the symbiosis, resulting in the term "mycorrhizosphere" (Rambelli 1973). Mycorrhizosphere refers to the zone of soil influenced by both the root and the mycorrhizal fungus (Oswald and Ferchan 1968) and includes the more specific term "hyphosphere". Hyphosphere refers only to the zone surrounding individual fungal hyphae. In mycorrhizosphere the appearance of mycorrhizae exert strong influence on the microbiota in the rhizosphere.

The mycorrhizosphere has been hypothesized to constitute an environment conducive to microorganisms antagonistic to soil-borne pathogen proliferation. Indeed, co-culture of the non-mycorrhizal species *Dianthus caryophyllus* with mycorrhizal species *Tagetes patula* in presence of *G. intraradices* clearly reduced the disease caused by *Fusarium oxysporium dianthi* in *D. Caryophyllus* in a manner clearly unrelated to plant nutrition which suggests a reduction in the pathogen development within the mycorrhizosphere (St-Arnaud et al., 1997). Moreover, a reduction in the number of infection loci of tomato roots pre-colonized with *G. mosseae* and inoculated with *P. nicotianae* zoospores infers that the pathogens may be affected prior to root penetration in the mycorrhizosphere (Vigo et al., 2000).

Different factors such as altered exudation, direct AM fungal effects, root size and architecture, altered physiology, etc., may contribute to quantitative and qualitative microbial community changes in mycorrhizosphere caused by AMF (Andrade et al., 1998; Hodge 2000; Johansson et al., 2004; Artusson et al., 2006; Toljander et al., 2007; Finlay et al., 2008). Changes in mycorrhizosphere microbial communities due to the influence of AMF may lead directly to the diminution of root pathogens. These changes may lead to stimulation of the microbiota which may be antagonistic to root pathogens. Actinomycetes antagonistic to pathogen were more prevalent in the rhizosphere of AM plants than non-mycorrhizal plants (Secilia and Bagyaraj 1987). Mycorrhization can modify both total microbial populations and specific functional groups of microorganisms in mycorrhizosphere (Meyer and Linderman 1986; Lindrman 1994; Citernesi et al., 1996; Filion et al., 1999; Larsen et al., 2003). Other studies indicate that pathogen suppression by AM fungi involves changes in mycorrhizosphere microbial populations. Caron et al. (1985, 1986a, b, c) showed reduction in *Fusarium* populations in mycorrhizosphere of tomatoes and corresponding reduction in root-rot in AM plants compared to non-mycorrhizal plants, probably due to the increased antagonism in the mycorrhizosphere. Recently, Siasou et al. (2009) provided evidence that rhizosphere fluorescent *Pseudomonas* strains, which produce the antibiotic 2, 4-diacetyphloroglucinol (DAPG) that confers plant protection against *Gaeumannomyces graminis* var *tritici*, produced significantly larger amounts of DAPG in the presence of soluble carbon from soil containing *G. intraradices*. Even though majority of studies provide support for this mechanism of plant pathogen protection (Whipps 2004), it is not universal. Li et al. (2007) found that bacterial strains associated with AM fungi were equally effective at providing pathogen protection as their counterparts isolated from non-mycorrhizal soils.

5.1 Alterations of Root Exudation

Alterations of root exudation have been suggested to be involved in the expression of the mycorrhizal bio-protection. These changes have been suggested to be at least partially involved in the altered susceptibility of arbuscular mycorrhizal plants towards root-borne pathogens (Vierheilig and Piche 2002; Vierheilig 2004a; Vierheilig et al., 2008). Analogous inhibitory effects were observed with root exudates from strawberry roots colonized by *G. etunicatumn* and *G. monosporum* on pathogen *P. fragariae* sporulation (Norman and Hooker 2000). The germination of *Fusarium oxysporium* f. sp. lycopersici was inhibited in the presence of root exudates from tomato (Scheffknecht et al., 2006). Meanwhile, depending upon the harvest time, exudates from *in vitro* grown tomato roots colonized with *G. intraradices* were repulsive or more attractive than exudates from non-mycorrhizal roots to *P. nicotianae* zoospores (Lioussanne et al., 2008). Root exudates of *G. mosseae* treated plants reduced hatching of *Rotylenchulus reniformis* juveniles. The reduction was concentration dependent (Subramanian and Senthamizh 2006). Differential growth of *Fusarium oxysporium* f. sp. *chrysanthemi, Trichoderma harzianum, Clavibacter michiganesis* and *Pseudomonas chlororaphis* was explained by substances released from *G. intraradices* under *in vitro* culture conditions (Filion et al. 1999). Grandmaison et al. (1993) suggested that phenolic compounds bound to the cell wall could be indirectly responsible for the resistance of AMF roots to pathogenic fungi since they increased the resistance of cell wall to the action of digestive enzymes.

5.2 Nematode Parasitism by Arbuscular Mycorrhizal Fungi

Parasitism of nematode eggs with AMF has been demonstrated, but the level of parasitism was not considered sufficient to negatively affect nematode activities (Francl and Dropkin 1985). Arbuscular mycorrhizal fungal spores have been reported to occupy seed and dead insects in soil (Rabatin and Rhodes 1982; Taber 1982) and have limited saprophytic capabilities (Harley and Smith 1983). It appears that the AMF colonize only stressed or weakened nematode eggs. The nematode parasitism by AM fungi is opportunistic and depends on carbon nutrition from autotrophic symbionts, rather than being representative of true host parasite relationship.

5.3 Physiological and Biochemical Changes

Physiological and biochemical alterations of the host after mycorrhization have been reported. Some of these changes are possibly linked with a protective efficacy of the mycorrhizal plant against root pathogens (fungi, bacteria, nematodes, etc.), e.g., enhanced level of PR-1 proteins (pathogenesis related proteins type-1, Cordier et al., 1998), accumulation of phytoalexins (Kaplan et al., 1980; Sylvia and Sinclair 1983; Harrison and Dixon 1993; Morandi 1996; Larose et al., 2002), callose (Cordier et al., 1998b; Lee et al., 2005), salicylic acid (Biliou et al., 2000a, b; Medina et al., 2003) and reactive oxygen species (Salzer et al., 1999), and induction of hydrolytic enzymes (Pozo et al., 1999), new isoforms of superoxide dismutases and peroxidises (Pozo et al., 2002; Garmendia et al., 2006), higher concentration of phenolic acids (Singh et al., 2004; Zhu and Zao 2004), accumulation of jasmonic acid (JA, Hause et al., 2002; Isayenkov et al., 2005). Further, synthesis of constitutively and additional isoforms of defence related enzymes such as chitinases, chitosanases, β-1, 3-glucanases, peroxidases and superoxide dismutase has been detected in mycorrhizal roots (Pozo et al., 1996, 1998, 1999). There is strong evidence that during AM development these defensive responses occur (Garcia-Garrido and Ocampo 2002) and that they are strongly stimulated when challenged by root pathogens. Possibly the mechanisms of plant defence are activated faster and to a greater extent in mycorrhizal plants when challenged by a pathogen compared to non-mycorrhizal plants. Formation of phytoalexins was higher on mycorrhizal roots than on non-mycorrhizal roots (Morandi 1987). Phytoalexins are assumed to play a major role in the host defense system against root pathogens (Kaplan et al., 1987). Phytoalexins and associated molecules accumulate in roots after mycorrhizal infection but less extensively and more slowly than in pathogenic interactions (Morandi 1996). Mycorrhization also amplified the accumulation of phytoalexins, rishitin and solavetivone in *Rhizoctonia*-infected potato plantlets, while AMF themselves did not alter the levels of these compounds (Yao et al., 2003).

It has been suggested that AM colonization acts as a priming system for the process of pathogen resistance (Azcon-Aguilar et al., 2002; Pozo and Azcon-Aguilar 2007, Pozo et al., 2009). In this respect, elevated JA levels occurring upon mycorrhization, likely associated with a fully established mycorrhiza and may mediate the enhance defense status of the Mycorrhizal plant (Vierheilig and Piche 2002; Hause et al., 2007, Vierheilig 2004a). Mycorrhizal protection of grape vine against *Meloidogyne incognita* has also been associated with primed expression, ubiquitously throughout the whole root system, of a chitinase gene, VCH3, in response to the nematode (Li et al., 2006).

6 CONCLUSION

Maintenance of soil fertility and management of plant diseases have become a crucial issue in meeting the ever increasing biomass demand for food, fodder and fuel as well as safeguarding the environment. An ideal agricultural soil is characterized not only by optimum physical and chemical constituents conducive for plant growth but also by microbial processes that is maintained in equilibrium. The symbiotic association between the mycorrhizal fungi and plant roots is widespread in the natural environment. There are numerous different types of fungi that form these associations, but for agricultural production system it is arbuscular mycorrhizal fungi (AMF) of the phylum Glomeromycota that are most important. The importance of AMF in recent years received considerable attention owing to their beneficial response. Besides improving the nutrient uptake, AMF improves plant health through increased resistance to various biotic and abiotic stresses. Of particular importance is its protective role against root-associated pathogens. Various direct and indirect mode of action have been suggested to be involved in the bioprotective effect of mycorrhization against pathogens but the exact mechanism by AM colonization confers that the protective effects are not completely understood. A greater understanding of the mechanism involved will permit the performance of an adequate management of such ecosystem and permit the maximization of AM benefits. Although further research is needed, the exact knowledge suggests the exploitation of AM fungi for bioprotection of target disease in sustainable production system. Appropriate AM fungi must be used against target root pathogens preferably in combination with other pathogen antagonistic microbes.

References

Akthar M.S and Siddiqui ZA (2008). Arbuscular Mycorrhizal fungi as potential bioprotectants shainst plant pathogens. In: Mycorrhizae: Sustainable Agriculture and Forestry (Siddiqui ZA, Akthar MS and Futai K (eds) Springer, Netherlands, Dordrecht, The Netherlands. 61–97.

Al-Karaki GN, Hammad R and Rusan M (2001). Reponse of two tomato cultivars differing in salt tolerance to inoculation with mycorrhizal fungi under salt stress. *Mycorrhiza* 11: 43–47.

Anjos ECT, Cavalcante UMT, Goncalves DMC, Pedrosa EMR, Santos VF and Maia LC (2010). Interaction between an arbuscular mycorrhizal fungus (*Scutellospora heterogama*) and the root-knot nematode (*Meloidogyne incognita*) on sweet passion fruit (*Passiflora alata*). *Braz. Arch. Biol. Technol.* 53: 801–809.

Artursson V, Finlay RD and Jansson JK (2006). Interaction between arbuscular mycorrhizal fungi and bacteria and their potential for stimulating plant growth. *Environ. Microbiol.* 8: 1–10.

Auge RM, Duan X, Ebel RC and Stodola AJW (1994). Non-hydraulic signalling of soil drying in mycorrhizal maize. *Planta* 193: 74–82.

Azaizeh HA, Marschner H, Romheld V and Wittenmayer L (1995). Effects of a vesicular-arbuscular mycorrhizal fungus and other soil microorganisms on growth, mineral nutrient acquisition and root exudation of soil grown maize plants. *Mycorrhiza* 5: 321–327.

Azcon-Aguilar C and Barea JM (1996). Arbuscular mycorrhizas and biological control of soil borne plant pathogen—an overview of mechanism involved. *Mycorrhiza* 6: 457–464.

Azcon-Aguilar C, Jaizme-Vega MC and Clavet C (2002). The contribution of arbuscular mycorrhizal fungi to the control of soil borne pathogens. In: Mycorrhizal technology in agriculture: From Genes to Bioproducts

(Gianinazzi S, Schuepp H, Barea JM and Haselwandter K and Barea JM eds) ALS Birkhauser, Basel, Switzerland 187–197.

Avis TJ, Gravel V, Antoun H and Tweddell RJ (2008). Multifacet beneficial effects of rhizosphere microorganisms on plant health and productivity. *Soil Biol Biochem* 40: 1733–1740.

Baath E and Hayman DS (1983). Plant growth responses to vesicular-arbuscular mycorrhizae XIV. Interactions with Verticillium wilt on tomato plants. *New Phytol* 95: 419–426.

Baltruschat H and Schonbeck F (1975). Studies on the influence of endotrophic mycorrhiza on the infection of tobacoo by *Thielaviopsis basicola*. *Phytopath.* 84: 172–188.

Becker WN (1976). Quantification of onion vesicular-arbuscular mycorrhizae and their resistance to *Pyrenochaeta terrestris*. Ph.D dissertation, University of Illinios, Urbana, IL.

Berta G, Fusconi A and Trotta A (1993). VA mycorrhizal infection and the morphology of root systems. *Environ Expt Bot* 33: 159–173.

Bethlenfalvay GJ and Schüepp H (1994). Arbuscular mycorrhizas and agrosystem stability. In: Impact of Arbuscular mycorrhizas on sustainable and Natural Ecosystems. Gianinazzi S and Schuepp H (eds), Birkhauser Verlag, Basel, Switzerland, 117–131.

Blilou I, Bueno P, Ocampo JA and Garcia-Garrido JM (2000a). Induction of catalase and ascorbate peroxidise activities in tomato roots inoculated with the arbuscular Mycorrhizal fungus *Glomus mosseae*. *Mycol Res* 104: 722–725.

Blilou I, Ocampo JA and Garcia-Garrido JM (2000b). Induction of Ltp (Lipid transfer protein) and Pal (Phenylalanine ammonia-lyase) gene expression in rice root colonized by the arbuscular fungus *Glomus mosseae*. *J Exp Bot* 51: 1969–1977.

Bodker L, Kjoller R and Rosendahl S (1998). Effect of phosphate and arbuscular Mycorrhizal fungus *Glomus intraradices* on disease severity of root rot of peas (*Pisum sativum*) caused by Aphanomyces euteiches. *Mycorrhiza* 8: 169–174.

Bago B, Pfeffer PE and Shachar-Hill Y (2000). Carbon metabolism and transport in arbuscular mycorrhizas. *Plant Physiol* 124: 949–957.

Boomsma CR and Tony J Vyn (2008). Maize drought tolerance: Potential improvements through arbuscular mycorrhizal symbiosis? *Field Crops Research* 108: 14–31.

Calvet C, Pera J and Barea JM (1993). Growth response of marigold (*Tagetes erecta* L.) to inoculation with *Glomus mosseae, tricoderma aureoviride* and *Pythium ultimum* in a peat-perlite mixture. *Plant Soil* 148: 1–6.

Calvet C, Pinochet J, Hernandez-Dorrego A, Estaun V and Camprubi A (2001). Field microplot performance of the peach-almond GF-677 after inoculation with arbuscular mycorrhizal fungi in a replant soil infested with root-knot nematode. *Mycorrhiza* 10: 295–300.

Cameron GC (1986). Interactions between two vesicular-arbuscular mycorrhizal fungi, the soybean cyst nematode and phosphorus fertility on two soybean cultivars. M.S. Thesis University of Georgia, Athens.

Camprubi A, Pinochet J, Calvet C and Estaun V (1993). Effects of root-lesion nematode *Pratylenchus vulnus* and vesicular arbuscular mycorrhizal fungus *Glomus mosseae* on the growth of three plum rootstocks. *Plant and Soil* 153: 223–229.

Cardoso IM and Kuyper TW (2006). Mycorrhizas and tropical soil fertility. *Agric Ecosyst Environ* 116: 72–84.

Caron M, Fortin JA and Richard C (1985). Influence of substrate on the interaction of *Glomus intraradices* and *Fusarium oxysporum* f sp. radicis-lycopersici on tomatoes. *Plant Soil* 87: 233–239.

Caron M, Fortin JA and Richard C (1986a). Effect of inoculation sequence on the interaction between *Glomus intraradices* and *Fusarium oxysporium* f. sp *radicis-lycopersici* in tomatoes. *Can J Plant Pathol* 8: 12–16.

Caron M, Fortin JA and Richard C (1986b). Effect of phosphorus concentration and *Glomus intraradices* On *Fusarium* crown and root rot tomatoes. *Phytopathology* 76: 942–946.

Caron M, Fortin JA and Richard C (1986c). Effect of pre-infection of the soil by a vesicular-arbuscular mycorrhizal fungus, *Glomus intraradices* on Fusarium crown and root-rot of tomatoes. *Phytoprotec.* 67: 15–19.

Citernesi AS, Fortuna P, Filippi C, Bagnoli G and Giovannetti M (1996). The occurrence of antagonistic bacteria in *Glomus mosseae* pot cultures. Agronomie 16: 671–677.

Cook RJ and Baker KF (1982). The nature and practice of biological control of plant pathogens. *APS St Paul,* MN.

Cordier C, Gianinazzi S and Gianinazzi-Pearson V (1998a). Colonization pattern of root tissue by *Phytophthora nicotiannae* var parasitica related to reduced disease in mycorrhizal tomato. *Plant Soil* 185: 223-232

Cordier C, Pozo MJ, Barea JM, Gianinazzi S and Gianinazzi-Pearson V (1998b). Cell defense responses associated with localized and systemic resistance to *Phytophthora parasitica* induced in tomato by an arbuscular mycorrhizal fungus. *Mol Plant Microb Interact* 11: 1017–1028.

Davis RM (1980). Influence of *Glomus fasciculatus* on *Thielaviopsis basicola* root rot of citrus. *Phytopathology* 70: 447–452.

Davis RM and Menge JA (1980). Influence of *Glomus fasciculatus* and soil phosphorus on Phytophthora root-rot of citrus. *Phytopathology* 70: 447–452.

Dehne HW (1982). Interaction between vesicular-arbuscular mycorrhizal fungi and plant pathogens. *Phytopathology* 72: 1115–1119.

Dehne HW, Schonbeck F and Baltruschat H (1978). Untersuchungen zum einfluss der endotrophen Mycorrhiza auf Planzenkrankheiten: 3. Chitinase-aktivitat und ornithinzyklus (The influence of endotrophic Mycorrhiza on plant disease: 3 chitinase-activity and ornithine cycle). *J. Plant Dis. Protec.* 85: 666–678.

Diaz S (1996). Effects of elevated [CO] at the community level mediated by root symbionts. *Plant Soil* 187: 309–320.

Diedhiou PM, Hallmann J, Oerke EC and Dehne HW (2003). Effects of arbuscular mycorrhizal fungi and non-pathogenic *Fusarium oxysporium* on *Meloidogyne incognita* infestation of tomato. *Mycorrhiza* 13: 199–204.

Elsen A, Beeterens R, Swennen R and Waele D (2003). Effects of an arbuscular mycorrhizal fungus and two plant parasitic nematodes on Musa genotypes differing in root morphology. *Biol Fertil Soils* 38 (6) 367–376.

Elsen A, Gervacio D, Swennen R and De Waele D (2008). AMF-induced biocontrol against plant parasitic nematodes in *Musa* sp.: asystemic effect. *Mycorrhiza* 18: 251–256.

Feng G, Zhang FS, Li XL, Tian CY, Tang C and Rengel Z (2002). Improved tolerance of maize plants to salt stress by arbuscular mycorrhiza is related to higher accumulation of soluble sugars in roots. *Mycorrhiza* 12: 185–190.

Filion M, St-Arnaud M and Fortin JA (1999). Direct interaction between the arbuscular mycorrhizal fungus *Glomus intraradices* and different rhizosphere microorganisms. *New Phytol* 141: 525–533.

Finlay RD (2008). Ecological aspects of mycorrhizal symbiosis: with special emphasis on the functional diversity of interactions involving the extraradical mycelium. *J. Exp. Bot.* 59: 1115–1126.

Fitter AH (2001). Specificity links and networks in the control of diversity in plant and microbial communities. In: Ecology, Achievement and Challenge. M.C., Huntly, N.J and Levin, S. Blackwell (Eds.), Oxford, UK, 95–114.

Fitter AH and Grabaye J (1994). Interaction between mycorrhizal fungi and other soil organisms. *Plant Soil* 159: 123–132.

Francl IJ and Dropkin VH (1985). *Glomus fasciculatum* a week pathogen of *Heterodera glycines*. *J Nematol* 17: 470–475.

Fritz M, Jakobsen I, Lyngkjaer MF, thordal-Christensen H and Pons-Kuhnemann J (2006). Arbuscular mycorrhiza reduces susceptibility of tomato to *Alternaria solani*. *Mycorrhiza* 16: 413–419.

Gange AC and West HM (1994). Interactions between arbuscular mycorrhizal fungi and foliar feeding insects in *Plantago lanceolata* L. *New Phytol.* 128: 79–87.

Garcia-Garrido JM and Ocampo JA (1989). Effect of VA Mycorrhizal infection of tomato on damage caused by *Pseudomonas syringae*. *Soil Biol Biohem.* 21: 65–167.

Garcia-Garrido JM and Ocampo JA (2002). Regulation of the plant defence response in arbuscular mycorrhizal symbiosis. *J Exp Bot* 53: 1377–1386.

Garmendia I, Aguirreolea J and Goicoechea N (2006). Defence related enzymes in pepper roots during interactions with arbuscular mycorrhizal fungi and/or *Verticillium dahlia*. *Biocontrol* 51: 293–310.

George E, Marchner H and Jakobsen I (1995). Role of arbuscular Mycorrhizal fungi in uptake of phosphorus and nitrogen from the soil. *Crt Rev Biotechnol* 15: 257–270.

Gosling P, Hodge A, Goodlass G and Bending GD (2006). Arbuscular Mycorrhizal fungi and organic farming. *Agric Ecosyst Environ* 113: 17–35.

Graham JH (1982). Effect of citrus root exudates on germination of chlamydospores of vesicular-arbuscular mycorrhizal fungus *Glomus epigaeum*. *Mycologia* 74: 831–835.

Graham JH (2001). What do root pathogen see in mycorrhiza? *New Phytol.* 149: 357–359.

Graham JH and Egel DS (1988). Phytophthora root rot development on mycorrhizal and phosphorus fertilized non-mycorrhizal sweet orange seedlings. *Plant Dis.* 72: 611–614.

Graham JH and Menge JA (1982). Influence of vesicular-arbuscular mycorrhizal fungi and soil phosphorus on take all disease of wheat. *Phytopathology* 72: 95–96.

Grandmaison J, Olah GM, Van Calsteren MR and Furlan V (1993). Characterization and localization of plant phenolics likely involved in the pathogen resistance expressed by endomycorrhizal roots. *Mycorrhiza* 3: 155–164.

Gutlahr C, Casieri L and Paszkowski U (2009). *Glomus intraradices* induces changes in root system architecture of independently of common symbiosis signalling. *New Phytol.* 182: 829–837.

Habte M, Zhang YC and Schmitt DP (1999). Effectiveness of *Glomus* species in protecting white clover against nematode species damage. *Can J Bot* 77: 95–98.

Harley JI and Smith SE (1983). Mycorrhizal symbiosis, Academic , New York.

Harrier LA and Watson CA (2004). The potential role of arbuscular mycorrhizal (AM) fungi in the bioprotection of plants against soil-borne pathogens in organic and or other sustainable farming systems. *Pest Manage Sci* 60: 149–157.

Harrison M and Dixon R (1993). Isoflavonoid accumulation and expression of defense gene transcripts during the establishement of vesicular arbuscular Mycorrhizal association in roots of *Medicago truncatula. Mol Plant-Microbe Interact* 6: 643–659.

Hause B, Cornelia M, Stanislav I and Strack D (2007). Jasmonate in arbuscular mycorrhizal interactions. *Phytochemistry* 68: 101–110.

Hiltner L (1904). Uber neuere erfahrungen und problem auf dem gebiet der bodenhakteruiligie und unter besonderer berucksichtiguang der grundungung und brache. *Arb Deutsch Landwirt* 98: 59–78 (In German).

Hodge A (2000). Microbial ecology of arbuscular mycorrhiza. *FEMS Microbiol Ecol* 32: 91–96.

Hooker JE, Jaizme-Vega M and Alkinson D (1994). Biocontrol of plant pathogen using arbuscular Mycorrhizal fungi, In: Impact of Arbuscualr Mycorrhizas on Sustainable Agriculture and Natural Ecosystems. Gianinazzi S and Schhepp H (eds) Birkhauser Verlag, Basle, Switzerland 191–209.

Jaizme-Vega MC and Pinochet J (1997). Growth response of banana to three mycorrhizal fungi in *Pratylenchulus goodeyi* infested soil. *Nematropica* 27: 69–76.

Johansson JF, Paul LR and Finlay RD (2004). Microbial interactions in the mycorrhizosphere and their significance for sustainable agriculture. *FEMS Microbiol. Ecol.* 48: 1–13.

Kaplan DT, Keen NT and Thompson IJ (1980). Association of glyceollin with incompatible response of soybean roots to Meloidogyne incognita. *Physiol. Plant Pathol,* 16: 309–318.

Karagiannidis N, Bletsos F and Stavropoulos N (2002). Effect of *Verticillium* wilt (*Veticillium dahlia* Kleb.) and mycorrhiza (*Glomus mosseae*) on root colonization, growth and nutrient uptake in tomato and egg plant seedlings. *Scientia Horticulturae* 94: 145–156.

Khaosaad T, Garcia-Garrido JM, Steinkellner S and Vierheilig H (2007). Take all disease is systemically reduced in roots of mycorrhizal barley plants. *Soil Biol Biochem* 39: 727–734.

Kim JS, Dungan RS, Kwon SW and Weon HY (2006). The community composition of root-associated bacteria of the tomato plant. *WJ Microbial Biotechnol* 22: 1276–1273.

Koide R (1991). Nutrient supply, nutrient demand and plant-response to mycorrhizal infection. *New Phytol* 117: 365–386.

Krishna KR and Bagyaraj DJ (1983). Interaction between *Glomus fasiculatum* and *Sclerotium rolfsii* in peanut. *Can J Bot* 61: 2349–2351.

Larose G, Chenevert, Moutoglis P, Gagne S, Piche Y and Vierheilig H (2002). Flavonoid levels in roots of *Medicago sativa* are modulated by the developmental stage of the symbiosis and the root colonizing arbuscular mycorrhizal fungus. *J Plant Physiol* 159: 1329–1339.

Larsen J and Bodker L (2001). Interactions between pea root-inhabiting fungi examined using signature fatty acids. *New Phytol* 149: 487–493.

Larsen J, Ravnskov S and Jakobsen I (2003). Combined effect of an arbuscular mycorrhizal fungus and a biocontrol bacterium against *Pythium ultimum* in soil. *Folia Geobotanica* 38: 145–154.

Lee CS, Lee YJ and Jeun YC (2005). Observation of infection structures on the leaves of cucumber plants pre-treated with arbuscular mycorrhiza *Glomus intraradices* after challenge inoculation with *Colletotrichum orbiculare*. *Plant Pathol J* 21: 237–243.

Lerat S, Lapointe L, Gutjahr S, Piche Y and Vierheilig H (2003a). Carbon partitioning in a split-root system of arbuscular Mycorrhizal plants is fungal and plant species dependent. *New Phytol* 157: 589–595.

Lerat S, Lapointe L, Piche Y and Vierheilig H (2003b). Variable carbon sink strength of different *Glomus mosseae* strains colonizing barley roots. *Can J Bot* 81: 886–889.

Li B, Ravnskov S, Xie GL and Larsen J (2007). Biocontrol of Pythium damping-off in cucumber by arbuscular mycorrhiza-associated bacteria from the genus Paenibacillus. *Biocontrol* 52: 863–875.

Linderman RG (1994). Role of VAM fungi in biocontrol. In: Mycorrhizae and Plant Health (Eds. Pfleger FL and Linderman RG) The American Phytopathological Society, St Paul, MN., USA, ISBN:0-89054-158-2, 1–27.

Lingua G, D'Agostino G, Massa N, Antosiano M and Berta G (2002). Mycoorhiza-induced differential response to a yellows disease in tomato. *Mycorrhiza* 12: 191–198.

Liu A, Hamel C, Hamilton RL and Smith DL (2000). Mycorrhizae formation and nutrient up take of new corn (*Zea mays* L.) hybrids with extreme canopy and leaf architecture as influenced by soil N and P levels. *Plant Soil* 221: 157–166.

Liu J, Baylock L, Endre G, Cho J, Town CD, Vanden Bosch K and Harrison MJ (2003). Transcript profiling coupled with spatial expression analyses reveals genes involved in distinct developmental stages of the arbuscular mycorrhizal symbiosis. *Plant Cell* 15: 2106–2123.

Liu J, Maldonado-mendoza I, lopez-Meyer M, Cheung F, Town CD and Harrisom MJ (2007). Arbuscular mycorrhizal symbiosis is accompanied by local and systemic alterations in gene expression and an increase in disease resistance in the shoots. *Plant J* 50: 529–544.

Lioussanne L, Jolicoeur M and St-Arnaud M (2008). Mycorrhizal colonization with *Glomus intraradices* and development stage of transformed tomato roots significantly modify the chemotactic response of zoospores of the pathogen *Phytophthora nicotianae*. *Soil Biol and Biochem* 40: 2217–2224.

Maherali H and Klironomos J (2007). Influence of phylogeny on fungal community assembly and ecosystem functioning. *Science* 316: 1746–1748.

Mark GL and Cassels AC (1996). Genotype-dependence in the interaction between *Glomus fistulosum, Phytophthora fragariae* and the wild strawberry (*Fragaria vesca*). *Plant Soil* 185: 233–239.

Marschner H (1995). Mineral nutrition of higher plants. Academic Press, New York.

Marschner H and Dell B (1994). Nutrient uptake in mycorrhizal symbiosis. *Plant Soil* 159: 89–102.

Martin F, Perotto S and Bonfante P (2000). Mycorrhizal fungi: A fungal communityat the interface between soil and roots. In: The Rhizosphere: Biochemistry and organic substances at soil-plant interface. (Eds) Pinton R, Varanini Z and Nannipieri P) New York/Basel: Marcel Dekker, 263–296.

Matsubara Y, Tamura H and Harada T (1995). Growth enhancement and Verticillium wilt control by vesicular-arbuscular mycorrhizal fungus inoculation in eggplant. *Journal of Japanese Society of Horticultural Science* 64: 55–561.

Medina HMJ, Gagnon H, Piche Y, Ocampo JA, GarciaGarrido JM and Vierheilig H (2003). Root colonization by arbuscular mycorrhizal fungi is effected by the salicylic acid content of the plant. *Plant Sci* 164: 993–998.

Meyer JR and Linderman (1986). Selective influence on population of rhizosphere or rhizoplane bacteria and actinomycetes by mycorrhizas formed by *Glomus fasciculatum*. *Soil Biol Biochem* 18: 191–196.

Miller RM and Jastrow JD (2000). Mycorrhizal fungi influence soil structure. In: Arbuscular mycorrhizas: physiology and function (Eds. Kapulnik Y and Douds DD) Kluwer Academics, Dordrecht, The Netherlands, 3–18.

Mohammad MJ, Malkawi HI and Shibli R (2003). Effects of mycorrhizal fungi and phosphorus fertilization on growth and nutrient uptake of barley grown on soils with different levels of salts. *J Plant Nutr* 26: 125–137.

Morandi D (1987). VA Mycorrhizae, nematode, phosphorus, and phytoalexins on soybean. In: Mycorrhizae in next decade, practical application and research priorities (Eds. Slyvia DM, Hung LL and Graham DH) Proc. of 7th North. Am. conference on Mycorrhizae, Institute of Food and Agriculture Sciences, University of Florida, Gaineville, GA.

Morandi D (1996). Occurrence of phytoalexins and phenolic compounds on endomycorrhizal interactions, and their potential role in biological control. *Plant Soil* 185: 241–251.

Newsham KK, Fitter AH and Watkinson AR (1995a). Arbuscular mycorrhiza protect an annual grass from root pathogenic fungi in the field. *Journal of Ecology* 83: 991–1000.

Newsham KK, Fitter AH and Watkinson AR (1995b). Multifunctionality and biodiversity in arbuscular mycorrhizas. *Trend in Ecology and Evolution* 10: 407–411.

Norman JR, Atkinson D and Hooker JE (1996). Arbuscular mucorrhizal fungal induced alteration to root architecture in strawberry and induced resistance to root pathogen *Phytophthora fragariae. Plant Soil* 185: 191–198.

Norman JR and Hooker JE (2000). Sporulation of *Phytophthora fragaria* shows greater stimulation by exudates of non-mycorrhizal than by mycorrhizal strawberry roots. *Mycol Res* 104: 1069–1073.

Olah B, Briere C, Becard G, Denarie J and Gough C (2005). Nod factors and a diffusible factor from arbuscular mycorrhizal fungi stimulate lateral root formation in *Medicago truncatula* via the DM11/DM12 signalling pathway. *Plant Journal* 44: 195–207.

Oswald ET and Ferchau HA (1968). Bacterial associations of coniferous Mycorrhizae. *Plant Soil* 28: 187–192.

Pacovsky RS, Bethelenfalvay GJ and Paul EA (1986). Camparisons between P-fertilized and mycorrhizal plants. *Crop Sci* 16: 151–156.

Paszkowski U, Kroken S, Roux C and Briggs SP (2002). Rice phosphate transporters include an evolutionarily divergent gene specifically activated in arbuscular mycorrhizal symbiosis. *Proceedings of the National Academy of Sciences* 99: 13324–13329.

Pearson JN and Jakobsen I (1993). Symbiotic exchange of carbon and phosphorus between cucumber and three arbuscular mycorrhizal fungi. *New Phytol* 124: 480–488.

Picone C (2003). Managing mycorrrhizae for sustainable agriculture in the tropics. In: Tropical Agroecosystems. Vandermeer JH (ed), CRC Press, Boca Raton, FL, 95–132.

Pozo MJ and Azcon-Aguilar C (2007). Unravelling Mycorrhiza-induced resistance. *Curr Opinion Plant Biol* 4: 393–398.

Pozo MJ, Azcon-Aguilar C, Dumas-Gaudot E, and Barea JM (1999). B-1, 3-glucanase activities in tomato roots inoculated with arbuscular mycorrhizal fungi and/or *Phytophthora parasitica* and their possible involvement in bioprotection. *Plant Sci* 141: 149–157.

Pozo MJ, Cordier C, Dumas-Gaudot E, Gianinazzi S, Barea JM and Azcon-aguilar C (2002). Localized verus systemic effect arbuscular mycorrhizal fungi on defense responses to *Phytophthora* infection in tomato plants. *J Expt Bot* 53: 525–534.

Pozo MJ, Dumas-Gaudot E, Azcon-Aguilar C and Barea JM (1998). Chitosanase and chitinase activities in tomato roots during interactions with arbuscular Mycorrhizal fungi or *Phytophthora parasitica. J Expt Bot* 49: 1729–1739.

Pozo MJ, Verhage A, Garcia-Andrade J, Garcia JM and Azcon-Aguilar C (2009). Priming plant defence against pathogens by arbuscular mycorrhiza fungi. In: Mycorrhizas-Functional Processes and Ecological Impact. (Eds. Azcon-Aguilar C, Barea JM, Gianinazzi S, and Gianinazzi-Pearson V). *Springer Verlag, Berlin* 123–136.

Price N, Roncardori R and Hussey R (1989). Turgor, solute import and growth in maize roots treated with galactose. *New Phytol* 31: 1095–1103.

Rabatin SC and Rhodes IH (1982). *Acaulospora bireticulatata* inside oribatid mites. *Mycologia* 74: 859–861.

Remy W, Taylor TN, Hass H and Kerp H (1994). Four hundred million year old vesicular arbuscular Mycorrhizae. *Proceedings of the National Academy of Sciences, USA* 91: 11841–11843.

Rilling MC (2004). Arbuscular mycorrhizae, glomalin and soil aggregation. *Can J Soil Sci* 84: 355–363.

Rosendahl S (1985). Interactions between the vesicular-arbuscular mycorrhizal fungus *Glomus intraradices* and *Aphanomyces euteiches* root rot of peas. *Phytopathol Z* 114: 31–40.

Saleh H and Sikora RA (1984). Relationship between *Glomus fasiculatum* root colonization on cotton and its effect on *Meloidogyne incognita. Nematologica* 30: 230–237.

Salzer P, Corbiere H, and Boller T (1999). Hydrogen peroxide accumulation in Medicago truncatula roots colonized by the arbuscular mycorrhiza-forming fungus *Glomus mosseae. Planta* 208: 319–325.

Sanchez-Daiz M and Honrubia M (1994). Water relations and alleviation of drought stress in mycorrhizal plants. In: impact of arbuscular mycorrhizas on sustainable agriculture and natural ecosystems. Gianinazzi S and Schüepp H (eds) Birkhauser, Basel, Switzerland, 167–178.

Sanders FE and Tinker PB (1971). Mechanism of absorption of phosphate from soil Endogone mycorrhizas. *Nature* 232: 2678–279.

Sankaranarayanan C and Sundarababu R (1994). Interaction of *Glomus fasciculatum* with *Meloidogyne incognita* inoculated at different timings on black gram (*Vigna mungo*). *Nematol medit* 22: 35–36.

Scheffknecht S, Mammerler R, Steinkellner S and Vierheilig H (2006). Root exudates of Mycorrhizal tomato plants exhibit a different effect on microconidia germination of *Fusarium oxysporium* f. sp. *lycopersici* than root exudates from non-mycorrhizal tomato plants. *Mycorrhiza* 16: 365–370.

Schellenbaum L, Berta G, Ravolanirina F, Tisserant B, Gianinazzi S and Fitter AH (1991). Influence of endomycorrhizal infection on root morphology in micropropagated woody plant species (Vitis vinifera L.). *Ann Bot* 68: 135–141.

Schippers B, Bakker AW and Bakker PAHM (1987). Interactions of deleterious and beneficial rhizosphere microorganisms and the effect of cropping practices. *Ann Rev Phytopath* 25: 339–358.

Schussler A, Schwarzott D and Walker C (2001). A new fungal phylum, the Glomeromycota, phylogeny and evolution. *Mycol Res* 105: 1413–1421.

Secilia J and Bagyaraj DJ (1987). Bacteria and actinomycetes associated with pot cultures of vesicular-arbuscular mycorrhizas. *Can J Microbiol* 33: 1069–1073.

Sharma MP, Gaur A and Mukerji KG (2007). Arbuscular mycorrhiza mediated plant pathogen interaction and the mechanisms involved. In: Biological control of plant diseases (Sharma MP, Gaur A and Mukerji KG (eds). Haworth Press, Binghamton, USA, 47–63.

Shaul O, Galili S, Volpin H, Ginzberg IElad Y, Chet I and Kapulnik Y (1999). Mycorrhiza induced changes in disease severity and PR protein expression tobacco leaves. *Mol Plant Microbe Inter* 12: 1000–1007.

Shetty KG, Hetrick BAD and Schwab AP (1995). Effects of mycorrhizae and fertilizer amendments on zinc tolerance of plants. Environ Pollut 88: 307-314

Siasou E, Standing D, Killham K and Johnson D (2009). Mycorrhizal fungi increase biocontrol potential of *Pseudomonas flouorescens. Soil Biology and Biochemistry* 41: 1341-1343.

Sieverding E and Oehl F (2006). Revision of Entrophospora and description of Kuklospora and Intraspora, two new genera in the arbuscular mycorrhizal Glomermycetes. *J Applied Bot Food Qual* 80: 69–81.

Sikotra RA and Sitaramaiah K (1980). Antagonistic interactions between the endotropic mycorrhizal fungus *Glomus mosseae* and *Rotylenchulus reniformis* on cotton. *Nematropica* 10: 72–73.

Singh R, Adholeya A and Mukerji KG (2000). Mycorrhiza in control of soil-borne pathogens. In: Mycorrhizal biology. Mukerji K.G, Chamola BP and Singh J (eds) Kluwer, New York 173–196.

Singh DP, Srivastava JS, Bahadur A, Singh UP and Sink SK (2004). Arbuscular mycorrhizal fungi induced biochemical changes in pea (*Pisum sativum*) and their effect on powdery mildew (*Erysiphe pisi*). *J Plant Dis Protect* 111: 266–272.

Smith GS (1988). The role of the phosphorus nutrition in interactions of vesicular-arbuscular mycorrhizal fungi with soilborne nematodes and fungi. *Phytopathology* 78: 371–374.

Smith SE and Giananizzi-Pearson V (1988). Physiological interactions between symbionts in vesicular-arbuscular mycorrhizal plants. *Ann Rev Plant Physiol Mol Boil* 39: 221–224.

Smith SE and Read DJ (1997). Mycorrhizal symbiosis. Academic, London

Smith, SE and Read DJ (2008). Mineral nutrient, toxic element accumulation and water relations of arbuscular mycorrhizal plants. In: Mycorrhizal Symbiosis, 3rd Ed., Academic Press, London, ISBN-10: 0123705266, 145–18.

Smith SE, Smith FA and Jokobsen I (2003). Mycorrhizal fungi can dominate phosphate supply to plants irrespective of growth responses. *Plant Physiol* 133: 16–20.

St-Arnaud M, Hamel C, Vimard B, Caron M and Fortin JA (1997). Inhibition of *Fusarium oxysporium* f. sp. dianthi in the non-VAM species Dianthus caryophyllus by co-culture with *Tagetes patula* companion plants colonized by *Glomus intraradices*. *Can J Bot* 75: 998–1005.

St-Arnaud M and Vujanovic V (2007). Effect of the arbuscular Mycorrhizal symbiosis on plant disease and pests. In: Arbuscular mycorrhizae in crop production. Hamel C and Plenchette C (eds), Haworth's Food Products Press, NY 67–122.

St-Arnaud M, Hamel C, Vimard B, Caron M and Fortin JA (1995). Altered growth of *Fusarium oxysporum* f. sp. *chrysanthemi* in an *in vitro* dual culture system with the vesicular arbuscular mycorrhizal fungus *Glomus intraradices* growing on *Daucus carota* transformed root. *Mycorrhiza* 5: 431–438.

Subramanian S and Senthamizh K (2006). Effect of culture filtrates of *Pseudomonas flourescens* and *Trichoderma viride* and root exudates of *Glomus mosseae* on hatching of Rotylenchulus reniformis. *Indian J Nematol* 36: 133–135.

Sylvia DM and Sinclair WA (1983). Phenolic compounds of resistance to fungal pathogens induced in primary roots of Douglas-fir seedlings by ectomycorrhizal fungus Laccaris laccata. *Phytopathology* 73: 390–397.

Sylvia DM, Wilson DO, Graham JH, Maddox JJ, Millner P, Morton JB, Skipper HD, Wright SF and Jarafster G (1993). Evaluation of vesicular-arbuscular mycorrhizal fungi in diverse plants and soils. *Soil Biol Biochem* 25: 705–713.

Taber HA (1982). Occurrence of glomus spores in weed seeds in soil. *Mycologia* 74: 515–520.

Tahat MM, Kamaruzaman S and Othman O (2010). Mycorrhizal fungi as a biocontrol. *Plant Pathol J* 9: 198–207.

Tahat MM, Kamaruzaman S and Othman O, Kadir J and Masdek HN (2008). Response of (Lycopersicum esculentum Mill.) to different arbuscular mycorrhizal fungi species. *Asian J Plant Sci* 7: 479–484.

Talavera M, Itou K and Mizukubo T (2001). Reduction in nematode damage by root colonization with arbuscular Mycorrhizal (Glomus spp.) in tomato *Meloidogyne incognita* (Tylenchida: Meloidogynidae)

and carrot *Pratylenchuus penetrans* (Tlenchida: Pratylenchidae) pathosystem. *Appl Entomol Zool* 36: 387–392.

Teljander JF, Lindahl BD, Paul LR, Elfstrand M and Finlay RD (2007). Influence of arbuscular mycorrhizal mycelia exudates on soil bacterial growth and community structure. *FEMS Microbiol Eco* 61: 295–304.

Tobar RM, Azcon R and Barea JM (1994a). Improved nitrogen uptake and transport from 15N-labeled nitrate by external hyphae of arbuscular mycorrhizae under water-stress conditions. *New Phytol* 126: 119–122.

Tobar RM, Azcon R and Barea JM (1994b). The improvement of plant N acquisition from an ammonium-treated, drought-stressed soil by the fungal symbiont in arbuscular mycorrhizae. *Mycorrhiza* 4: 105–108.

Trotta A, Vanese E, Gnavi a Sampo, S and Berta, G (1996). Interaction between the soilborne root pathogen *Phytophthora nicotianne* var *parasitica* and the arbuscular mycorrhizal fungus *Glomus mosseae* in tomato plant. *Plant Soil* 185: 199–209.

Umesh KC, Krishnappa K and Bagyaraj DJ (1988). Interaction of burrowing nematode, *Radophilus similis* (Cobb, 1938) Thorne, 1949, and VA Mycorrhiza, *Glomus fasciculatum* (Thaxt.) Ger. And Trappe, in banana (*Musa acuminate* Colla.). *Indian J Nematol* 18: 6–11.

Vaast PH, Caswell-Chen EP and Zasoski RJ (1998). Influences of a root-lesion nematode, Pratylenchus coffeae, and two arbuscular mycorrhizal fungi, *Acaulospora mellea* and *Glomus clarum* on coffee (Coffea arabica L.). *Biol Fert Soils* 26: 130–135.

Van der Heijden MGA and Sanders IR (2002). Mycological ecology: Synthesis and perspectives. In: Mycological Ecology. Van der Heijden MGA and Sanders IR (eds) studies in Ecology 157, Heidelberg: Springer-Verlag, 441–456.

Vierheilig H (2004a). Regulatory mechanisms during the plant-arbuscular mycorrhizal fungus interaction. *Can J Bot* 82: 1166–1176.

Vierheilig H (2004b), Further root colonization by arbuscular mycorrhizal in already mycorrhizal plants is suppressed after critical level of root colonization. *J Plant Physiol* 161: 339–341.

Vierheilig H, Alt M, Boller T and Wiemken A (1994). Ethylene biosynthesis and activities of chitinase and β-1, 3-glucanase in the roots of host and non-host plants of vesicular-arbuscular Mycorrhizal fungi after inoculation with *Glomus mosseae*. *J. Plant Physiol* 143: 337–343.

Vierheilig H and Piche Y (2002). Signalling in arbuscualr mycorrhiza: fact and hypotheses. In: Flavonoids in cell function. Buslig B and Manthey J (eds) Kluwer, New York, 23–39.

Vierheilig H, Steinkellner S, Khaosaad T and Garcia-Garrido JM (2008). The biocontrol effect of mycorrhization on soilborne fungal pathogens and autoregulation of the AM symbiosis: One Mechanism, Two effects? In: Varma A (ed) Mycorrhiza. Springer-Verlag, Berlin Heidelberg.

Vigo C, Norman JR and Hooker JE (2000). Biocontrol of pathogen *Phytophthora parasitica* by arbuscular mycorrhizal fungi is a consequence of effects on infection loci. *Plant Pathol* 49: 509–514.

Whipps JM (2004). Prospects and limitations for mycorrhizas in biocontrol of root pathogens. *Can J Bot* 82: 1198–1227.

Wick Rl and Moore LD (1984). Histologiy of mycorrhizal and non-mycorrhizal Ilex crenata 'Helleri' challenged by *Thievlaviopsis basicola. Can J Plant Pathol* 6: 146–150.

Wright SF (2005). Management of arbuscular mycorrhizal fungi. In: Roots and Soil Management: Interactions between Roots and Soil. Zobel RW and Wright SF (eds), American Society of Agronomy, Crop Science Society of America, Soil Science Society of America, Madison, WI, 183–197.

Wyss P, Boler T and Wiemken A (1991). Phytoalexin response is elicited by a pathogen (*Rhizoctonia solani*) but not my a mycorrhizal fungus (*Glomus mosseae*) in bean roots. *Experiential* 47: 395–399.

Xavier LJC and Boyetchko SM (2004). Arbuscular mycorrhizal fungi in plant disease control, Fungal Biotechnology in agricultural, food and environment applications. (Arora DK (ed) Dekker, New York 183–184.

Yano K, Yamauchi A and Kono Y (1996). Localized alteration in lateral root development in roots colonized by arbuscular mycorrhizal fungus. *Mycorrhizas* 6: 409–415.

Zhang L, Zhang J, Christle P and Li X (2008). Pre-inoculation with arbuscular Mycorrhizal fungi suppresses root knot nematode (*Meloidogyne incognita*) on cucumber (*Cucumis sativus*). *Biol Fertil Soils* 45: 205–211.

Zhu HH and Zao Q (2004). Localized and systemic increase of phenols in tomato roots induced by *Glomus versiforme* inhibits *Ralstania solanacearum. J Phytopathol* 152: 537–542.

14

Resistance: Origin, Evolution and Challenges in Future

Ramesh Kumar*

Virology Section, Department of Microbiology, All India Institute of Medical Sciences, New Delhi-110029, India

**Email: ramesh.aiims07@gmail.com*

1 INTRODUCTION

Infections are very common and responsible for a large number of diseases, adversely affecting human health. Most of the infectious diseases are caused by bacteria. Infections caused by bacteria can be prevented, managed and treated through anti-bacterial group of compounds known as antibiotics.

Antibiotics can be defined as the variety of substances derived from bacterial sources (microorganisms) that control the growth of or kill other bacteria. However, synthetic antibiotics, usually chemically related to natural antibiotics, have since been produced that accomplish comparable tasks.

Antibiotics were a breakthrough in medicine. They changed the method of infections treatment. Antibiotics considerably increased life expectancy, thanks to antibiotics, there was a possibility to successfully treat many dangerous infections.

Antibiotics are used around the world in all countries, at hospitals and at home. Antibiotics save lives of people.

2 HISTORY

The word antibiotic comes from the Greek *anti* meaning 'against' and *bios* meaning 'life' (a bacterium is a lifeform).' Antimicrobials are probably one of the most successful forms of chemotherapy in the history of medicine. It is not necessary to reiterate here how many lives they have saved and how significantly they have contributed to the control of infectious diseases that were the leading causes of human morbidity and mortality for most of human existence. Contrary to the common belief that the exposure to antibiotics is confined to the modern "antibiotic era," research has revealed that this is not the case. The traces of tetracycline, for example, have been found in human skeletal remains from ancient

Sudanese Nubia dating back to 350–550 CE (Bassett et al., 1980; Nelson et al., 2010). The distribution of tetracycline in bones is only explicable after exposure to tetracycline-containing materials in the diet of these ancient people. Another example of ancient antibiotic exposure is from a histological study of samples taken from the femoral midshafts of the late Roman period skeletons from the Dakhleh Oasis, Egypt (Cook et al., 1989). These samples showed discrete fluorochrome labeling consistent with the presence of tetracycline in the diet at that time (Cook et al., 1989). The postulated intake of tetracycline in these populations possibly had a protective effect because the rate of infectious diseases documented in the Sudanese Nubian population was low, and no traces of bone infection were detected in the samples from the Dakhleh Oasis (Armelagos, 1969; Cook et al., 1989). Therefore the history of antibiotics can be described in two segments as under.

2.1 Early History

While our scientific knowledge of antibiotics has only recently been developed, the practical application of antibiotics has existed for centuries. The first known use was by the Chinese about 2500 years ago. During this time, they discovered that applying the mouldy curd of soybeans to infections had certain therapeutic benefits. It was so effective that it became a standard treatment. Evidence suggests that other cultures used antibiotic-type substances as therapeutic agents. The Sudanese-Nubian civilization used a type of tetracycline antibiotic as early as 350 A.D. In Europe during the Middle Ages, crude plant extracts and cheese curds were also used to fight infection. Although these cultures used antibiotics, the general principles of antibiotic action were not understood until the twentieth century.

During ancient times, Greeks and Indians used moulds and other plants to treat infections. In Greece and Serbia, mouldy bread was traditionally used to treat wounds and infections. Warm soil was used in Russia by peasants to cure infected wounds. Sumerian doctors gave patients beer soup mixed with turtle shells and snake skins. Babylonian doctors healed the eyes using a mixture of frog bile and sour milk. Sri Lankan army used oil cake (sweetmeat) to serve both as desiccant and antibacterial.

2.2 Modern History

We usually associate the beginning of the modern "antibiotic era" with the names of Paul Ehrlich and Alexander Fleming. Ehrlich's idea of a "magic bullet" that selectively targets only disease-causing microbes and not the host was based on an observation that aniline and other synthetic dyes, which first became available at that time, could stain specific microbes but not others. Ehrlich argued that chemical compounds could be synthesized that would "be able to exert their full action exclusively on the parasite harbored within the organism." This idea led him to begin a large-scale and systematic screening programme (as we would call it today) in 1904 to find a drug against syphilis, a disease that was endemic and almost incurable at that time.

Sir Alexander Fleming, a Scottish biologist, defined new horizons for modern antibiotics with his discoveries of enzyme lysozyme (1921) and the antibiotic substance penicillin (1928). The discovery of penicillin from the fungus Penicillium notatum perfected the treatment of bacterial infections such as, syphilis, gangrene and tuberculosis. He also contributed immensely towards medical sciences with his writings on the subjects of bacteriology, immunology and chemotherapy.

The period between the 1950s and 1970s was indeed the golden era of discovery of novel antibiotics classes, with no new classes discovered since then.

3 CLASSIFICATION

Although there are several classification schemes for antibiotics, based on bacterial spectrum (broad versus narrow) or route of administration (injectable versus oral versus topical), or type of activity (bactericidal vs. bacteriostatic), the most useful is based on chemical structure.

Antibiotics within a structural class will generally show similar patterns of effectiveness, toxicity, and allergic potential. Antibiotics can also be classified based on their chemical structure. A similar level of effectiveness, toxicity and side-effects is rendered by the antibiotics of same structural group. Broad spectrum antibiotics are effective against a broad range of microorganisms in comparison to narrow spectrum antibiotics. Bactericidal antibiotics kill the bacteria whereas bacteriostatic antibiotics halt the growth of bacteria.

4 HOW ANTIBIOTICS WORK?

Antibiotics differ chemically so it is understandable that they also differ in the types of infections they cure and the ways in which they cure them. Certain antibiotics destroy bacteria by affecting the structure of their cells. This can occur in one of the two ways. First, the antibiotic can weaken the cell walls of the infectious bacteria, which causes them to burst. Second, antibiotics can cause the contents of the bacterial cells to leak out by damaging the cell membranes. Another way of antibiotic's function is by interfering with the bacteria's metabolism. Some antibiotics such as tetracycline and erythromycin interfere with protein synthesis. Antibiotics like rifampin inhibit nucleic acid biosynthesis. Still other antibiotics, such as sulfonamide or trimethoprim have a general blocking effect on cell metabolism.

Various types of antibiotics work in either of the following two ways:

- A bacteriocidal antibiotic kills the bacteria generally by either interfering with the formation of the bacterium's cell wall or its cell contents. Penicillin, daptomycin, fluoroquinolones, metronidazole, nitrofurantoin and co-trimoxazole are some examples of bactericidal antibiotics.
- A bacteriostatic antibiotic stops bacteria from multiplying by interfering with bacterial protein production, DNA replication, or other aspects of bacterial cellular metabolism. Some bacteriostatic antibiotics are tetracyclines, sulphonamides, spectinomycin, trimethoprim, chloramphenicol, macrolides and lincosamides.

5 PRODUCTION AND USES OF ANTIBIOTICS

Although most antibiotics occur in nature, they are not normally available in the quantities necessary for large-scale production. For this reason, a fermentation process was developed. The process of antibiotic production basically involves fermentation of the medium by some specific micro-organism. Hence, (i) the medium should be ideal for the microbial action and (ii) the selected strain of the microorganism should give the maximum yield.

6 ANTIBIOTICS IN ENVIRONMENT

Understanding the fate and transport of antibiotics in the environment is essential to assess their impact and subsequent risks to ecosystems. Antibiotics may enter the environment through a number of sources. A 1999 study by the Toxic Substances Hydrology Programme of the US Geological Survey found a range of chemicals in residential, industrial and agricultural wastewaters and at low concentrations in surface waters. The chemicals included human and veterinary pharmaceuticals (including antibiotics), natural and synthetic hormones, detergent metabolites, plasticizers, insecticides, and fire retardants. Measured concentrations of pharmaceuticals in wastewaters were much lower than would be found if a person or animal were consuming the chemical. Most are detected in water at concentrations less than 1 microgram per liter. These amounts are relatively small compared to the dosage provided to humans and animals to treat disease. For example, the antibiotic tetracycline may be given to humans at a rate of 2 grams per day, usually in 2 to 4 pills taken orally. If given to pigs as a growth promoter, the dosage is 0.05 grams per day. Higher concentrations occurred in sediments because of sorption. In Huron County, Michigan, a USGS study found both human-use and veterinary-use antibiotics in stream (surface) water, but not in groundwater (Duris and Haack, 2004). Whether these findings are of biological, environmental or health consequence are currently unknown. Research is being proposed and conducted.

Up to 40% or more of the antibiotic dose may be excreted, especially for antibiotics given at therapeutic doses (Boxall et al., 2004). This is true for both humans and animals. However, different classes of antibiotics are more or less metabolized. Antibiotics are excreted in urine and faeces either unchanged or metabolized in the form of the conjugated, oxidized or hydrolyzed products of parent compounds.

Antibiotics may enter the environment in wastewater, or when human waste solids and animal manures are applied to cropland as plant fertilizer. Some antibiotics degrade quite slowly, possibly surviving the processes of storage and handling, and may be present in land-applied bio-solids. The continual land application of bio-solids could cause the rates of antibiotic accumulation in soils to exceed the rates of degradation. However, accumulation in soils is less probable as environmental regulations and voluntary generally accepted management practices limit manure application so that applied nutrients meet the requirement of the crop being grown. Sorption (absorption or to take up; and adsorption or to hold) by soil plays a determinant role in controlling transport, bioavailability and hence fate of antibiotics in the environment.

The complicated chemical structures of antibiotics lead to multiple interactions with soils (Tolls, 2001). In general, soil organic matter and minerals are the two soil components responsible for holding antibiotics.

Unmetabolized antibiotic substances are often passed into the aquatic environment in wastewater. Antibiotics used for veterinary purposes or as growth promoters are excreted by the animals and end up in manure. Manure is used as an agricultural fertilizer; thus, the antibiotics seep through the soil and enter groundwater. However, very little is known about the occurrence, fate and risks associated with antibiotics entering the environment after being used in human and veterinary medicine and as growth promoters.

7 ANTIBIOTICS RESISTANCE

Antibiotics are probably one of the most successful forms of chemotherapy in the history of medicine. They have saved many millions of lives and placed the majority of infectious diseases that plagued human history for many centuries under control. Initially, on their introduction into clinical practice in the 1940s, antibiotics were extremely efficient in clearing pathogenic bacteria leading many to believe that infectious diseases would become a problem of the past and would be wiped out from all human populations eventually.

However, the emergence and rapid dissemination of antibiotic-resistant pathogens, especially multidrug-resistant bacteria, during recent decades, exposed our lack of knowledge about the evolutionary and ecological processes taking place in microbial ecosystems. Antibiotics are extremely important in medicine, but unfortunately bacteria are capable of developing resistance to them. Antibiotic-resistant bacteria are germs that are not killed by commonly used antibiotics. When bacteria are exposed to the same antibiotics over and over, the bacteria can change and are no longer affected by the drug.

Bacteria have number of ways how they become antibiotic-resistant. For example, they possess an internal mechanism of changing their structure so the antibiotic no longer works, they develop ways to inactivate or neutralize the antibiotic. Also bacteria can transfer the genes coding for antibiotic resistance between them, making it possible for bacteria never exposed to an antibiotic to acquire resistance from those which have. The problem of antibiotic resistance is worsened when antibiotics are used to treat disorders in which they have no efficacy (e.g., antibiotics are not effective against infections caused by viruses), and when they are used widely as prophylaxis rather than treatment.

Resistance to antibiotics poses a serious and growing problem, because some infectious diseases are becoming more difficult to treat. Resistant bacteria do not respond to the antibiotics and continue to cause infection. Some of these resistant bacteria can be treated with more powerful medicines, but there are some infections that are difficult to cure even with new or experimental drugs.

Even before the extensive use of penicillin, some observations suggested that bacteria could destroy it by enzymatic degradation (Abraham and Chain, 1940). In general, though, the outlook was more or less optimistic. One of the earlier studies of possible resistance emergence under laboratory conditions concluded that: "Syphilis has now been treated with arsenicals for about 40 years without any indications of an increased incidence of arsenic-resistant infections, and this work gives grounds for hoping that the widespread use of penicillin will equally not result in an increasing incidence of infections resistant to penicillin" (Rollo et al., 1952). Surprisingly, this is still true for *T. palladium* (Cha et al., 2004), but not for many other pathogenic bacteria, including the *Enterobacteriaceae*, which have become resistant not only to the original penicillin but also to semi-synthetic penicillins, cephalosporins, and newer carbapenems (Kumarasamy et al., 2010). The mortality rates due to multidrug-resistant bacterial infections are high.

The issue of antibiotic resistance has received considerable attention due to the problem of the emergence and rapid expansion of antibiotic-resistant pathogenic bacteria. The common perception, however, is that the problem is seen as exclusively associated with the use and overuse/misuse of antibiotics in humans and animals.

It is now evident that microbial populations possess enormous metabolic diversity, from which they may deploy protective mechanisms allowing them to withstand the selective pressures imposed by their natural environment as well as human interventions such as antibiotics. Revealing the nature and functional role played by antibiotics and antibiotic resistance in various natural ecosystems may help to understand the processes leading to the emergence of antibiotic-resistant pathogens. Finally, armed with knowledge accumulated through many years of genetic, genomic and metagenomic studies and with new concepts about antibiotics and antibiotic resistance, can we now predict the emergence and dissemination of resistance to newly introduced antibiotics?

8 CHALLENGES AND IMPACT OF ANTIBIOTIC RESISTANCE

The emerging problem of antibiotic resistance is a serious threat to global public health. The situation is aggravated by a substantial decline in the research and development of antibacterial agents. Hence, very few new antibacterial classes are brought to market when older classes lose their efficacy.

The main problem we are facing with antibiotic therapy is that after a new antibiotic is introduced, resistance to it will, sooner or later, arise. This scenario has been witnessed on multiple occasions, and thus, there is a continuing race between the discovery and development of new antibiotics and the bacteria that will respond to this selective pressure by the emergence of resistance mechanisms. So, how to protect the power of antibiotics and extend their lifespan? There are many factors contributing to the emergence and dissemination of antibiotic resistance and, as mentioned before, the problems require a complex approach.

Antibiotic resistance is now a global healthcare threat and today's armoury of antibiotics is increasingly limited. For some pathogens, the choice of available drugs is now greatly reduced. Increasing mortality from infections caused by resistant strains, and the strong link between resistant pathogens and increasing levels of hospital-acquired infections, together with escalating healthcare costs, have put antibiotic resistance at the top of the healthcare agenda.

Despite its importance, patho-genomics (genome research on pathogenic microorganisms) is still at an early stage in its development, compared to human genome research. While bacterial genomes and phenotypes are being mapped, much less information is available on the horizontal spread of virulent genes across bacterial populations, an area that has fundamental relevance to keeping pace with the emergence of new resistant strains and the therapeutic strategies that can be used to target them. However, important initiatives are moving forward, notably the ERA-NET Patho-genomics programme and the development of the LLNL database.

According to the US Centres for Disease Control and Prevention, 1.7 million patients per year in the US acquire an infection while in hospital, resulting in 99,000 (5.8%) deaths. In 1992, deaths from hospital-acquired infections in the US were 13,300, showing a 670% increase over a decade, equivalent to around a 20% annual growth during that time. The CDC also reported that 70% of bacteria responsible for hospital-acquired infections are resistant to at least one of the antibiotics that were once used to treat them.

In a report from the US state of Pennsylvania, state-wide hospitals reported 19,154 cases in which patients acquired an infection while staying in hospital, a rate of 12.2 cases per 1000 patients. The hospital costs associated with this patient group averaged $185,260 per patient, giving overall hospital costs of around $3.5 billion. By comparison, hospital costs for patients who did not acquire an infection were $32,389, around 17% of that seen in the case of patients who acquired an infection.

In the US, it is reported that between 50–60% of all hospital-acquired infections are caused by antibiotic resistant bacteria. While little has been published on the relationship between antibiotic resistance and subsequent infection rates and prevalence, increases in both are an inevitable consequence of longer treatment times (durations of infection), due to antibiotic resistance. Also, the more time that is required to eradicate a pathogen from a patient, the higher the probability the pathogen will spread, particularly within the hospital environment, where patients and healthcare workers are in close proximity.

Based on these figures, estimates of the cost of hospital-acquired infections, as a proportion of the general population, suggest a figure of $28.0 million/million of general population in the UK and $22.3 million/ million of population in the US. The average of these two figures is approximately $25 million/million of general population. Based on this average figure, hospital-acquired infections in the USA, Canada, Australia, Japan and Europe (population = 1.3 billion) is approximately $32.5 billion.

9 CONCLUSION

On the global evolutionary and human history scales, the antibiotics era went through many ups and downs, providing us valuable lessons on many aspects of how the microbial world around us functions. While antibiotics are effective for most infections, increasing resistance in some pathogens threatens to undermine the few remaining drugs that are still effective against them. The rate of growth of antibiotic resistant bacteria over the last two decades, coupled with an evident fall in the numbers of new antibiotics being approved during this time, has long raised concerns over the future treatment of infections. Whilst antibiotics that offer different bactericidal mechanisms have been developed over the last 60 years, the increasing difficulty of treating infections today is defined by the resistance attributes of bacterial pathogens and it is here that discovery and innovation are now focusing. This strategy is generating new molecular classes to which pathogens have shown little resistance, alongside others (such as lactamases inhibitors), that directly target resistance mechanisms. We need to learn to be more precise in targeting the pathogens and limit the indiscriminate use of antimicrobials and other practices that accelerate the emergence of novel resistance mechanisms. Investigation of the microbial world around us for potential mechanisms of antibiotic resistance and dissemination may help to design the early warning and preventive measures to sustain the efficacy of antimicrobials and chemotherapy.

References

Dorlands Medical Dictionary: antibiotic. Archived from the original on 2010-11-17. Retrieved 2010-10-29.

SA Waksman (1947). What is an antibiotic or an antibiotic substance? *Mycologia* 39 (5): 565–569.

von Nussbaum F. et al. (2006). Medicinal chemistry of antibacterial natural products – exodus or revival?. *Angew. Chem. Int. Ed.* 45 (31): 5072–5129.

Lindblad WJ (2008). Considerations for determining if a natural product is an effective wound-healing agent. *International Journal of Lower Extremity Wounds* 7 (2): 75–81

Forrest R.D. (March 1982). Early history of wound treatment. *J R Soc Med* 75 (3): 198–205.

M. Wainwright (1989). Moulds in ancient and more recent medicine. *Mycologist* 3 (1): 21–23.

Kingston W (2008). Irish contributions to the origins of antibiotics. *Irish Journal of Medical Science* 177 (2): 87–92.

Harley JP, Klein DA and Prescott LM (1993). Microbiology. Wm. C. Brown Publishers, Dubuque, Iowa.

Hogberg LD, Heddini A and Cars O (2010). The global need for effective antibiotics: challenges and recent advances. *Trends Pharmacol Sci;* 31(11): 509–15.

Antibiotic Resistance (2012). The antibiotic development pipeline and strategies to combat antibiotic resistance.

Abraham EP and Chain E (1940). An enzyme from bacteria able to destroy penicillin. *Nature* 146: 837.

Armelagos GJ (1969). Disease in ancient Nubia. *Science* 163: 225–258.

Bassett EJ, Keith MS, Armelagos GJ, Martin DL and Villanueva AR (1980). Tetracycline-labeled human bone from ancient Sudanese Nubia (A.D. 350). *Science* 209: 1s532–1534.

Cha JY, Ishiwata A and Mobashery S (2004). A novel β-lactamase activity from a penicillin-binding protein of Treponema pallidum and why syphilis is still treatable with penicillin. *J. Biol. Chem.* 279: 14917–14921.

Cook M, Molto E and Anderson C (1989). Fluorochrome labelling in Roman period skeletons from Dakhleh Oasis, Egypt. *Am. J. Phys. Anthropol.* 80, 137–143.

Kumarasamy KK, Toleman MA, Walsh TR, Bagaria J, Butt F, Balakrishnan R, Chaudhary U, Doumith M, Giske CG, Irfan S, Krishnan P, Kumar AV, Maharjan S, Mushtaq S, Noorie T, Paterson DL, Pearson A, Perry C, Pike R, Rao B, Ray U, Sarma JB, Sharma M, Sheridan E, Thirunarayan MA, Turton J, Upadhyay S, Warner M, Welfare W, Livermore DM and Woodford N (2010). Emergence of a new antibiotic resistance mechanism in India, Pakistan, and the UK: a molecular, biological, and epidemiological study. *Lancet Infect. Dis.* 10: 597–602.

Nelson ML, Dinardo A, Hochberg J and Armelagos GJ (2010). Brief communication: mass spectroscopic characterization of tetracycline in the skeletal remains of an ancient population from Sudanese Nubia 350–550 CE. *Am. J. Phys. Anthropol.* 143: 151–154.

Rollo IM, Williamson J and Plackett RL (1952). Acquired resistance to penicillin and to neoarsphenamine in *Spirochaeta recurrentis*. Br. *J. Pharmacol. Chemother.* 7: 33–41.

15

Algae and Wastewater Remediation: Current and Future Scenario

Priyanku Teotia[1] and Vivek Kumar[2*]

[1]Department of Biotechnology, Division of Biosciences, College of Applied Education & Health Sciences, Roorkee Road, Meerut-250003, India

[2]Amity Institute of Microbial Technology, Amity University Noida-201303, India

**Email: vkumar51@amity.edu; vivekbps@gmail.com*

1 INTRODUCTION

Increasing environmental deterioration and the need for energy and food are forces driving man to explore more feasibility of biological wastewater treatment coupled to resource recovery. Large amounts of water used for domestic, agricultural, municipal and industrial purposes result in the generation of large volumes of wastewater. Municipalities and industries the world over are exploring bioremediation as an important route by which to treat wastewater. Bioremediation uses naturally occurring microorganisms (microalgae) to treat wastewater of its nutrients. This method provides an economical and environmentally sustainable and effective treatment method. Algae are an important bioremediation agent, and are already being used by many wastewater facilities. Wastewater treatment eliminates easily settled materials (primary treatment) and oxidizes the organic material present in wastewater (secondary treatment). As the resulted wastewater is apparently clean, it is discharged into natural water bodies. However, this effluent contains large amounts of inorganic nitrogen and phosphorus which cause eutrophication of the water. Also, the presence of heavy metals and organic compounds can cause long-term problems. Wastewater treatment system with algae has been particularly attractive, because of their photosynthetic capabilities, transforming sunlight into useful biomass. Under appropriate conditions, the waste-borne algae develop biomass equivalent or superior to higher plants. Therefore, one of the possible solutions for wastewater treatment could be algal cultivation. Many species of microalgae are able to effectively grow in wastewater because of their ability to utilize organic carbon and inorganic nitrogen and phosphorus. The growth of microalgae results in the production of large amounts of biomass which can be used for the production of different products as biodiesel, biomethane and bioethanol.

Dual use of microalgae for wastewater treatment coupled with biofuel generation is an attractive option for reducing energy, fertilizer and freshwater costs, as well as reducing green-house gas emissions. The aim of this paper is to review the present status of algal wastewater treatment systems with regard to achievements made in harvesting algal biomass and its potential uses and the possibility of cultivation of algae in the wastewater for bioremediation including CO_2 mitigation and uptake of pollutants and production of beneficial components of high value energy products such as biogas, bioethanol, hydrogen with respect to present and future status.

2 PREREQUISITES FOR ALGAL WASTEWATER TREATMENT

In wastewater treatment system, to be functional, an alga must be provided with sufficient incident light and an optimum temperature, which thereby means that depending upon the latitude, algae should be productive and functionally active round the year or only during a part of the year. Depending on the latitude, or light or temperature or both, may be the major limiting factor in treating wastewater. Another condition for algal water treatment is the need for space to capture necessary sunlight energy to enable the algae to be functional. Therefore, algal wastewater treatment system will not be suitable in crowded urban areas, where space is limited and economically costly. Tropical and subtropical countries with ample sunshine and high temperature will be better in comparison to countries with low temperature and at high altitudes. Another precondition is that the wastewater to be treated should not be toxic to the algae and sufficient quantities of nutrients should be present. To allow optimum growth of algae in wastewater treatment plant, retention time of wastewaters should not be too short or too long. To maintain the aerobic conditions, too much organic matter should not be present in the wastewater. Finally, the population dynamics and proliferation of algae involved in wastewater remediation will be influenced by, pH, temperature, turbulence, aeration, sunlight, space, organic matter, retention time, bacteria-algae interactions, salinity, predation, recirculation seeding and toxicants present.

3 COMPOSITION OF WASTEWATER

Water courses receive pollution from many different sources, which vary both in strength and volume. The composition of wastewater is a representation of the lifestyles and methods opted in the generating culture (Gray 1989). It is a composite blend of usual organic and inorganic materials as well as man-made compounds. Nearly seventy per cent of organic carbon in sewage is present as fats, proteins, carbohydrates, amino acids, and volatile acids. This water has high chemical oxygen demand (COD), biological oxygen demand (DOD) and low dissolved oxygen (DO), which represents a poor quality of water for consumption. The inorganic constituents include large amount of elements like Na, K, HCO_3, P, Ca, Cl, S, Mg, NH_4 salts and heavy metals (Lim et al., 2010). There are many sources of toxic pollutants such as unwanted chemicals and pesticides, which include domestic and industrial discharge from big cities and small towns, leached and runoff water from cultivable lands. Need of water for consumption, irrigation and industrial use force us to treat and recycle the wastewater.

4 WASTEWATER MICROBIOLOGICAL COMPOSITION

Water quality has inspired development of tests designed to measure its suitability for drinking, bathing, and release back to the environment. Water that looks clear and pure may be contaminated with pathogenic microorganisms. Even water that appears "pure" must be tested to ensure that it contains no microorganisms that might cause disease. The environment of wastewater is highly suitable for many pathogenic and non-pathogenic microorganisms, such as virus, bacteria, algae, fungi and protozoa. The pathogenic microbes are responsible for many water-borne diseases and non-pathogenic microbes such as algae are used for treatment. The common pathogens of wastewater are examples of viruses commonly found in wastewater are Hepatitis and Norwalk virus and a common fungus is *Candida*. Common bacteria, such as *Salmonella* can cause food poisoning whereas *Vibrio cholerae* is the pathogen that causes cholera. Parasites such as *Cryptosporidium* and *Schistomsoma* can cause diarrhea, and just about all wastewater pathogens can lead to serious gastrointestinal illness in addition to other, more specific implications, which is a significant cause of deaths worldwide. Another very common bacterium is *E. coli*, which is usually harmless but its pathogenic strains may cause serious health problems (Mishra et al., 2012).

5 WASTEWATER TREATMENT

The role of microalgae in the treatment of wastewater is in the release of free oxygen thus promoting aerobic degradation processes by other microorganisms. Also, microalgae have a major role in the accumulation and conversion of wastewater nutrients to biomass and lipids (Rawat et al., 2011). These processes, called phytoremediation, have several applications: removal of nutrients from municipal wastewater and effluents rich in organic matter; removal of nutrient and xenobiotic compounds with the aid of algae based biosorbents; treatment of acidic and metal wastewaters; CO_2 sequestration; transformation and degradation of xenobiotics; and detection of toxic compounds with the aid of algae-based biosensors (Levoie and Noue 1985). Microalgae can assimilate nitrogen in the form of nitrate, nitrite and ammonium, while the preferable forms of phosphorus are H_2PO_4 and HPO_4 (Pittman et al., 2011; Rawat et al., 2011). Nitrogen in sewage effluent comes primarily from metabolic interconversions of extra derived compounds, whereas 50% or more of phosphorus comes from synthetic detergents. The forms in which they occur in wastewater are ammonia, nitrite, nitrate and orthophosphate (Cai et al., 2013). Typical concentration of total nitrogen and phosphorus in different types of wastewater are in the ranges of 11–2700 mgl^{-1} and 4–324 mgl^{-1}, respectively (Christenson and Sims 2011). These amounts can support high productivities during the cultivation of most fresh water microalgae strains (McGinn et al., 2011).

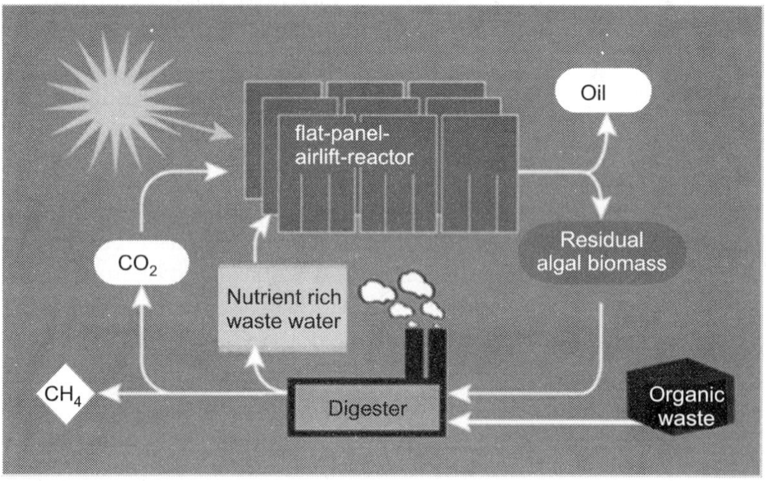

Fig. 1 Recycling of wastewater by algae

The removal of the nutrients depends on the species used for remediation and the properties of the wastewater itself (Cai et al., 2013). The presence of toxic metals as cadmium or mercury, or organic chemicals is critical factor of algal growth in wastewater. This is particularly an issue in industrially derived wastewaters. Biotic factors that may have adverse impact on algal growth include pathogenic bacteria or predatory zooplankton. In addition, other microorganisms in the wastewater might out compete with the microalgae for essential nutrients (Lau et al., 1995). Selection of the appropriate algal strain for wastewater treatment depends on various factors like the characteristics of the wastewater, the original habitat of the strain and the climatic conditions in the treatment plant (Olguin 2012). One of the major practical limitations in algal treatment systems is harvesting or separation of algal biomass from the treated water discharge. An efficient removal of algal biomass is essential for wastewater recycling. Harvesting of algal biomass can be performed by centrifugation, filtration or sedimentation. These processes are often followed by flocculation. Separation of the biomass is a problem because of the small size of algal cells and relatively high dilutions (Molina Grima et al., 200).

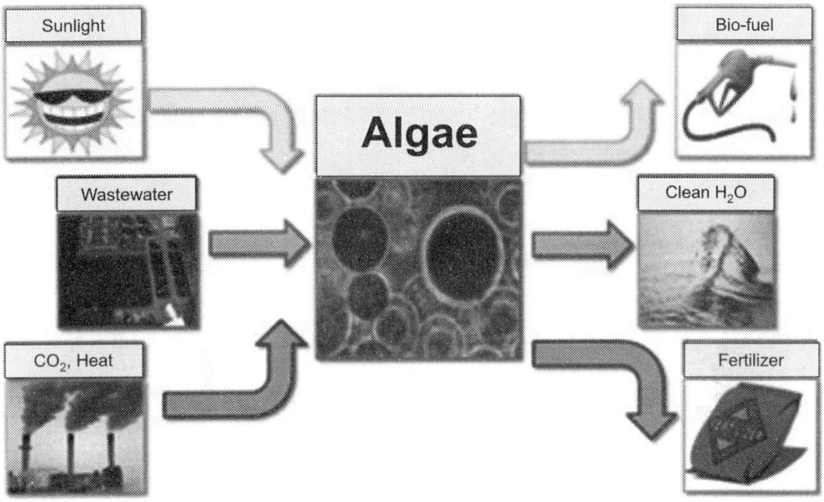

Fig. 2 Application of algae in different fields

6 BIODIESEL PRODUCTION

Simple unicellular structure of microalgae allows high rate of photosynthesis, effective consumption of carbon and rapid accumulation of oil. Microalgae can accumulate oil up to 77% dw. They have the possibility to produce 100 times more biodiesel per unit of cultivation area compared to oil crops with much lower costs (Mata et al., 2010). Restrictions of the cultivation of microalgae refer mostly to the intensity of the light and temperature (Pan et al., 2011). During the initial stages of cultivation large amounts of polar lipids and polyunsaturated fatty acids are being generated, but in the stationary stage lipids consist of neutral saturated long chain fatty acids (Rawat et al., 2013). Oil extraction is usually performed by the use of organic solvents suitable for dry biomass or with supercritical fluids which are suitable for the use on wet algal biomass (Lam and Lee 2012). Ideal method for the extraction of oil has to be specific for the extraction of lipids, so that the concentration of non-lipid component would be minimal. Also, a method has to be selective for some fractions of lipids, for example neutral lipids (Macias-Sanchez et al., 2007). Mixture of solvents chloroform/methanol is often used in the ratios 1/2 v/v, 2/1 v/v 17-19 or 1/1 v/v (Rodolfi et al., 2008). Water represents tertiary component of the mixture and allows complete extraction of neutral and polar lipids (Medina et al., 1998). Less toxic alternative with similar properties is the mixture of hexane/iso-propanol (Halim et al., 2012).

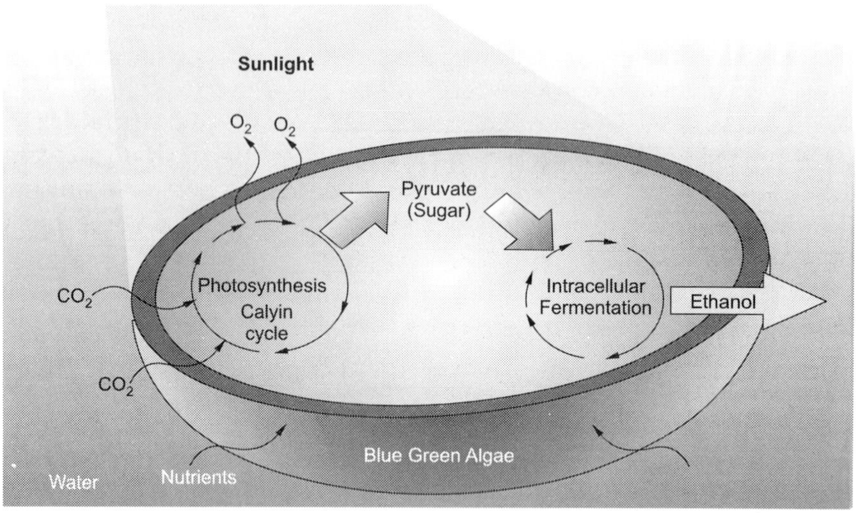

Fig. 3 An algal cell at work

The most effective method of improving microalgae lipid accumulation is nitrogen limitation, which not only results in the accumulation of lipids, but also results in a gradual change of lipid composition from free fatty acids to triacylglycerol (TAG) (Widjaja et al., 2009). TAGs are more useful for conversion to biodiesel (Meng et al., 2009). Lipid accumulation in microalgae occurs when a nutrient (typically nitrogen, but can be silicate for diatoms) is exhausted from the medium or becomes the growth limiting factor. Cell proliferation is prevented but carbon is still assimilated by the cell and converted to TAG lipids that are stored within existing cells thereby increasing the concentration (Meng et al., 2009). Wu and Hsieh (2008) investigated the effects of salinity, nitrogen concentration and light intensity on lipid productivity, and recorded up to 76% increase in production of lipids for specific growth conditions when compared to more typical growth processes.

Another possibility is the use of supercritical fluids which improve the mass transfer during the process. Compared to conventional solvent extraction, supercritical fluids give higher yield and higher purity of the product (Tabernero et al., 2012). The production of biodiesel can be performed by direct transesterification of algal biomass or by transesterification of oil previously extracted from biomass (Mulumba and Farag 2012). Homogenous alkaline catalysts usually used in the transesterification of crops oil cannot be applied to the transesterification of microalgal oil because of the high amounts of free fatty acids (FFA) (Leung et al., 2010). The use of acid catalysts is more acceptable since they are not sensitive to the presence of FFA and at the same time they catalyze the esterification of FFA and transesterification of triacilglycerols. Another possibility is the use of two-stage process with acid pretreatment and alkaline catalyzed reaction (Lam and Lee 2012). The advantages of applying enzymatic transesterification process are high selectivity, lower energy consumption and smaller production of by-products and wastes (Shah et al., 2003). The enzymatic transesterification process is influenced by the optimum pH range of the enzyme itself, the substrates concentration, and the spacing between the enzyme molecules and the substrate (Suali and Sarbatly 2012).

7 BIOETHANOL PRODUCTION

Ethanol production from microalgae is obtained by the fermentation of starch, sugar and cellulose contained in microalgal biomass (Kim et al., 2011). *Enteromorpha* species has been employed as the starting material for the bioethanol production. This algae is very rich in carbohydrate source ranging from 70–72%.The percentage of fermentable sugar was obtained by such hydrolysis ranges from 0.75 mg gm^{-1} to 0.940 mg gm^{-1} in dry algae and 0.21 to 0.52 mg gm^{-1} in case of fresh algae as partial hydrolysis of different polysaccharides. The content of carbohydrates in microalgae is up to 72% (Nahak et al. 2011). Since the high concentration of carbohydrates is entrapped within the cell wall of alga (John et al., 2011). This can be accomplished by ultrasound or by hydrolytic enzymatic conversion (Suali and Sarbatly 2012). Fermentation of the microalgal biomass is catalyzed by microbes such as bacteria, yeast and fungi and the main by-products are CO_2 and water (Harun et al., 2010). Bioethanol is usually produced by the action of yeasts *Saccharomyces cerevisiae, S. uvarum* and *S. bayanus* (John et al., 2011). The spent biomass after fermentation process can be used for anaerobic digestion and the production of biomethane. Simultaneous biodiesel and bioethanol production from microalgae is also possible. Microalgal oil can be extracted by supercritical CO_2 prior to fermentation process and be used for biodiesel production. Supercritical CO_2 breaks down microalgae cell wall causing the release of oil and carbohydrates and increasing the concentration of bioethanol (Harun et al., 2010).

8 BIOGAS PRODUCTION

Wastewater grown algae are a potentially important organism for biogas or methane production. Biogas is a mixture of methane (55–75%) and CO_2 (25–45%) produced during anaerobic digestion by anaerobic microorganisms. Most of the wastewater treatment ponds or plants do not employ algae harvesting. Any one that do, generally return the algal biomass to the ponds, in which algal biomass decompose on the bottom of pond, results in producing CH_4 to the atmosphere and pollute the aquatic environment (Chaiprasert 2011). Waste grown algae has vast amount of lipid contents and it can be extracted (Woertz et al., 2009) or anaerobic digestion of algal biomass could be an appropriate utilization at wastewater treatment plants. Residual biomass can further be processed into fertilizer making this process cost effective. Microalgae contain almost no lignin and lower cellulose, so they represent good substrate for anaerobic digestion (Song et al., 2011; Vergara-Fernandez et al., 2008). Biogas production is affected by organic composition of the feedstock, temperature, pH and retention time in reactors (Harun et al., 2010). Methane production was doubled when equal masses of wastewater sludge and *Spirulina* biomass were co-digested (Samson and LeDuy, 1983). The use of waste paper (50%w/w) to aquacultural microalgal slude to adjust the C:N ratio to 20–25:1 which resulted in production of double methane from 0.6L day^{-1} to 1.2L day^{-1} at 35°C with a hydraulic retention time of 10 days (Yen and Brune 2007).

9 ORGANIC CHEMICALS RECOVERY

Microalgae can represent an ideal feedstock for the production of fine chemicals such as biobutanol, acetone, *n*-butanol or propandiol since they have a carbohydrate fraction (Harun et al., 2010). Biobutanol can be used as organic solvent, but also as a renewable transport fuel. It is produced by the fermentation with the bacteria *Clostridium* spp. The yield of the solvents, among others, depends on the used strain and pre-treatment method (Efremenko et al., 2012).

10 ALGAE FOR NUTRIENTS HARVESTING

Wastewater contains large amount of chemicals like nitrogen, phosphorus, sulphur, etc. Processing is the main step to remove these unwanted materials from wastewater. The type of processing, in turn, depends on the characteristics of wastewater. For nutrient removal from the wastewater, single microalgae strain or consortium of microalgae along with microalgae growth-promoting bacteria are used. Several studies were performed to analyze the capability of microalgae consortium along with symbiotic bacteria for nutrient removal capacity from wastewater. One such study was the analysis of euglenophyta, cyanobacterium, green microalgae and two types of indigenous microalgae consortium along with symbiotic bacteria for the nutrient removal capability from diluted piggery wastewater. The unialgal cultures like *Euglena viridis, Chlorella sorokiniana* were able to grow in diluted wastewater, while *Scenedesmus obliquus* and consortium were able to grow in eight times diluted wastewater. In case of phosphorus and nitrogen removal capacity, *E. viridis* and *C. sorokiniana* showed more nitrogen removal in both the dilutions; on the other hand, *C. sorokiniana*, *S. obliquus* and *E. viridis* showed phosphorus (20–65%) removal in eight times diluted wastewater (de-Godos et al., 2010).

One of the largest groups of microalgae is chlorophyta, which is present all over the world. The *Chlorella* sp. has been used in most studies and proved to be potential candidate for harvesting N and P from the different aquatic bodies. The N and P harvesting efficiency ranged from 8–100%. *C. vulgaris* proved to be most efficient in removing nutrients compared to *C. kessleri* in laboratory conditions (Lee and Lee 2001). Some species of *Chlorella* are mixotrophic of heterotrophic and can utilize organic carbon along with inorganic nutrients as part of their metabolic activities. While using wastewater, this could be an advantage for utilizing carbon residues in case of dairy waste (Wang et al., 2010). One more non-motile microalgae populary used from nutrients harvesting is *Scenedesmus* sp., which removes N and P from wastewater up to 30–100% (Gonzalez et al., 1997). *Botryococcus braunii* utilized No_3 and NO_2 but not NH_4, ammonium ion was inhibitory to its cell growth (Sawayama et al., 1992). On the other hand, cyanobacteria are considered the commonest and trouble creating algae, which is responsible for algal blooms.

Many species of cyanobacteria are serious environmental threat, toxic and may cause health concern. But few genera like *Arthrospira* contains high amount of essential amino acids and fatty acids, therefore, could be utilized as nutritional requirement for animals. *Arthrospira* sp. has been reported the removing of 84–96% of NH_4 and 72–87% of P from pig wastewater. Ammonium removed at a rate of 13.6 mgl^{-1} d^{-1} corresponded to 93% removal (Olguin et al., 2003). Various environmental factors affect the nutrients harvesting by microalgae such as the nutrients concentration itself, light, temperature, pH and the inoculum density. The light intensity and nutrients concentration causes a direct effect on the proliferation of algae; therefore, overall concentration and toxicity of nutrients and specially concentration of N and P affect the nutrients removal by algae.

11 TOXIN BIOREMEDIATION

Bioaccumulation of toxins is a separate issue to environmental toxicity, though there are likely to be direct relationships between toxin accumulation and the nature and strength of toxic effects. Bioaccumulation occurs where tissue-based concentrations exceed those available in the environment:, the latter may include algal toxins available through drinking, dietary and or through direct contact routes. Algae comprise the greatest abundance of plant biomass in aquatic environments and are a rational preference for aquatic toxicological studies, yet have been underutilized in this capacity. The lipid content of many algal species provides a point of entry for trophic transfer of lipophilic organic contaminants.

12 HEAVY METAL REMEDIATION

Accumulation of toxic metals, e.g., Hg, Cu, Cd, Cr and Zn in humans has several consequences such as growth and developmental abnormalities, carcinogenesis, neuromuscular control defects, mental retardation, renal malfunction and wide range of other illnesses. Elevated levels of such metal ions are generally toxic and cause major damage to cell (Inouhe et al., 1996). Metals are taken up by algae through adsorption. At first, the metal ions are adsorbed over the cell surface very quickly just in a few seconds or minutes; this process is called physical adsorption. Then, these ions are transported slowly into the cytoplasm in a process called chemisorption.

Several researchers have established that metals such as Ti, Pb, Mg, Zn, Cd, Sr, Co, Hg, Ni and Cu are sequestered in polyphosphate bodies in green algae. These bodies perform two different functions in algae; provide a "storage pool" for metals and act as a "detoxification mechanism". The alga *Scenedesmus obliquus* was also found to accumulate some metals on increasing the amount of phosphorus in the media. It was able to accumulate increased Cd and Zn with higher phosphorus concentrations, whereas Selenium (Se) accumulation was found to be inhibited. Shehata and Badr (1980) cultured *Scenedesmus* in different concentrations of Cu, Cd, Ni, Zn and Pb to evaluate their effects on the growth of algae. The concentration of metal that reduced *Scenedesmus* growth was 0.5 mg l^{-1} for Cu, 0.5 mg l^{-1} for Ni and 2 mg l^{-1} for Cd, 2 mg l^{-1} for Zn. The nickel solution was less toxic than copper for *Scenedesmus* growth. The alga tolerated high lead concentrations up to 30 mg l^{-1}. Cd^{2+} contamination in surface water comes mainly from phosphatic fertilizers used in agricultural operations, which is reflected in municipal water supplies drawing water from river sources.

Toxic heavy metals often vie with essential trace metals for binding to and uptake into cells. Transgenic approaches have been developed to further enhance the heavy metal specificity and binding capacity of microalgae with the objective of using these microalgae for the treatment of heavy metal contaminated wastewaters and sediments. These transgenic strategies have included the overexpression of enzymes whose metabolic products ameliorate the effects of heavy metal-induced stress, and the expression of high-affinity, heavy metal binding proteins on the surface and in the cytoplasm of transgenic cells. The most effective strategies have substantially reduced the toxicity of heavy metals allowing transgenic cells to grow at wild-type rates in the presence of lethal concentrations of heavy metals. In addition, the metal binding capacity of transgenic algae has been increased fivefold relative to wild-type cells. Recently, fluorescent heavy metal biosensors have been developed for expression in transgenic *Chlamydomonas* (Rajamani et al., 2007). Algae possess the ability to take up toxic heavy metals from the environment, resulting in higher concentrations than those in the surrounding water. Bio-accumulation studies reveal the accumulation of the contaminant in the organism via uptake of food or water containing the contaminant. Algae the species of *Chlorella, Anabaena inaequalis, Westiellopsis prolifica, Stigeoclonium tenue, Synechococcus* sp. tolerate heavy metals. However, several species of *Chlorella, Anabaena* and marine algae have been utilized for harvesting of heavy metals. But the operating condition limits the practical application of these organisms (Muhaemin 2004).

13 PESTICIDE BIOREMEDIATION

A wide variety of organochlorine compounds are introduced into aquatic environments via domestic sewage, pesticide runoff from agricultural lands and industrial effluents. Recent studies have indicated that different microalgal species apparently have different sensitivity to pesticides (Solomon 1996). The response of pesticides vary widely depending upon concentrations used, duration of exposure, and algal species tested. Generally, there was a broad range of sensitivity among species and between divisions of algae tested. Biodegradation of pesticides is determined by two groups of factors, the first relates to microbial consortium and the optimum condition for their survival and activity while the second relates to the chemical structure of the pesticides. Factors related to microorganisms including the presence and number of appropriate algae, the contact between employed algae and the substrate (pesticide), pH, temperature, salinity, nutrients, light quality and intensity, available water, oxygen tension and redox potential, surface binding, presence of alternative carbon substrates and alternative electron acceptors. Other factors including chemical structure, molecular weight and functional groups of the applied pesticides, their concentration and toxicity and their solubility in water. Some information on the interactions between pesticides and algae was compiled by (Kobayashi and Rittam 1982), showing that not only were microalgae capable of bioaccumulating pesticides, but they were also capable of bio-transforming some of these environmental pollutants.

14 IMMOBILIZED ALGAE AND WASTEWATER REMEDIATION

The major difficulty in algal bioremediation identified is the harvesting of microalgae after treating the wastewater (Mallick, 2002; Aslan and Kapdan, 2006). Immobilization of microalgae is one of the solutions for harvesting problem (Vılchez et al., 2001; Jimnez-Perez et al., 2004). Carrageenan, chitosan and alginate are generally employed polymers for the microalgae immobilization. Immobilized microalgae beads are easy to harvest and reuse. There are quite a few reasons affecting the nutrient harvesting potential of immobilized microalgae. Number of microalgae immobilized beads inoculated to treat wastewater, gel permeability for the nutrients present in the wastewater, cell density present in each bead, environmental factors, chemical components present in wastewater, type of gel matrix used for immobilization.

Some studies were done to identify the key factor affecting the immobilized microbe's nutrient removal efficiency. In one study, the effect of two key factors such as gel thickness and cell concentration in the gel were analyzed for the nutrient removal efficiency. Effect of immobilization on the growth and nutrient removal of *Scenedesmus obliquus* and *Chlorella vulgaris* in artificial and urban wastewater was analyzed. Both the cultures were immobilized in alginate beads. Immobilized cultures showed immediate growth when inoculated into the medium than the free cell, which showed lag period. *C. vulgaris* had higher cell densities within the beads than *S. obliquus*. In case of nitrogen uptake, immobilized microalgae had higher nitrogen uptake than the free cell in both the types of wastewater but in phosphorus removal efficiency, immobilized cultures removed more phosphorus in artificial wastewater than in urban wastewater (Ruiz-Marin et al., 2010).

15 RESTRAIN FACTORS IN ALGAL REMEDIATION

15.1 Abiotic Factor

Many substances can act inhibitorily on photosynthesis and algal growth if present in too high concentrations. Examples of such substances are heavy metals, herbicides, pesticides, substances in detergents, household cleaning products and personal care products. High concentrations of ammonia act inhibitory on algal growth at high pH, and this toxicity is intensified at higher temperatures when a higher proportion of the ammonia occurs as free ammonia which may freely diffuse over membranes into the cells (Borowitzka, 1998; Ogbonna et al., 2000). Total ammonium concentration should not exceed 20 mg $NH4^+$-N. Too high levels of organic compounds can inhibit the nutrient uptake by microalgae (Ogbonna et al., 2000), and acetate can be toxic to some species, due to the unionized molecule, which can penetrate the cell membrane and damage the cell interior by ionization (Wood et al., 1999). Some algae also produce substances toxic to themselves in the course of their metabolism. These eventually accumulate to concentrations high enough to inhibit growth; a phenomenon called auto-inhibition (Fogg 1975).

15.2 Biotic Factors

Not only physical and chemical factors affect algal growth, in nature, species have to compete with each other for space and nutrients, and this can be reflected in algal cultures as well. Some species inhibit the growth of others in mixed culture; e.g., some cyanobacteria can produce inhibitory substances to the growth of eukaryotic algae, and some eukaryotic algae can produce antibacterial substances (Borowitzka 1998; Fogg 1975). The later may indirectly affect competition among algal species by their effects on associated bacteria. In pure monocultures, infections by parasites, predators or competing species can be deleterious, and protozoa and rotifers present the greatest threat (Grobbelaar, 1982; Mostert and Grobbelaar 1987). In open wastewater treatment systems, infections are hard to avoid, but by keeping optimal conditions for the algae the cultures are less susceptible. Methods to avoid infections can be acidification of the cultures to pH 2 for a short period or by daily removal of particulate matter larger than 100 μ with a small porosity screen, which removes mainly zooplankton but not the algae (Grobbelaar, 1982). The acidification is adequate to kill most rotifers and protozoa; however, this is difficult in large ponds. Establishing a pond regime that leads to diurnal anaerobic conditions for a short period can also prevent the development of animal and fungal populations (Borowitzka 1998), and short periods of high ammonia concentration can eliminate contamination by zooplankton (Becker 1994). Alternatively, biocides can be applied, but this can spoil the quality of the product, is expensive and is not environmentally sound (Soeder 1981).

16 CONCLUSION

Domestic and industrial wastewater is very nutritionally rich, and basically all nutrients needed for algal growth are present. The factors limiting algal growth, and hence treatment efficiency is therefore more likely to be light and carbon. Light is the most important parameter to optimize, and hence, culture depth and turbulence are vital for good performance. To avoid temperature limitation in northern climate, greenhouses would be recommended in order to have functioning treatment during longer periods than just during summer. To increase the performance during winter, artificial light may also be needed, however that demand extra costs for energy. The easiest way to start a microalgal wastewater treatment process is to inoculate with water containing a large variety of algae, e.g., water from outdoor ponds. This will create a mixture of algae and other organisms, where the best suited species will grow fastest and dominate the treatment step. Other microalgae will also be introduced eventually, partly from the wastewater itself, partly from algal particles in the air dust. This approach requires less supervision and operation than if a particular alga is chosen to be cultivated for any purpose. A drawback is that the fastest growing microalgae are most often unicellular green algae (Chlorophyceae) which are difficult to harvest. Harvesting of algal biomass accounts for the highest proportion of energy input during production, but currently, there are no standard harvesting techniques.

Adaptation of technologies already in use in the food, biopharmaceutical and wastewater treatment sector may provide possible solutions. Lipids are the most readily extractible biofuel feedstock from algae, but potential storage is hindered by the presence of polyunsaturated fatty acids (PUFAs) causing oxidation reactions and high moisture content of algal feedstock. Therefore, large-scale production of microalgae for biofuels will increase the availability of these products. Overall, with the current demand for renewable fuels, especially for use in the transportation sector, there is a need to develop a range of sustainable biofuels resources as the combined mix will be a significant step towards the replacement of fossil fuels. Continued development of technologies to optimize the microalgae production, oil extraction and biomass processing for biogas and other chemicals recovery has the capacity to make significant contributions towards this goal. Therefore, application of algae in wastewater treatment is cost effective, low energy requirement, reductions in sludge formation, GHG emission reduction and production of useful algae biomass.

17 FUTURE FACETS

Algae are tiny biological factories that use photosynthesis to transform carbon dioxide and sunlight into energy so efficiently that they can double their weight several times in a day. As part of the photosynthesis process algae produce oil and can generate 15 times more oil per acre than other plants used for biofuels, such as corn and switchgrass. Algae can grow in salt water, freshwater or even contaminated water, at sea or in ponds, and on land not suitable for food production.

Algae can provide a number of different products or services. The product market is segmented into the large commodity market like the polluted environment management, energy generation, and niche higher value markets that can be easily saturated. Products for a large market with low revenue on individual unit sales include biofuels (biodiesel, green diesel or renewable diesel, ethanol, jet fuel, etc.), and high protein biomass (animal feed). Products for the higher value markets include nutraceuticals (beta-carotene, omega oils, etc.), pharmaceuticals, pigments, or bioproducts. Services fall into current services such as wastewater remediation and future services such as carbon dioxide recycling, mitigation, and sequestration. Mitigation and sequestration all depend on the final use of the biomass. It is possible and likely to use a combination of algal products and services to generate revenue in near future.

Nutrient pollution (eutrophication) can lead to undesirable changes in ecosystem structure and function. The impact of algal aquaculture could be positive or negative. Negative impacts could occur if residual nutrients in spent culture medium are allowed to leach into local aquatic systems. On the other hand, positive impacts could occur if algae production were to be integrated into the treatment of water bodies already suffering from excess nutrient supply. For example, remediation of polluted water bodies suffering from algal blooms may also provide locally significant amounts of free waste biomass, and this could be used for biofuel production on a small scale. In the search for algae that can deliver high biomass productivity and lipid content simultaneously, genetic modification is one possible option. Applications of molecular genetics range from speeding up the screening and selection of desirable strains, to cultivating modified algae on a large scale. Traits that might be desirable include herbicide and heavy metal resistance to prevent contamination of cultures by wild-type organisms and increased tolerance to high light levels. Containment of genetically modified algae poses a major challenge. In open pond systems, culture leakage and transfer (e.g., by waterfowl) is unavoidable. Closed bioreactors appear more secure but as far as containment is concerned, photo bioreactors are only cosmetically different from open ponds and some culture leakage is inevitable.

Large-scale algae cultivation systems are required for the production of sufficient amount of biomass, which can be converted into biodiesel for the replacement of fossil fuels. Ideally, this would be met by growing algae in areas with lots of sunshine and surface that are not used for agriculture i.e., wastelands and deserts. The availability and supply of wastewater having nutrients in the form of phosphate and nitrogen, and delivery of CO_2 for these kinds of remote areas will not be a problematic issues, if handled by technically feasible methods in case of large-scale production. The ultimate aim should be to design a algal wastewater cultivation system that utilizes CO_2 from the air and without any extra input to make it cost effective and performance efficient.

Microalgae culture can have a diverse range of environmental impacts, many of which are location specific. Depending on how the system is configured the balance of impacts may be positive or negative. Impacts such as the use of genetic engineering are uncertain, but may affect what systems are viable in particular legislatures. Possibly the most important environmental aspect of micro-algae culture that needs to be considered is water management: both the water consumed by the process, and the emissions to water courses from the process. In any algal cultivation scheme, it should be anticipated that environmental monitoring will play an important role and will be an ongoing requirement. Finally, the application of genetic engineering in potential algal strains to improve energy production phenotypes seems to be in its infancy, significant advances in the development of genetic manipulation tools have recently been achieved with microalgal model systems and are being used to manipulate central carbon metabolism in these organisms. It is likely that many of these advances can be extended to industrially relevant organisms in near future. Another important aspect of algal bioremediation could be selection and modification of potential strains to resist change in temperature, pH, salinity and toxicity and to work optimally in any location and altitude for a cleaner and sustainable environment.

References

Abou-Shanab RAI, Matter IA, Kim SN, Oh YK, Choi J and Jeon BH (2011). Characterization and identification of lipid-producing microalgae species isolated from a freshwater lake. *Biomass and Bioenergy* 35(7): 3079–3085.

Aslan S and Kapdan IK (2006). Batch kinetics of nitrogen and phosphorus removal from synthetic wastewater by algae. *Ecol. Eng.* 28: 64–70.

Becker EW (1994). Microalgae, Biotechnology and Microbiology. Cambridge: Cambridge University Press.

Borowitzka MA (1998). Limits to growth, in Wastewater treatment with algae, Y.-S. Wong and N.F.Y. *Tam, Editors. Springer Verlag.* 203–226.

Bravo-Linares CM, Mmudge SM and Loyola-Sepulveda RH (2010). Production of volatile organic compounds (VOCS) by temperate macroalgae. The use of solid phase microextraction (SPME) coupled to GC-MS as method of analysis. *J. Chile Chem. Soc.* 55(2): 272–232.

Cai T, Park SY and Li Y (2013). Nutrient recovery from wastewater streams by 368 microalgae: status and prospects. *Renew. Sust. Energ. Rev.* 19: 360–369.

Christenson L and Sims R (2011). Production and harvesting of microalgae for wastewater treatment, biofuels, and bioproducts. *Biotechnol. Adv.* 29: 686–702.

de-Godos I., Vargas VA, Blanco S, Garcia-Gonzalez MC, Soto R, Garcia-Encina PA, Becares E and Muioz R. 2010. A comparative evaluation of microalgae for the degradation of piggery wastewater under photosynthetic oxygenation. *Bioresour. Technol.* 101: 5150–5158.

Efremenko EN, Nikolskaya AB, Lyagin IV, Senko OV, Makhlis TA, Stepanov NA, Maslova OV, Mamedova F and Varfolomeev SD (2012). Production of biofuels from pretreated microalgae biomass by anaerobic fermentation with immobilized *Clostridium acetobutylicum* cells. *Bioresour Technol.* 114: 342–348. doi: 10.1016/j.biortech.2012.03.049.

Fogg GE (1975). Algal cultures and phytoplankton ecology. Second edition 1975 ed. Wisconsin: The University of Wisconsin Press.

González LE, Cañizaresb RO and Baena S (1997). Efficiency of ammonia and phosphorus removal from a Colombian agro industrial wastewater by the microalgae *Chlorella vulgaris* and *Scenedesmus dimorphus*. *Biores. Technol.* 60: 259–262.

Sawayama S, Minowa T, Dote Y and Yokoyama S (1992). Growth of hydrocarbon rich microalgae *Botryococcus braunii* secondarily treated sewage. *Appl. Environ. Microbiol.* 29: 145–148.

Gray NF (1989). Biology of wastewater treatment. Oxford University Press, Oxford.

Grobbelaar JU (1982). Potential of algal production. *Water SA* 8(2): 79–85.

Harun R, Danquah MK and Forde GM (2009), Microalgal biomass as a fermentation feedstock for bioethanol production. *J. Chem. Biotechnol.* 85 (2): 199–203.

Harun, R, Singh M, Forde GM and Danquah MK (2010). Bioprocess engineering of microalgae to produce a variety of consumer products *Renewable Sustainable Energy Rev.* 14: 1037–1047.

Jimenez-Perez, M.V., Sanchez-Castillo P, Romerab O, Fernandez-Moreno D and Perez-Martınez C (2004). Growth and nutrient removal in free and immobilized planktonic green algae isolated from pig manure. Enzyme Microb. *Technol.* 34: 392–398.

John RP, Anisha GS, Nampoothiri KM and Pandey A (2011). Micro and macroalgal biomass: a renewable source for bioethanol. *Bioresource Technol.* 102(1): 186-93. doi: 10.1016/j.biortech.2010.06.139.

Kaur S, Sarkar M, Srivastava RB, Gogoi HK and Kalita MC (2012). Fatty acid profiling and molecular characterization of some freshwater microalgae from India with potential for biodiesel production. *New Biotechnol.* 29(3): 332-344. doi: 10.1016/j.nbt.2011.10.009

Kim NJ, Li H, Jung K, Chang HN and Lee PC (2011). Ethanol production from marine algal hydrolysates using *Escherichia coli*. *Bioresource Technol.* 102: 7466-7469.

Kobayashi H and Rittman BE (1982). Microbial removal of hazardous organic compounds. *Environ. Sci. Technol.* 16: 170–183.

Lam MK and Lee KT (2012). Microalgae biofuels: A critical review of issues, problems and the way forward. *Biotechnol Adv.* 30(3): 673–90. doi: 10.1016/j.biotechadv.2011.11.008.

Lau PS, Tam, NFY and Wong YS (1995). Effect of Algal Density on Nutrient Removal from Primary Settled Wastewater. *Environ. Pollu.* 89: 59–66.

Lavoie A and De la Noüe J (1985). Hyperconcentrated cultures of *Scenedesmus obliquus*. A new approach for wastewater biological tertiary treatment?. *Water Res.*, 19, 1437–1442.

Lee K and Lee CG (2001). Effect of light/dark cycles on wastewater treatments by micro algae. *Biotechnology and Bioprocess Engneering.* 6: 194–199.

Leung DYC, Wu X and Leung MKH (2010). A review on biodiesel production using catalyzed transesterification. *Applied Energy*, 2010, 87: 1083-1095. DOI: http://dx.doi.org/10.1016/j.apenergy.2009.10.006

Macías-Sánchez MD, Mantell C, Rodríguez M, Martínez de la Ossa E, Lubián LM and Montero O (2007). Supercritical Fluid Extraction of Carotenoids and Chlorophyll a from *Synechococcus* sp., *Journal of Supercritical Fluid*, 39: 323–329. doi:10.1016/j.supflu.2006.03.008.

Mallick N (2002). Biotechnological potential of immobilized algae for wastewater N, P and metal removal: A review. *BioMetals.* 15: 377–390.

Mata TM, Martins AA and Caetano NS (2010). Microalgae for biodiesel production and other applications: A review. *Renewable Sustain Energy Rev.* 14-2017-232.

McGinn PJ, Dickinson KE, Bhatti S, Frigon JC, Guiot SR and O'Leary SJ (2011). Integration of microalgae cultivation with industrial waste remediation for biofuel and bioenergy production: opportunities and limitations. *Photosynth. Res.* 109(1-3):231-247. doi: 10.1007/s11120-011-9638-0.

Meng X, Yang J, Xu X, Zhang L, Nie Q and Xian M (2009). Biodiesel production from oleaginous microorganisms. *Renewable Energy.* 34 (1): 1–5.

Mishra S, Bisht S, Malik R, Singh J, Teotia P, Sharma S, Kela R and Kumar V (2012). Occurrence and distribution of microbiological and physico-chemical indicators in ground water contaminated by drainages, North India. *Env. Asia.* 6(1): 27–35.

Molina Grima E, Belarbi EH, Acién Fernández FG, Robles Medina A and Chisti Y (2003). Recovery of microalgal biomass and metabolites: process options and economics. *Biotechnol. Adv.* 20: 491–515.

Mostert ES and Grobbelaar JU (1987). The influence of nitrogen and phosphorus on algal growth and quality in outdoor mass algal cultures. *Biomass* 13: 219–233.

Muhaemin M (2004). Toxicity and bioaccumulation of lead in *Chlorella* and *Dunaliella*. *J. Coastal Develop.* 8 (1): 27–33.

Mulumba N and Farag IH (2012). Tubular photobioreactor for microalgae biodiesel production. *Int. J. Eng. Sci. Technol.* 4 (2): 703–709.

Nahak S, Nahak G, Pradhan I and Sahu RK (2012). Bioethanol from marine algae: A solution to global warming problem. *J. Appl. Environ. Biol.* Sci. 1(4): 74-80.

Ogbonna JC, Yoshizawa H and Tanaka H (2000). Treatment of high strength organic wastewater by a mixed culture of photosynthetic microorganisms. *J. Appl. Phycol.* 12: 277–284.

Olguín EJ, Galicia S, Mercado G and Perez T (2003). Annual productivity of *Spirulina* (*Arthrospira*) and nutrient removal in a pig wastewater recycle process under tropical conditions. *J. Appl. Phycol.* 15: 249–257.

Olguín EJ (2012). Dual purpose microalgae-bacteria-based systems that treat wastewater and produce biodiesel and chemical products within a biorefinery. *Biotechnol. Adv.* 30: 1031-1046. doi: 10.1016/j. biotechadv.2012.05.001.

Pan YY, Wang ST, Chuang LT, Chang YW and Chen CN (2011). Isolation of thermo-tolerant and high lipid content green microalgae: oil accumulation is predominantly controlled by photosystem efficiency during stress treatments in *Desmodesmus*. *Bioresour Technol.* 102(22): 10510-10517. doi: 10.1016/j.biortech.2011.08.091.

Pittman JK, Dean AP and Osundeko O (2011). The potential of sustainable algal biofuel production using wastewater resources. *Bioresour Technol.* 102(1): 17–25. doi: 10.1016/j.biortech.2010.06.035.

Pruvost J, Van Vooren G, Le Gouic B, Couzinet-Mossion A and Legrand J (2011). Systematic investigation of biomass and lipid productivity by microalgae in photobioreactors for biodiesel application. *Bioresource Technol.* 102(1): 150–158. doi: 10.1016/j.biortech.2010.06.153.

Rajamani S, Siripornadulsil S, Falcao V, Torres M, Colepicolo P, Sayre R (2007). Phycoremediation of heavy metals using transgenic microalgae. *Adv Exp Med Biol.* 616 : 99–109.

Rawat I, Kumar RR, Mutanda T and Bux F (2013). Biodiesel from microalgae: A critical evaluation from laboratory to large scale production. *Applied Energy*. 103: 444–467.

Rawat R, Kumar R, Mutanda T and Bux F (2011). Dual role of microalgae: Phycoremediation of domestic wastewater and biomass production for sustainable biofuels production. *Applied Energy*. 88: 3411–3424.

Rodolfi L, Zittelli GC, Bassi N, Padovani G, Biondi N, Bonini G and Tredici MR (2009). Microalgae for oil: Strain selection, induction of lipid synthesis and outdoor mass cultivation in a low-cost photobioreactor. *Biotechnol Bioeng*. 102: 100–112.

Samson R and LeDuy A (1983). Improved performance of anaerobic digestion of *Spirulina* maxima algal biomass by addition of carbon rich wastes. *Biotechnol. Lett.* 5: 677-682.

Shehata SA and Badr SA (1980). Growth response of *Scenedesmus* to different concentrations of copper, cadmium, nickel, zinc and lead. Environ Intern. 4: 431–434.

Soeder CJ (1981). Productivity of microalgal systems. In: Wastewater for aquaculture. University of the OFS, Bloemfontein: University of the OFS Publication, Series C, No. 3.

Solomon KR (1996). Ecological risk assessment of atrazine in North America surface waters. Environ. Toxicol. *Chem.* 15: 31–76.

Song W, Rashid N, Choi W and Lee K (2011), Biohydrogen production by immobilized *Chlorella* sp using cycles of oxygenic photosynthesis and anaerobiosis. Bioresource Technol. 102: 8676–8681.

Tabernero A, Martín del Valle EM and Galán MA (2011). Evaluating the industrial potential of biodiesel from a microalgae heterotrophic culture: Scale-Up and Economics. *Biochem. Eng. J.* 63: 104–115. doi: 10.1016/j.bej.2011.11.006.

Vergara-Fernandez A, Vargas G, Alarcon N and Velasco A (2008). Evaluation of marine algae as a source of biogas in a two-stage anaerobic reactor system. *Biomass Bioenergy*. 32: 338–344.

Vílchez C, Garbayo I, Markvicheva E, Galván F and León R (2001). Studies on the suitability of alginate-entrapped *Chlamydomonas reinhardtii* cells for sustaining nitrate consumption processes. Bioresource *Technol.* 78: 55–61.

Wang L, Li Y, Chen P, Min M, Chen Y, Zhu J and Ruan RR (2010). Anaerobic digested dairy manure as a nutrient supplement for cultivation of oil-rich green microalgae *Chlorella* sp. Bioresource *Technol.* 101: 2623–2628.

Widjaja A, Chien CC and Ju YH (2009). Study of increasing lipid production from fresh water microalgae *Chlorella vulgaris*. *J. Taiwan Inst. Chem. Eng.* 40(1): 13–20.

Wood BJB, Grimson PHK, German JB and Turner M (1999). Photoheterotrophy in the production of phytoplankton organisms. *J. Biotechnol.* 70(1–3): 175–183.

Wu WT and Hsieh CH (2008). Cultivation of microalgae for optimal oil production. *J. Biotechnol.* 136: 1521–1524.

Yen HW and Brune DE (2007). Anaerobic co-digestion of algal sludge and waste paper to produce methane. *Bioresource Technol.* 98: 130–134.

16

Screening of Thiosulphate Utilizing Bacteria from Mining Site of Ambaji, Gujarat, India

Sunita J. Varjani*

Science and Technology Department, Gujarat National Law University, Gandhinagar-382007, Gujarat, India

**Corresponding author E-mail: drsvs18@gmail.com*

ABSTRACT

Metallurgy is one of the old sciences. Bioleaching is a potential route for valorization of waste materials and may serve as a pre-treatment stage to mineral carbonation and/or sorbent synthesis processes. In bioleaching processes, a native flora of microorganisms consisting of a mixture of acidophilic, autotrophic, heterotrophic, or mixotrophic bacteria, fungi, and protozoa is active. Isolation of heterotrophs from mining site water sample of Ambaji was carried out using high plate count (HPC) agar and nutrient agar plates. Screening of thiosulphate utilizers was performed using modified 9 K medium. Fifteen pure heterotrophic isolates were obtained on HPC agar among which only two isolates were found to be thiosulphate utilizers. These two isolates along with six isolates obtained on thiosulphate agar plates are referred as thiosulphate utilizers. Cultural, morphological and biochemical characteristics of these organisms will be discussed. Final screening led to selection of one Gram-positive rod belonging to *Bacillus* sp. as potent thiosulphate utilizer.

Keywords: Ambaji, *Bacillus*, Bioleaching, Thiosulphate utilizer, 9 K medium

1 INTRODUCTION

Metals are a class of chemical elements with very useful properties such as strength, malleability, and conductivity of heat and electricity. As the population increases and our standard of living advances, so does our need for metals. We now use three times as much copper and four times as much lead and zinc as we did 75 years ago. The increasing need for metals is a need shared throughout the world. The desire to raise global living standards, coupled with a growing world population, will increase worldwide demand for metals in the future. Metal mining industry is responsible for extracting metals from the earth for use in our daily life (Hudson et al., 1999).

Mining is one of the oldest industrial processes in which mankind has been involved for thousands of years (Saxena, 2002). Therefore, it is important to make mining environmentally sustainable or at least minimize the environmental impact and increase the productivity of the metal recovery process (Dold 2007). A number of metal sulphides are formed as a result of biological activity or degraded due to microbial action; hence they are of geo-microbiological interest (Varjani 2010). Nowadays, microbes are used in mining industries as they have benefits over traditional physico-chemical methods (Rawlings 2002; Dave et al., 2012). Olson et al. (2003) reported that application of microbial processes by the mineral industry - biohydrometallurgy - predates the understanding of the role of microorganisms in metal extraction, commonly referred to as "bioleaching". Bioleaching is the extraction of metals from ores using the principal components of water, air and microorganisms, all of which are found readily within the environment (Jonglertjunya, 2003). *Acidithiobacillus ferrooxidans* and *Leptospirillum* sp. are chemolithoautotrophic acidophilic iron oxidizers whereas species of genera *Acidiphilium, Delftia* and *Bacillus* are able to grow autotrophically with sulphur and iron (II) compounds and heterotrophically with glucose or yeast extract and mixotrophically with all of these substrates (Brandl 2001; Paknikar and Pathkar, 2002; Kamal et al., 2010; Jablonovská et al., 2012). These microorganisms play a vital role in acid mine drainage (AMD) generation (Dave et al., 2012).

The present study demonstrates screening and primary characterization of thiosulpahte utilizers obtained from the water sample collected from multimetal mine of Ambaji, Gujarat, India.

2 MATERIALS AND METHODS

2.1 Sample Collection

Under the technical assistance of mining authorities water sample was collected in sterile carboy from multimetal mine situated at Ambaji, Gujarat, India. Sample was labeled and transported to laboratory and stored in refrigerator at 4°C before analysis.

2.2 Physico-Chemical Characterization of Sample

Physico-chemical characteristics were analysed in terms of pH, redox potential, total dissolved solids (TDS), conductivity and salinity, using Eu-tech multimeter, Singapore. Concentration of metal ions (Zn, Cu and Fe) was analysed by using standard atomic absorption spectrophotometer (Elico, Sl 194, India).

2.3 Growth Media and Chemicals

The medium used for cultivation and isolation of thiosulphate utilizers was consisting of (g/l): $(NH_4)_2SO_4$, 3.0; KCl, 0.1; K_2HPO_4, 0.5; $MgSO_4.7H_2O$, 0.5; $Na_2S_2O_3$, 2.5. The medium pH was adjusted to 4.5 with 10% v/v H_2SO_4.

All chemicals for isolation, screening and cultural, morphological as well as biochemical tests were purchased from CDH, Merck and Himedia.

2.4 Enumeration of Microorganisms

Most Probable Number (MPN) technique was initially used for enumeration of total thiosulphate utilizers using modified 9 K medium (basal medium supplemented with thiosulphate as energy source) (Varjani, 2010). Results were observed in terms of colour change from colourless to brick red by using methyl orange as indicator. The results thus obtained were adapted as number of organisms per 100 ml of water sample.

2.5 Isolation of Microorganisms

Isolation of heterotrophs was carried out on high plate count (HPC) agar plates. The inoculated plates were incubated at 30 ± 1°C for 48-72 h. Colony characteristics as well as cellular characteristics were observed after incubation period. All colonies from an entire plate or a representative sector were picked up and transferred on other HPC agar plates. After 3-4 successive transfers pure isolated colonies were transferred on HPC agar slants and on nutrient agar slants and in refrigerator.

For the isolation of thiosulphate utilizers media with specific nutrient constituents and conditions were employed, as this group of microorganisms requires special techniques for their isolation. In this study modified 9K medium was used for isolation of thiosulphate utilizers. The inoculated plates were incubated at 30 ± 1°C for 20-22 days. All colonies from an entire plate or a representative sector were picked up and transferred on other TS (thiosulphate) agar plates. After 3-4 successive transfers, pure isolated colonies were transferred on TS agar and TSB (tryptone soya broth) agar slants and preserved on TS agar slants and TSB agar slants in refrigerator.

2.6 Screening of Potential Thiosulphate Utilizers

2.6.1 Primary screening

All heterotrophic isolates were streaked on TS agar plates, to check whether they are utilizing thiosulphate as energy source or not. For this TS agar plates were inoculated with pure cultures of heterotrophs and were incubated for 48-72 h at 30 ± 1°C. Colonies of heterotrophs which were able to grow on TS agar plates were used for further studies along with isolates obtained on TS agar plates. Also isolates obtained on TS agar plates were streaked on HPC agar plates, to check their ability to grow on organic media.

2.6.2 Evaluation of thiosulphate utilization in liquid medium

All the selected thiosulphate utilizing bacterial isolates were inoculated in flask(s). Evaluation of thiosulphate utilization by the isolates was checked by determining thiosulphate activity. All experiments were carried out in 90 ml TS broth in 250 ml Erlenmeyer flasks. Medium was sterilized by autoclaving. 10% (v/v) inoculum of active culture of respective isolates was added to each flask. The pH of medium was adjusted to 4.5 ± 0.1 using 10% H_2SO_4. All flasks were incubated on Newtronics orbital shaker at 30 ± 1°C rotating at 150 rpm. At regular interval of time sample was withdrawn and analysis was carried out. Thiosulphate activity was determined by estimating utilized thiosulphate by standard iodimetric method (Vogel, 1962).

2.7 Cultural and Morphological Studies of Selected Thiosulphate Utilizing Isolates

Morphological characteristics were studied using Gram's staining technique and cultural characteristics were studied on HPC agar plates having pH 6.5. Colony morphology on HPC agar (pH: 6.5) was studied under 100 × magnification using microscope.

2.8 Biochemical Characterization of Selected Thiosulphate Utilizing Isolates

Biochemical characterization of the selected thiosulphate utilizers was done by performing various tests described in Bergey's manual of systematic bacteriology (Holt, 1984). Media for each test were inoculated as per requirement. Liquid media were inoculated by 1% (v/v) exponential phase culture of each isolate. Solid media in plate were spot inoculated and solid media in slopes were streaked from the exponential phase culture suspension. Incubation of all inoculated media was done at 30 ± 1°C for 24-48 h. When sufficient growth was obtained after incubation, suitable reagents were added wherever necessary and results were noted.

3 RESULTS AND DISCUSSION

Physico-chemical characterization and metal ion concentration of water sample are represented in Table 1. pH, redox potential, conductivity, TDS and salinity of the sample were 3.5, 360 mV, 2.10 mS, 1.7 ppt and 2.2 ppt, respectively. The observed results indicate extreme physico-chemical characteristics of the sample. Major metallic constituents of the sample in ppm were Zn (55.4), Cu (60.2) and Fe (1.2). On the basis of the data obtained in MPN test cell load of thiosulphate utilizers in studied sample was 2.45 × 10^3 cells/mL. Presence of both motile as well as non-motile microorganisms with dominance of Gram-positive organisms was observed.

Table 1 Physico-chemical characteristics of water sample from multi-metal of Ambaji

pH	*Redox potential (mv)*	*Conductivity (mS)*	*TDS (ppt)*	*Salinity (ppt)*	*Metal ion compsition (ppm)*		
					Zn	*Cu*	*Fe*
3.5	360	2.10	1.7	2.2	55.4	60.2	1.2

In laboratory conditions fifteen pure heterotrophic bacterial cultures were successfully isolated from the sample. Based on their ability to grow on modified 9 K medium and TS agar plates (containing 0.25% thiosulphate) two organisms were screened. These two heterotrophic isolates along with six isolates obtained on TS agar plates were used for further study. Together all these eight isolates are now referred to as thiosulphate utilizers. Results for thiosulphate utilization by the selected eight thiosulphate utilizing bacterial isolates are shown in Table 2. Depending on their utilization of thiosulphate in percentage one potent thiosulphate utilizer was screened as best thiosulphate utilizer. Among these organisms isolate SVTH6 which showed 82.50 % thiosulphate utilization was selected as a potent thiosulphate utilizer.

Table 2 Comparison for thiosulphate utilization by selected thiosulphate utilizing isolates

Sr. No.	*Isolate No.*	*Thiosulphate utilized (%)*
		192 h
1.	SVTH1	42.00
2.	SVTH 2	38.64
3.	SVTH3	63.12
4.	SVTH 4	40.15
5.	SVTH 5	35.43
6.	SVTH 6	82.50
7.	SVTH 7	41.31
8.	SVTH 8	38.00

At 0 h of incubation thisulphate utilized was in range 0.2 to 0.5% for all the eight (08) isolates.

The cultural and morphological characteristics of selected eight thiosulphate utilizing isolates are presented in Table 3. Microscopic photograph of SVTH6 colony, a potent thiosuphate utilizer on HPC agar plate is shown in Fig. 1. The results for various biochemical tests and sugar utilization profile of the selected isolates are presented in Tables 4 and 5, respectively. Isolates SVTH6, SVTH1, SVTH5, SVTH7 and SVTH8 were

Table 3 Cultural and morphological characteristics of selected thiosulphate utilizers on HPC agar plate (pH 6.5)

Isolate	*Size*	*Shape*	*Elevation*	*Margin*	*Texture*	*Opacity*	*Pigment*	*Gram staining*
SVTH1	Small	Round-Irregular	Slightly raised	Irregular	Smooth	Transparent	Lemon yellow	Gram-positive, Cocci
SVTH2	Small	Round	Slightly raised	Undulate	Smooth	Translucent	-	Gram-positive, Rod
SVTH3	Medium	Irregular	Slightly raised	Undulate	Smooth	Transparent	-	Gram-positive, Rod
SVTH4	Small	Round	Slightly raised	Irregular	Smooth	Transparent	Lemon yellow	Gram-positive, Cocci
SVTH5	Small	Round	Slightly raised	Irregular	Smooth	Translucent	-	Gram-negative, short rod
SVTH6	Big	Irregular	Slightly raised	Irregular	Gummy	Opaque	Off white	Gram-positive, Rod
SVTH7	Big	Irregular	Slightly raised	Undulate	Smooth	Transparent	-	Gram-positive, Rod
SVTH8	Small	Round	Slightly raised	Irregular	Smooth	Translucent	-	Gram-positive, Rod

oxidase positive. All organisms showed catalase test positive. Gelatinase enzyme was produced by only one isolate, i.e., (SVTH7). Except isolate SVTH1 all organisms were able to reduce nitrate. Isolates SVTH6 and SVTH7 were able to produce ammonia. SVTH6 was the only isolate showed Voges–Proskauer test positive. All organisms showed negative results for Methyl Red test, H_2S production test, Indole production test, Phenyl alanine deamination test and Urease production test. Organism in presence of Andrade's indicator changes colour of the medium to dark pink, when acid is produced and gas is collected in the Durham's vial. All isolates produced acid and gas both when glucose was provided. Mannitol was utilized by isolates SVTH1, SVTH2, SVTH3, SVTH4, SVTH7 and SVTH8 but only acid was produced. Maltose was utilized by only isolates SVTH2 and SVTH4 and produced only acid. Xylose was not utilized by any of the isolates. Sucrose was utilized by isolates SVTH2, SVTH4, SVTH5, SVTH6 and SVTH8 (in case of isolates SVTH4, SVTH5 and SVTH6 both acid and gas were produced; but isolate SVTH2 and SVTH8 produced only acid). Lactose was utilized by only SVTH4 isolate and acid was produced. Based on cultural, morphological and biochemical characteristics isolate SVTH6 was characterized as *Bacillus* sp.

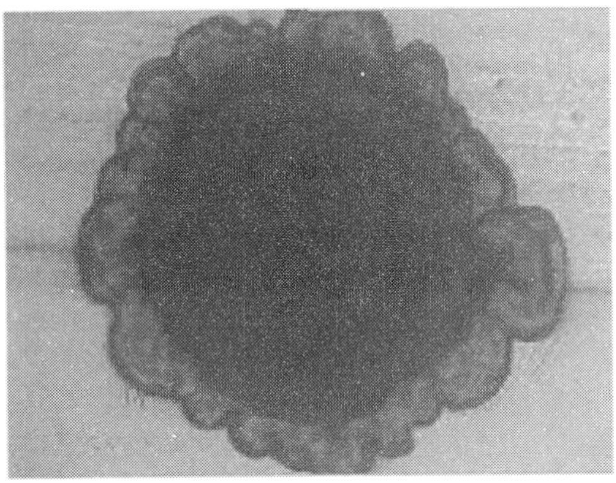

Fig. 1 Microscopic photograph (100X): Colony of thiosuphate utilizing SVTH6 isolate on HPC agar plate (pH 6.5).

Song et al. (2008) reported that bioleaching has been commercially applied in extraction of copper, gold and uranium from sulphides. Bioleaching of zinc, cobalt and nickel from sulphides has been extensively studied and commercially used in some cases. The ecosystems are dominated by acidophilic sulphur and/or iron oxidizing organism, which tolerated comparatively higher metal ion concentrations and acidity together with a variety of other metal and acid-tolerant heterotrophic bacteria, fungi, yeast, algae and protozoa. Most of the autotrophs isolated from Indian mine waters are included under the genus *Acidithiobacillus*. They are Gram-negative, small rods, 0.5 by 1-3 µm, occurring singly or

Table 4 Biochemical characterization of selected thiosulphate utilizers

Biochemical tests	*Isolates*							
	SVTH1	*SVTH2*	*SVTH3*	*SVTH4*	*SVTH5*	*SVTH6*	*SVTH7*	*SVTH8*
Oxidase test	+	–	–	–	+	+	+	+
Catalase test	+	+	+	+	+	+	+	+
Starch hydrolysis test	+	+	+	–	+	+	–	–
Milk hydrolysis test	+	–	+	–	–	+	+	–
Gelatin utilization test	–	–	–	–	–	–	+	–
Lipase production test	+	+	–	–	+	+	+	+
Methyl red test	–	–	–	–	–	–	–	–
Voges–Proskauer test	–	–	–	–	–	+	–	–
H_2S production test	–	–	–	–	–	–	–	–
Indoleproduction test	–	–	–	–	–	–	–	–
Citrate utilization test	–	+	–	–	+	+	+	–
Nitrate reduction test	–	+	+	+	+	+	+	+
Ammonia production test	–	–	–	–	–	+	+	–
Phenyl alanine deamination test	–	–	–	–	–	–	–	–
Urease production test	–	–	–	–	–	–	–	–

Negative: – ; Positive: +

occasionally in pairs (Natarajan 1998). However, in this study the potential thiosulphate utilizing isolate is gram positive in nature and belongs to genus *Bacillus*.

Farbiszewska-kiczma et al. (2004) reported presence of *Bacillus amyloliquefaciens* and *Bacillus cereus* from Cupriferous black shale ore, Lubin region in Poland. Bosecker (1997) has reviewed and reported presence of heterotrophic bacterial and fungus of the genera *Bacillus, Aspergillus* and *Penicillium* along with autotrophic bacterial sp. of genera *Thiobacillus* and *Leptospirillium*. A comparative phylogenetic analysis of 16S rRNA genes was performed with respect to large group of *Delftia* sp., *Oxalobacter* sp., *Bacillus* sp. and *Ferribacter* sp. Dave et al. (2012) reported 93.34% thiosulphate utilization by *Delftia* sp. (HQ398998) from polymetallic concentrate leachate sample. There are reports on isolation of the bacterial isolates of genera *Bacillus, Delftia, Ferribacte*r and *Oxalobacter*, etc. from mine water (Heinzel et al. 2009).

Table 5 Sugar fermentation test thiosulphate utilizers

Name of sugar	*Isolates*															
	SVTH1		*SVTH2*		*SVTH3*		*SVTH4*		*SVTH5*		*SVTH6*		*SVTH7*		*SVTH8*	
	A	G	A	G	A	G	A	G	A	G	A	G	A	G	A	G
Glucose	+	+	+	+	+	+	+	+	+	+	+	+	+	+	+	+
Mannitol	+	-	+	-	+	-	+	-	-	-	-	-	+	-	+	-
Maltose	-	-	+	-	-	-	+	-	-	-	-	-	-	-	-	-
Xylose	-	-	-	-	-	-	-	-	-	-	-	-	-	-	-	-
Sucrose	-	-	+	-	-	-	+	+	+	+	+	+	-	-	+	-
Lactose	-	-	-	-	-	-	+	-	-	-	-	-	-	-	-	-

Negative: – ; Positive: +; A: Acid; G: Gas; Expt. conducted using sugar broth medium

4 CONCLUSION

Under laboratory experimental conditions at the end of screening techniques eight organisms were selected as thiosulphate utilizers. Furhter study by measuring thiosulphate activity leads to the selection of one isolate belonging to the genera *Bacillus* as potent thiosulphate utilizer. This organism may be of great significance in tratment of AMD water. Optimatization of growth conditions as well as thisulphate utilization profiling will be envisaged.

Acknowledgements

I am grateful to representatives of mining authority at Ambaji multimetal mine for providing assistance in sample collection. I am thankful to all faculty members of the department, non-teaching staff and colleagues for their kind support.

References

Bosecker K (1997). Bioleaching: metal solubilisation by microorganisms. FEMS Microbiol. Rev. 20: 591–604.

Brandl H (2001). Microbial Leaching of Metals: Bitechnology. Rehm HJ, Reed G, (eds.).Vol. 10, 2nd edition. pp. 191–224.

Dave SR, Varjani SJ and Tipre D (2012). Isolation and characterization of thiosulphate utilizing *Delftia* sp. from bulk polymetallic concentrate leachate. Res. J. Biotech 7(4): 119–124.

Dold B (2007). Biogochemical processes in mine tailing with special focus on marine shore tailings deposits and their remediation: Biohydrometallurgy: From the single cell to the environment. Schippers A, Sand W, Glombitza F and Willscher S (eds.), IBS, Proceedings of 17th IBS held at Frankfurt, Germany, Trans Tech Publications, Germany. pp 177–188.

Farbiszewska-kiczma J, Farbiszewska T and Bąk M (2004). Bioleaching of metals from polish black shale in neutral medium. Physicochemical problems of mineral processing 38: 273–280.

Heinzel E, Hedrich S, Janneck E, Glombitza F, Seifert J and Schlomann M (2009). Bacterial diversity in marine water treatment plant. Appl. Environ. Microbiol. 75(3): 858–861.

Bergey (1984). Bergey's manual of systematic bacteriology. Holt JG (ed). Vol. 1, Williams and Wilkins, Baltimore, United States of America.

Hudson TL, Fox DF and Plumlee GS (1999). It helps to know: AGI environmental awareness series 3, Metal mining and the environment. CLB Printing Company. pp. 7–12.

Jablonovská K, Pallova Z and Styriakova I (2012). Bioleaching of Zn, Ni and Fe from contaminated sediments of water reservoir Ružín I with using heterotrophic bacterial strains. Acta Montanistica Slovaca 17: 47–50.

Jonglertjunya W (2003). Bioleaching of chalcopyrite. Ph.D. thesis submitted to the University of Birmingham, Edgbaston, United Kingdom. (http://etheses.bham.ac.uk/238/1/Jonglertjunya03PhD.pdf (Last accessed: 25th May 2015).

Kamal S, Prasad R and Varma A (2010). Soil microbial diversity in relation to heavy metals. In: Soil Heavy Metals (eds. Sherameti I and Varma A) Springer-Verlag, Germany 31–64.

Natarajan KA (1998). Biotechnological aspects of metal extraction and leaching bacteria: Microbes minerals and Environment. Geological Survey of India, India. pp. 43–56.

Olson GJ, Brierle JA and Brierley CL (2009). Bioleaching review part B: Progress in bioleaching: applications of microbial processes by the minerals industries. Appl. Microbiol. Biotechnol. 63: 249–257.

Paknikar KM and Pathkar AV (2002). Development of strain to enhance the bioleaching process: Brain storming session of the consortium of experts on biomining/biomineral processing to extract mineral values hel d at Dhanbad. Gupta A and Misra DD (eds.), Allied Publishers Pvt. Ltd., Dhanbad, India. pp. 26–30.

Rawlings DE (2002). Heavy metal mining using microbes. Annu. Rev. Microbiol. 56: 65–91.

Saxena RK (2002). Mineral processing: Brainstorming session of the consortium of experts on biomining/biomineral processing to extract mineral values held at Dhanbad. Gupta A and Misra DD (eds.), Allied Publishers Pvt. Ltd., Dhanbad, India. pp. 31-46.

Song C, Wen-qing Q and Guan-zhou Q (2008). Effect of Cu^{2+} ions on bioleaching of marmatite.Tras. Nonferrous Met. Soc., China, pp. 1518–1522.

Varjani SJ (2010). Microbial profiling of preserved polymetallic concentrate leachate sample. M. Phil. Thesis submitted to Gujarat University, Gujarat, India.

Vogel AI (1962). A Textbook of Quantitative Inorganic Analysis. 3rd ed., The English Language Book Society and Longmans, Green and Co. Ltd., London, United Kingdom.

17

Biodegradable Polymers from Microbes

Charu Gupta[1*], Dhan Prakash[1] and Sneh Gupta[2]

[1]*Amity Institute for Herbal Research and Studies, Amity University Uttar Pradesh, Noida-201303, India*

[2]*Department of Zoology, RGPG College, Meerut-250001 (UP), India*

**E-mail: charumicro@gmail.com*

ABSTRACT

Microbes are also the potential sources of biodegradable polymers besides plant carbohydrates. The main microbial polymers are polyhydroxyalkanoates (PHA) and polylactic acid (PLA). These polymers are biodegradable and have significant potential in biotechnology and bioengineering. *Azotobacter eutrophus, A. latus* and mutant strain of *Azotobacter vinelandii* are the major biopolymer accumulating bacteria. These biopolymers are produced by microbial fermentation whereas some are partially synthesized. The natural polymers are completely degraded by microorganisms by the process of enzymatic scission of the polymer chain. Besides these, some nylon eating bacteria like *Flavobacteria* and *Pseudomonas* are capable of breaking down nylon as they possess nylonase enzyme. Chitin is another biopolymer that finds applications in the cosmetic industry. The present chapter focuses on the biopolymers produced by microorganisms and their industrial importance.

Keywords: Biopolymers, Microorganisms, Biodegradable, Poly-hydroxy-alkanoates (PHA), Polylactic acid (PLA), Poly-3-hydroxybutyrate (PHB).

1 INTRODUCTION

In the last few years, there has been a gradual shift from the use of plastics to the biodegradable polymers. Plastic waste causes high degree of contamination, which requires high energy cost to recycle (Sanchez-Garcia et al., 2008). The plastic wastes are persistent in environment, causes dwindling of petroleum resources, shortage of landfill space, and emits poisonous gases during incineration thereby causing environmental pollution (Abdelwahab et al., 2012). Natural polymers are biodegradable as they can be disintegrated by the enzymatic action of microorganisms like bacteria, fungi, yeast into the carbon dioxide, water and biomass under the aerobic conditions and into hydrocarbons like methane and biomass under anaerobic conditions.

Microorganisms naturally produce a huge variety of biopolymers, such as polysaccharides, polyesters and polyamides. These range from viscous solutions to plastics and their physical properties are dependent on the composition and molecular weight of the polymer.

Bioplastics are chiefly made from a compound called polyhydroxyalkanoate (PHA) that are produced by microbial fermentation. The material properties exhibited by PHAs ranges from stiff, brittle to rubber-like that makes it a close substitute for the synthetic plastic. Bacteria accumulate PHA in the presence of excess carbon source, similar to how humans accumulate fat deposits on their bodies after consuming excess food. Thus, microbial PHA accumulators are microorganism's versions of Sumo wrestlers, accumulating storage biopolymers after consuming excess nutrients. Poly 3-hydroxy butyric acid (PHB) is the most common microbial PHA. Industrial interest in microbial polymers has been stimulated by their unique properties and the opportunity to develop new materials, which can be used for specific applications in medical and pharmaceutical industries, e.g., insulin producing cells and tissue engineering (Hernández et al., 2010). The newer applications of these polymers include their use as a source of soluble fiber, or in medical products. The intracellular polyester PHB and other PHAs have been drawing attention because they are biodegradable and biocompatible thermoplastics, which can be processed to create a wide variety of consumer products, including plastics, films, and fibers (Aldor and Keasling, 2003). There are new attractive applications for PHAs based on their properties of biocompatibility and biodegradability in the medical field where the chemical composition and product purity are critical (Williams and Martin, 2005). But the cost of production of PHAs is higher that has restricted its applications. Poly-lactic acid (PLA) is another 100% compostable biopolymer which can fully degrade above 60°C in an industrial composting facility. It is produced by microbial fermentation and then polymerized using chemical catalysis. PLA is used in biomedical applications such as implants, sutures, drug delivery, tissue engineering and stent development due to its biocompatibility properties.

2 HISTORICAL BACKGROUND

Poly-hydroxy-alkanoate (PHA) was first observed by Beijerinck in 1888 as refractile bodies inside bacterial cells. Lemoigne in 1926 first established PHA composition. The potential of these biopolymers was recognized in the 1960s, but their possible exploitation was seriously considered only in the late 1970s. In 1976, ICI of England explored if PHB could be satisfactorily produced by microbial fermentation. In 1993, Zeneca

Bioproducts took over ICI's activities and in 1996 Monsanto purchased the rights of bioplastics production business from Zeneca. Soon Monsanto closed down its activities in 1998 that were successfully taken up by many new players like Metabolix, Proctor & Gamble, DuPont, General Motors, and Toyota etc. BIOPOL is one of a marketed bioplastic product (Tokiwa et al., 2009).

Scientists have now developed genetically modified bacteria that can synthesize plastic polymers ("PHAs") using organic materials. PHAs are important because they are a biodegradable plastic source that does not require petroleum, and is renewable, as scientists can synthesize the plastics in a lab with the correct materials and bacterial strains. The PHA polymers can be extracted using a host of enzymes and solvents, and later refined into plastic products (Reemer 2009).

3 CLASSIFICATION OF BIOPOLYMERS

3.1 Renewable Resource-Based

It includes PLA Polymer (from corn), Cellulosic plastics, Soy-based plastics, Starch plastics.

3.2 Petro-bio (Mixed) Sources

It includes Sorona, Bio based Polyurethane, Bio based Epoxy Blends etc.

3.3 Petro-based Synthetic

It includes aliphatic polyester, Aliphatic-aromatic polyesters, Polyesteramides, Polyvinyl alcohols.

3.4 Microbial Synthesized

It includes Polyhydroxy alkanoates (PHAs), Polyhydoxybutyrate co-valerate (PHBV) (Reddy et al., 2003; see Fig. 1)

4 BIO-PLASTIC PRODUCING BACTERIA

PHB granules were first observed by Lemoigne in 1926 in *Bacillus megaterium* (Gram-positive). Since then several groups of archaebacteria, Gram-positive and Gram-negative bacteria have been shown to be the producers of PHAs. Few examples are *Ralstonia eutropha, Ralstonia* spp., *Pseudomonas putida,* other *Pseudomonas* spp., *Bacillus mycoides, Alcanivorax borkumensi, Rhodococcus ruber* etc (Tokiwa et al., 2009). Some of the properties of biodegradable plastic are given in Table 1.

Table 1 PHB (and its copolymers) biosynthesis from inexpensive substrates (Ceyhan et al., 2011)

Substrate	*Microorganisms*
Sugar cane liquor	*Ralstonia pickettii* 61A6 *Pseudomonas fluorescens* A2a5
Sugar cane and beet molasses	*Bacillus cereus* M5
Starch-based materials	*Azotobacter chroococcum* and Recombinant *Escherichia coli* strain
Molasses	Recombinant *Escherichia coli*
Cellulose hydrolysates, xylose, etc.	*Burkholderia cepacia* IPT 048 and B. sacchari IPT 101
Dairy whey	Recombinant *Escherichia coli* *Pseudomonas hydrogenovora* *Methylobacterium sp.* ZP24
Corn steep liquor	*Bacillus* sp. CFR 256
Corn steep liquor and molasses	*Bacillus megaterium* ATCC 6748
Corn syrup	*Cupriavidus necator**
Plant oils from biotechnological rhamnose production	*Alcaligenes eutrophus*
Crude palm kernel oil	*Cupriavidus sp.* USMAA2-4
Fatty acids and waste glycerol	*Ralstonia eutropha* *Pseudomonas aeruginosa* NCIB 40045 *Cupriavidus necator*
Paper mill wastewater	Activated sludge
Swine waste liquor	*Azotobacter vinelandii* UWD.
Petrochemical plastic waste	*Pseudomonas putida* CA-3
Sunflower cake, soy bran and olive mill	*Pseudomonas hydrogenovora*
Soy and malt wastes	Mixed bacteria
Waste activated sludge	*Bacillus sp.*
Dairy waste and seawater	*Bacillus megaterium*
Wastewater from olive oil mills (called alpechín)	*Pseudomonas putida* KT2442 *Azotobacter chroococcum* H23
Acid-hydrolysed malt waste	*Recombinant Bacillus subtilis* strains

* *Cupriavidus necator = Ralstonia eutropha = Alcaligenes eutrophus= Wautersia eutropha.*

4.1 Bacterial Polyesters

The bacteria produce natural polyesters as intracellular reserve materials. They can be utilized as biodegradable, melt processable polymers that

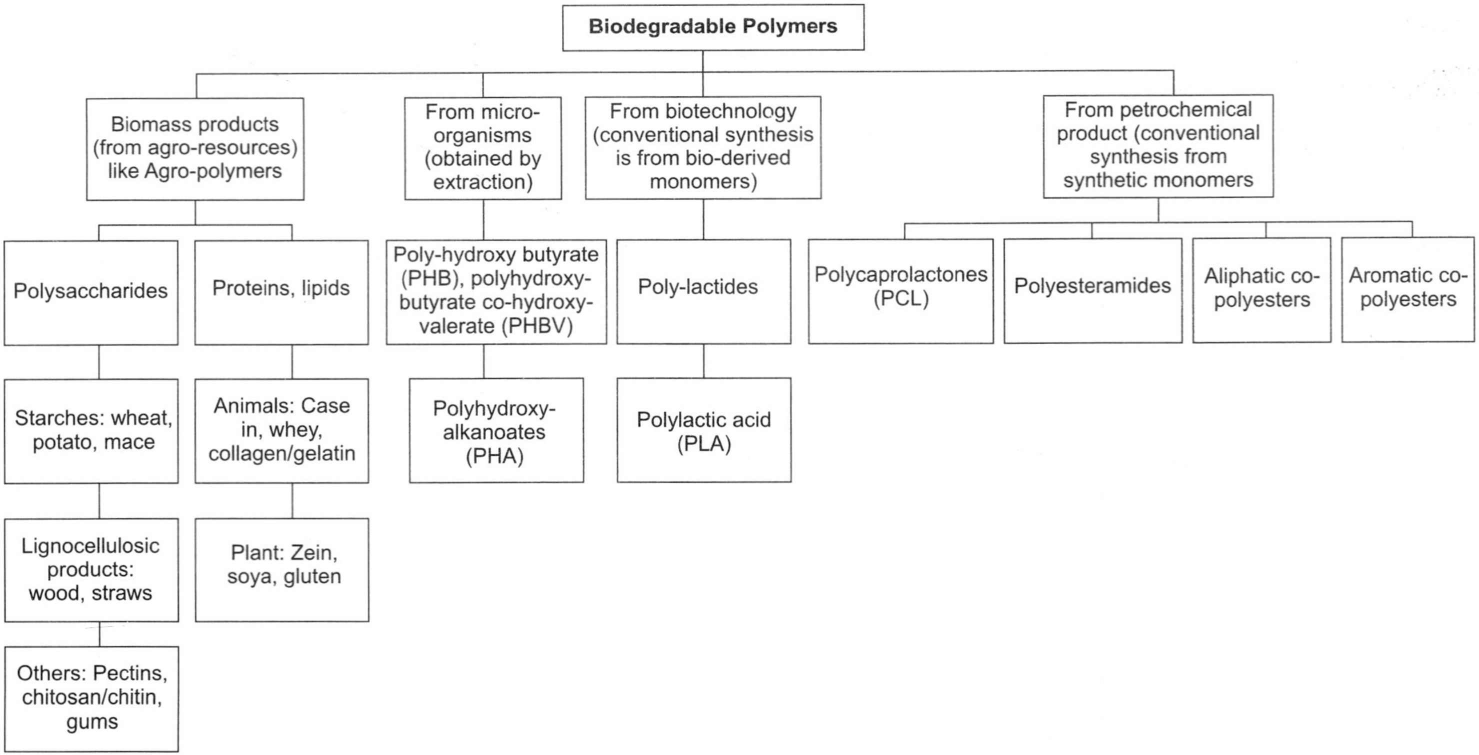

Fig. 1 Classification of biodegradable polymers (Reddy et al., 2003)

can be produced from renewable resources. These polyesters belong to the family of thermoplastic biopolymers. They show variation in their material properties from rigid brittle plastics to flexible plastics with good impact properties to strong tough elastomers depending on the size of the pendant alkyl group R and the composition of the polymer. These natural polyesters are made up of units that are 100% optically pure at the β-position; therefore all are 100% isotactic. The polymer with R=CH_3, poly-β-hydroxybutyrate (PHB), is highly crystalline with a melting temperature of 180°C and a glass transition temperature, T_g of approximately 5°C. This combination of very high crystallinity and relatively high T_g makes the films and plastics of PHB very brittle, so copolymers with units containing other alkyl groups, especially R=C_2H_5, are preferred. All of these materials are therefore inherently biodegradable. Bacteria generally produce polyesters with longer alkyl substituents with $x = 3 - 6$ or so in the form of copolymers which have lower degrees of crystallinity and lower T_m and T_g values. As a result, these longer alkyl chain polyesters are useful as thermoplastic elastomers, which can have excellent strength and toughness, and yet are also inherently biodegradable (Chandra and Rustgi, 1998).

Table 2 Properties of some biodegradable plastics (Verlinden et al., 2007)

Material	*Film preparation*	*Moisture barrier*	*Oxygen barrier*	*Mechanical properties*	*Cost, \$/kg*
Starch/PVOH	Extrusion	–	+	+	3-7
PHB/V	Extrusion	+	+	+/–	7-13
PLA	Extrusion	+/–	–	+	2-11

(NC-W = Nitrocellulose-wax; PVOH = Poly-vinyl-alcohol; PHB/V = Poly-hydroxy-butyrate/valerate).

Recently, a large scale controlled fermentation process was developed for the production of copolymers of PHB. The bacteria were fed with a variety of carbon sources that lead to the production of different copolymers with better mechanical properties than PHB.

In a study, the biodegradation of PHB and its copolymers has been studied in soil, activated sludge and seawater. Films (0.07 mm thick) of PHB (homopolymer), a copolymer of 91% 3HB and 9% 4HB and a copolymer of 50% 3HB and 50% 3HV were subjected to biodegradation in soil. The fastest biodegradation rate was obtained for P(3HB*co*-9% 4HB). In activated sludge P(3HB-*co*-9% 4HB) was completely decomposed after two weeks. The native polyesters are also hydrolyzed in water but at a very slow rate. The main degradation mechanism is chain scission of polymer under *in vivo* conditions. Studies have shown that the hydrolytic degradation of microbial polyesters occurs by homogeneous erosion over two stages: random hydrolytic chain scission of the ester group leading to a decrease in molecular weight followed by a second step in which more weight loss occurs (Chandra and Rustgi, 1998).

5 BIOPLASTIC TYPES

Some of the major types of bioplastics are soy-based bioplastics, starch-based bioplastics, microbial bioplastics, polylactides (obtained mostly from plants). The microbial based bioplastics include Poly 3-hydroxy butyrate (PHB), Poly 3-hydroxy alkanoate (PHA), Poly-lactic acid (PLA) and other miscellaneous bioplastics (Tokiwa et al., 2009).

5.1 Poly-Hydroxy-Alkanoates (PHAs)

Poly-hydroxy-alkanoates is exclusively produced by microorganisms chiefly bacteria. These are the best studied groups of biodegradable plastics (Reddy et al., 2003). They are type of microbial polyesters that has numerous applications such as bioplastics, fine chemicals, implant biomaterials, medicines and biofuels. They constitute an effective biocompatible substitute for unconventional plastics (Safak et al., 2002; Suriyamongkol et al., 2007). Although due to their high cost of large-scale production, their widespread use has been restricted.

Various bacterial strains, either wild-type or recombinant have been utilized with a wide spectrum of utilizable carbon sources. New fermentation strategies have been developed for the efficient production of PHAs at high concentration and productivity. Vegetable oils have been found to be possible substrates in the production of PHAs (Alias and Tan, 2005). Waste streams from oil mills or used oils, which are even cheaper than purified oils can also be used (Fernández et al., 2005; Mumtaz et al., 2010). The synthesis of bacterial polyhydroxyalkanoates (PHA) is very much dependent on the expression and activity of a key enzyme, PHA synthase (PhaC). With the current advances, PHAs can now be produced to a concentration of 80 gL^{-1} with productivities greater than 4 g PHA $L^{-1}h^{-1}$. These advances will further lower the production cost of PHAs and allow this family of polymers to become a leading biodegradable polymer in the near future (Akaraonye et al., 2010).

Polyhydroxyalkanoates (PHAs) are polyesters that are synthesized by various microorganisms such as *Cupriavidus necator, Alcaligenes latus, Aeromonas hydrophila, Pseudomonas putida* and *Bacillus* spp. (Khanna et al., 2005; Keshavarz et al., 2010). The biosynthesis of PHA is described in Fig. 2.

Out of the many different bacterial cultures that can produce PHAs, *Cupriavidus necator* has been most extensively studied (Volova and Kalacheva, 2005; Yan et al., 2005). Wild-type *C. necator* is known as a poor oil utilizer, but studies into novel plant oil media can increase this yield (Budde et al., 2011).

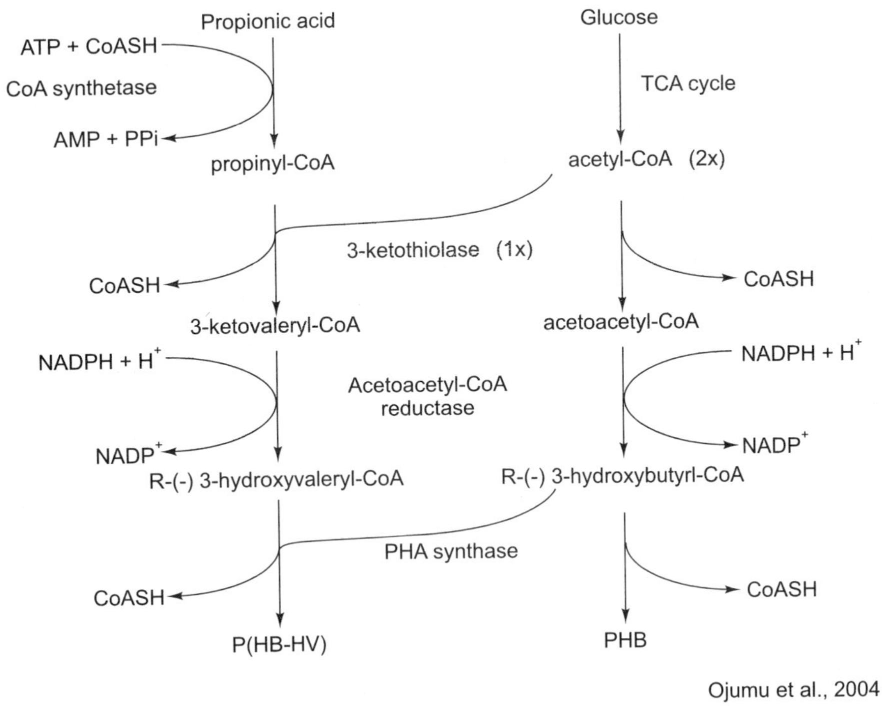

Fig. 2 The biosynthetic pathway of PHB and P(HB-HV) in © (Ojumu et al. 2004)

The microbes producing PHA also serve as hydrocarbon degraders, e.g., the PHA producers belong to the oil degrading group comprising genera *Pseudomonas, Acinetobacter, Sphingobacterium, Brochothrix, Caulobacter, Ralstonia, Burkholderia* and *Yokenella* (Fig. 3). All these PHA producers can be isolated from oil-contaminated sites (Dalal et al., 2010).

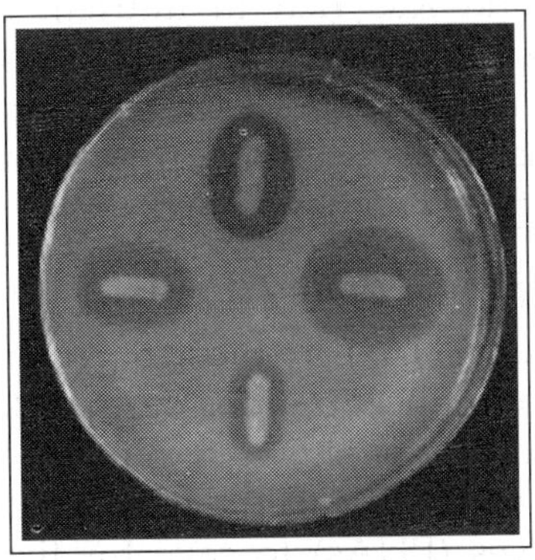

Fig. 3 Degradation by PHA depolymerases (Ojumu et al., 2004)

In a study on *Ralstonia eutropha* JMP 134, the bacteria accumulated PHB up to 50% of cell dry weight when grown in presence of substances like phenol or sodium benzoate as carbon source under nutrient limiting state (Maskow and Babel, 2000).

Some microbes on exhaustion of nutrients and in presence of excess of carbon sources also produce extracellular metabolites like rhamnolipids, alginate oligosaccharides, extracellular polymeric substances and biohydrogen along with intracellular PHA production stored as inclusion bodies (Marsudi et al., 2008; Guo et al., 2011).

There are various ecological niches to PHA producers like oil contaminated sites, hot springs, salt lakes, marsh lands. In fact the first case of polyhydroxy butyrate (PHB) accumulation by archae *Halobacterium marismortui* was reported in 1970 from the Dead Sea. Photosynthetic bacteria such as cyanobacteria have also been reported to produce PHA by oxygenic photosynthesis. *Synechococcus* sp. MA19, *Nostoc muscorum* and *Spirulina platensis* produce PHB under phosphate-limited conditions (Panda et al., 2005).

Plant growth promoting Rhizobia (PGPR) also possess hidden or untapped reservoirs for PHA accumulators. Some microorganisms like *Burkholderia terricola, Lysobacter gummosus, Pseudomonas extremaustralis, Pseudomonas brassicacearum* and *Pseudomonas orientalis* have been reported as PHA producers based on PCR technique having *pha*C as targeted gene (Gasser et al., 2009). It is believed that stored PHA fuels rapid colonization, efficient nodulation and nitrogen fixation.

Antibiotic producers like *Streptomyces* have also been reported for PHA production intracellularly in granular form working as supplier of carbon units for antibiotic synthesis and sporulation (Kannan and Rehacek, 1970).

Activated sludge also serves as PHA producer. Wastes are rich in organic contents and less rich in nutrient contents. This state of unbalanced nutrients supports growth of PHA producers but has minimal consumption rate of organic matter and storage rate (Beccari et al., 2002; Dircks et al., 2001; Saharan et al., 2014). The carbon cycle of biodegradable plastics is diagrammatically shown in Fig. 4.

5.1.1 *Conditions of production of PHAs*

Polyhydroxyalkonoates (PHAs) are those polymers that bacteria produce under conditions of low concentrations of important nutrients (typically nitrogen, but sometimes oxygen) and high concentrations of carbon sources. This process occurs because the excess carbon leads to bacteria creating carbon reserves (PHAs) to save for time with more plentiful nutrients in which they need energy to carry out regular functions (growing, reproducing, other biosynthesis, etc). Bacteria store PHAs

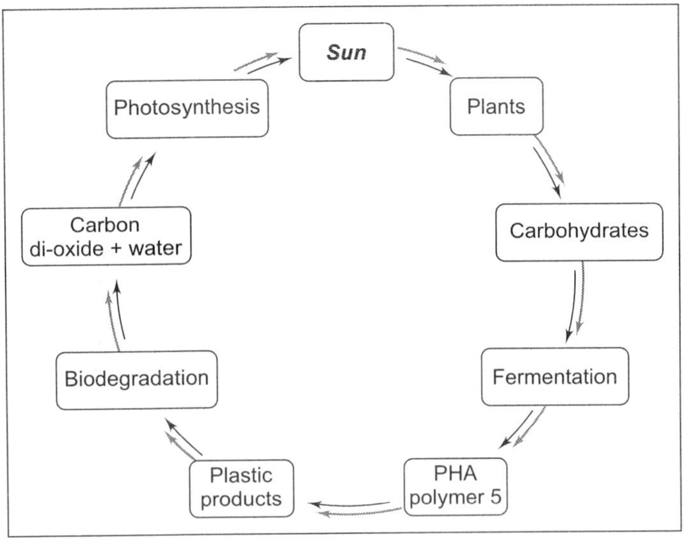

Fig. 4 Diagrammatic representation of carbon cycle of biodegradable plastics (Reemer 2009)

in granules for later use. Humans can harvest these PHAs to create bioplastics as the polymers produced by bacteria produce resemble the chemical structure of petroleum-produced plastics (Glimore, 2005).

5.1.2 Key enzymes of production of PHAs

It is through a biochemical process during fermentation, that the bacterial enzymes engage a cycle that stops protein synthesis and begins the production of PHAs when there is a high concentration of acetyl-CoA inside the cells, instead of regular CoA. This high concentration occurs when the bacterial cells are starved of essential nutrients (nitrogen, oxygen, phosphorus) because the pyruvate molecules do not enter its normal cycle, and is carboxylated to form acetyl CoA. The three key enzymes in P(3HB) (a specific form of PHAs) production are β-ketothiolase, acetoacetyl CoA reductase, and PHB reductase, successively (Fig. 1). β-ketothiolase begins by condensing two acetyl-CoA molecules. These products (acetoacetyl-CoA) are reduced to hydroxyalkanoic acids by acetoacetyl-CoA reductase. The last step is a PHA synthase (more than 50 different PHA synthases have been identified, and each designates which type of PHA will be the final products) polymerizes the production of a PHA, which is typically defined by its "R" group (Fig. 5) (Volova 2004).

5.1.3 Gram-negative bacteria for PHA production

In order to produce PHBs, the most commonly produced form of PHA, scientists typically use Gram-negative strains because they can utilize cheap carbon sources and produce up to 80% of the cell's dry weight in PHB. The disadvantage of this method is that PHB production through

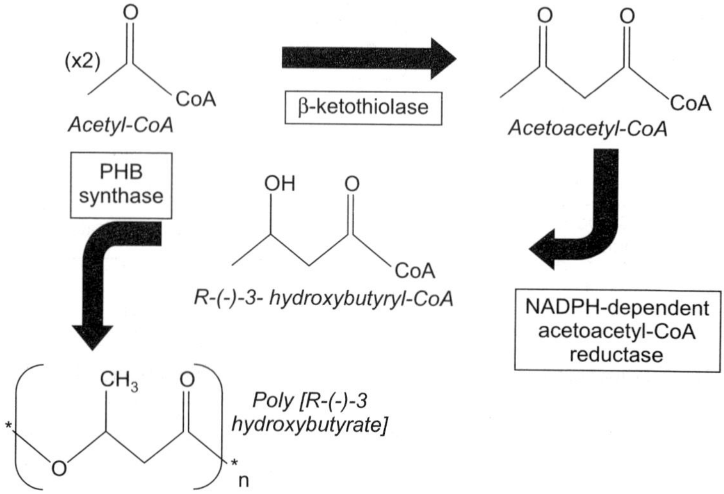

Fig. 5 Biochemical pathway and enzymes in PHA production (Reemer 2009)

Gram-negative strains require an extra step in order to eradicate the endotoxins in the additional outer membrane of these strains. Although this outer membrane is only typical of Gram-negative strains, PHB production through Gram-positive strains raises other issues. When put under the same sorts of conditions required for PHB production, Gram-positive strains tend to undergo sporulation, or the production of spores, which detracts from PHB productivity (Lopez et al., 2012).

Scientists have recently found a Gram-positive strain called *Bacillus megaterium* that can produce PHB under the same sort of methods and conditions that most Gram-negative strains use. According to experiments, the *Bacillus megaterium* strain uses the same cheap carbon sources like glucose and glycerol but can only produce 60% of its dry cell weight in PHB. This method avoids the extra process to remove endotoxins, but it is unclear whether the loss in PHB production is worthwhile economically. The results are promising, however, because the success with *Bacillus megaterium* does give potential for finding more useful Gram-positive strains or altering existing ones (Lopez et al. 2012). This ability to alter strains appears to be the hope for the future of PHB production. Some researchers have managed to create synthetic organisms by injecting a synthesized DNA code into a bacterial cell. The bacterial cells were engineered to grow by consuming carbon dioxide, which could be helpful for making not only plastics, but food and fuel also (Bates 2012).

5.1.4 Recovery of PHAs from bacterial cells

The PHA producing microorganisms are stained with Sudan black or Nile blue. Cells are then separated by centrifugation or filtration. PHA is recovered using solvents (chloroform) to break cell wall and extract polymer. Finally, the polymers are purified.

5.2 Polyhydroxy Butyrate (PHB)

Polyhydroxy butyrate is one of the most commonly used polyhydroxyalkanoates (PHAs) also known as bioplastic (bio-based polymer) that is synthesized and catabolized by numerous microorganisms (Akar et al., 2006; Verlinden et al., 2007). It was first discovered by French microbiologist *Maurice Lemoigne* in 1923. They have biodegradable and biocompatible properties. Several halophilic microbes (including *Haloferax mediterranei*, *Vibrio* spp., *V. natriegens*, *V. nereis* and *V. harveyi*) reportedly produce PHB (Weiner, 1997). The identification of the gene that is involved in PHA synthesis, polyhydroxyalkanoic acid synthase, was verified using *V. parahemolyticus* and *V. cholera*.

The biopolymer poly-3-hydroxybutyrate (PHB) is polyester produced by certain bacteria processing glucose, corn starch or wastewater. Its characteristics are similar to those of the petroplastic polypropylene. It produces transparent film at a melting point higher than 130°C and is biodegradable without residue. PHAs are accumulated as intracellular inclusion in some bacteria under unfavourable growth condition such as limitation of nitrogen, phosphorus, oxygen or magnesium in the presence of excess supply of carbon source.

In a scientific study by Kumar and Prabhakaran 2006, *Alcaligenes eutrophus* MTCC 1285 has been found to produce bioplastic on bioeffluents. The bacteria produced bioplastic PHB in nitrogen-deficient medium containing sago effluent, thippi and molasses as substrates.

PHB is linear thermoplastic polyester produced by controlled bacterial fermentation process. This polymer is primarily a product of carbon assimilation (from glucose or starch) and is employed by microorganisms as a form of energy storage material, accumulated intracellularly, and is metabolized when other common energy sources are not available (Lugg et al., 2008). The properties of PHB are similar to those of various synthetic thermoplastics such as polypropylene. Various microorganisms completely degrade PHB to water and carbon dioxide under aerobic conditions and to methane under anaerobic conditions.

The accumulation of PHB has been reported in both rhizobia associated with roots actively involved in nodulation (rhizobia-legume symbiosis) or free living nitrogen fixing organisms. The ability to accumulate PHB is directly correlated to type of nodule being formed, that is determinate or indeterminate differing in possession of persistent meristem (Trainer and Charles, 2006). Proteobacteria and Firmicutes were isolated from root nodules of eight leguminous plants that have been reported to accumulate PHB (Kumbhakar et al., 2012).

They are adopted in the biomedical field, for example, medical implants and drug delivery carriers. It is believed that next generation PHB will come from transgenic plants/microorganisms. Their typical cost ranges from 1-2 $/lb (Khazir and Shetty, 2014).

5.3 Alginates

Alginates are linear polysaccharides composed of variable amounts of (1-4)-β-D-mannuronic acid and its epimer, α-L-guluronic acid. Alginate, an extracellular polysaccharide is produced by *Azotobacter vinelandii* (Galindo et al., 2007). *Azotobacter* alginates are true block co-polymers, composed of homo-polymeric regions of M and G residues, separated by regions of alternating structure (Clementi, 1997). Unlike alginates produced by algae, bacterial alginates are acetylated to a variable extent at positions O-2 and/or O-3 of the mannuronate residues. The variability in molecular mass, monomer block structure and acetylation influence the physicochemical and rheological characteristics of the polymer. Alginate in marine brown alga (*Phaeophyceae*) is believed to act as structure forming component, providing mechanical strength and flexibility (Moe et al., 1995).

The capability of alginate to confer viscosity in solution is dependent on its molecular mass (MM). The MM of algal alginates has been found to range from 48 to 186 kDa. In contrast, some alginates isolated from *A. vinelandii* present MM in the range of 80 to 4,000 kDa (Galindo et al., 2007). Alginates present a wide range of applications, acting as stabilizing, thickening, gel or film-forming agents, in various industrial fields. Currently, new applications are being discovered for these polymers, such as their use as a source of soluble fiber, or in medical products.

5.4 Poly (Lactic Acid) or Poly-Lactide (PLA)

PLA ($[-O(CH^3)CHCO-]_n$) is a thermoplastic aliphatic polyester derived from renewable resources, such as corn starch (in the United States), tapioca products (roots, chips or starch mostly in Asia) or sugarcanes (in the rest of world). PLA belongs to the family of aliphatic polyesters commonly made from α-hydroxy acids, which also includes polyglycolic acid (PGA). PLA is a biodegradable and biocompatible thermoplastic which can be produced by fermentation from renewable resources like corn starch or cane sugar. It can also be synthesized either by condensation polymerization of lactic acid or by ring opening polymerization of lactide in the presence of a catalyst. It can biodegrade under certain conditions, such as the presence of oxygen and is difficult to recycle. This polymer exists in the form of three stereoisomers: poly (L-lactide) (L-PLA), poly (D-lactide) (D-PLA) and poly (DL-lactide) (DL-PLA). The manufacture of PLA from lactic acid was pioneered by Carothers in 1932 (Carothers and

Hill, 1932). However, lactic acid cannot be directly polymerized to a useful product, because each polymerization reaction generates one molecule of water, the presence of which degrades the forming polymer chain to the point that only very low molecular weights are observed. Instead, two lactic acid molecules undergo a single esterification and then catalytically cyclized to make a cyclic di-lactate ester. Although dimerization also generates water, it can be separated prior to polymerization due to a significant drop in polarity. PLA of high molecular weight is produced from the di-lactate ester by ring-opening polymerization using most commonly a stannous octoate catalyst. Polylactic acid can be processed like most thermoplastics into fiber and film (Tokiwa et al., 2009).

Studies have shown that the degradation of PLA is slow as their microbial degraders are not widely distributed and therefore they are less susceptible to microbial attack as compared to other microbial and synthetic aliphatic polymers. The degradation of PLA in soil is slow and that takes a long time for degradation to start (Ohkita and Lee, 2006). Many strains of genus *Amycolatopsis* and *Saccharotrix* are able to degrade PLA (Tokiwa et al., 2009).

Besides being biodegradable, PLA is also biocompatible when in contact with living tissues, e.g., biomedical applications such as implants, sutures, drug encapsulation, etc. PLA can be degraded by abiotic degradation, i.e., simple hydrolysis of ester bond without requiring the presence of enzymes to catalyze the reaction. During the biodegradation process, and only in the second step, the enzymes degrade the residual oligomers till final mineralization (biotic degradation). As long as the basic monomers (lactic acid) are produced from renewable resources (carbohydrates) by fermentation, PLA compiles with the rising worldwide concept of sustainable development and is classified as an environmentally friendly material.

5.5 Chitin

Chitin, the second most abundant natural biopolymer, is a linear aminoacetyl glucan that is present in the shell of crabs, lobsters, shrimp, and insect cuticle. It is a natural, biologically safe polymer synthesized from chitin by deacetylation (hydrolysis of acetamide groups). Recent advances in fermentation technology suggest that the cultivation of fungi can provide an alternative source of chitin (Teng et al., 2001). It is a tough, biodegradable, biocompatible, non-toxic (depends on the source, method of preparation, molecular weight and degree of acetylation) linear polysaccharide suitable for various applications in pharmaceutical drug delivery technology. Along with these properties it also possesses some medical applications such as analgesic, hypo-cholesterolemic, haemostatic, antitumor, spermicidal, CNS depressant, immuno-adjuvant properties (Balamurugan, 2012). The materials are biocompatible; possess antimicrobial activity and has the ability to absorb heavy metal ions. They also possess water-retaining and moisturizing ability. Chitin and its derivatives are also used as drug carriers, blood anticoagulants, anti-tumor products and immuno-adjuvants (Nair and Laurencin, 2007). More recently some studies have shown the anti-oxidative and radical scavenging activities of chitosans. Chitin, collagen and poly-L-leucine have been used to prepare skin substitutes or wound dressing (Khor and Lee, 2003).

From chitin, chitosan is obtained by deacetylation. Chitosan is not widespread in nature. It is found in some mushrooms (zygote fungi) and in the termite queen's abdominal wall. The chitosan is characterized by its acetylation degree and its molecular weight. These parameters influence its viscosity and solubility. Chitosan can be plasticized with glycerol to obtain a kind of thermoplastic like material, e.g., plasticized starch (Epure et al., 2011).

6 USING FERMENTATION TO PRODUCE PLASTICS

Fermentation is even more powerful when coupled with new biotechnology techniques. Fermentation is the use of microorganisms to break down organic substances usually in the absence of oxygen. Fermentation can be carried out with genetically engineered microorganisms, specially designed for the conditions under which fermentation takes place, and for the specific substance that is being broken down by the microorganism (Keshavarz and Roy, 2010). There are two ways fermentation can be used to create biopolymers and bio-plastics.

6.1 Bacterial Polyester Fermentation

Fermentation is the process by which bacteria can be used to create polyesters. Bacteria called *Ralstonia eutropha, Bacillus megaterium, Ralstonia* spp., *Pseudomonas putida, Pseudomonas* spp., *Bacillus mycoides, Alcanivorax borkumensi, Rhodococcus ruber* or other suitable bacterial species are used to accomplish this purpose. The bacteria use the sugar of harvested plants, such as corn, to fuel their cellular processes. The by-product of these cellular processes is the polymer. The polymers are then separated from the bacterial cells (Kahar et al., 2004; Khazir and Shetty, 2014).

6.2 Lactic Acid Fermentation

Lactic acid is fermented from sugar, much like the process used to directly manufacture polymers by bacteria. However, in this fermentation process, the final product of fermentation is lactic acid, rather than a polymer. After the lactic acid is produced, it is converted to polylactic acid using traditional polymerization processes.

7 GROWING PLASTICS IN PLANTS

Plants are becoming factories for the production of plastics. Researchers created *Arabidopis thaliana* plant (commonly referred as wall cress) through genetic engineering. The plant contains the enzymes used by bacteria to create plastics. It produces a flexible, moldable plastic called poly-hydroxy-butyrate-co-poly-hydroxy-valerate, or PHBV. Bacteria create the plastic through the conversion of sunlight into energy. The researchers have transferred the gene that codes for this enzyme into the plant; as a result the plant produces plastic through its cellular processes. The plant is harvested and the plastic is extracted from it using a solvent. The liquid resulting from this process is distilled to separate the solvent from the plastic (Khazir and Shetty, 2014).

A non-polluting, renewable plastic, PHBV is used in wide range of products like grocery bags and disposable razors. It can be easily degraded by bacteria in soil into water and carbon dioxide. Its production does not have to compete with corn or soybeans, whose starch and protein was previously used to make the plastic.

To produce higher quantities of PHBV, researchers have successfully transferred the biologically engineered reaction wherein bacterial enzymes combine with the plants own PHBV chemical precursors to other plants like switchgrass. However, without *Arabidopsis thaliana* the discovery may have never been made, and for that reason, this little mustard plant made the list (Griffen, 2009).

8 ADVANTAGES OF BIOPLASTICS

The major advantage of bioplastics is that they are biodegradable with eco-friendly synthesis. Besides this, they possess high processibility with good mechanical properties and they are derived from renewable resources.

There are some advantages of bioplastics over petroleum-based polymers. PHB and other bioplastics are natural polymers; they are thermoplastic with elastic properties with melting points ranging from 40° to 180°C; they can be degraded in microbe active environments in 5-6 weeks producing carbon dioxide and water, which are environmentally friendly byproducts. The released carbon dioxide and water are absorbed during photosynthesis in nature. The synthesis and biodegradation of biopolymers are totally compatible to the carbon cycle. Bioplastics can be produced from renewable carbon resources. As long as there is fuel shortage and rise in crude oil prices traditional energy sources are safe. Thus there can be a conservation of finite fossil resources like mineral oil and coal and their neutrality with regard to the emission of CO_2. The wider use of bioplastics in daily life will solve the increasing problem of organic waste. The use of biodegradable plastics will decrease the country's dependence on other countries for fossil fuels. They are biocompatible and hence are suitable for medical applications and thus can be implanted in the body without causing inflammations. Some other possible applications of bioplastics include they can be used as biodegradable carriers that demonstrate the ability to deliver drugs for a given time within the individual's body, surgical needles, suture materials, bone tissue replacement, etc. The advantage of using biodegradable plastics is that they do not require surgical removal. Due to their biocompatibility and low oxygen permeability, further allow applications for the production of films and coatings and special biomedical like patch materials, stents, bone implants, drug delivery systems, scaffolds for tissue engineering.

The main advantages of bacterial bioplastics are their biodegradability, ecofriendly synthesis, and excellent processability with good mechanical properties (Khazir and Shetty, 2014).

9 LIMITATIONS OF BIOPLASTICS

Some of the limitations of bioplastics are their poor interactions with fibers; their narrow processing window; lack of reactive groups; thermal degradation and brittleness. As discussed in the above paragraphs that bioplastics are economically feasible but their biodegradable nature may be undesirable in certain cases. The energy required for manufacture of bioplastics, which currently comes from fossil fuels, thereby compromising on the goal of bioplastics. Their bioplastic properties currently lack certain desirable properties, e.g., malleability and pliability.

10 POSSIBLE SOLUTIONS

These problems may be overcome partly through the development of blends and nanocomposites of bioplastics with other materials, e.g., copolymers of PHB and PHV (poly hydroxy valerate), blends of starch or other materials with bioplastics, etc. Hemp oil, agricultural and other wastes, urban wastes, crop residues are being investigated as raw materials for production of bioplastics using bacteria.

10.1 Occurrence of Polymer-Degrading Microorganisms

There are some microorganisms that are capable of degrading the polymers. Their biodiversity and occurrence depends on the environment, such as soil, sea, compost, activated sludge, etc. The major mechanism in the microbial degradation of plastics is first the adherence of microorganisms on the surface of plastics followed by the colonization of the exposed surface. The enzymatic degradation of plastics by microbial hydrolysis is a two-step process: first, the enzyme binds to the polymer substrate then subsequently catalyzes a hydrolytic cleavage. The polymers are degraded into low molecular weight oligomers, dimers and monomers and finally mineralized to CO_2 and H_2O.

The polymer degrading microbes can also be isolated and identified from the environment. The clear zone method with agar plates is a widely used technique for screening polymer degraders and for assessment of the degradation potential of different microorganisms towards a polymer. Agar plates containing emulsified polymers are inoculated with microorganisms and the presence of polymer degrading microorganisms can be confirmed by the formation of clear halo zones around the colonies. This happens when the polymer-degrading microorganisms excrete extracellular enzymes which diffuse through the agar and degrade the polymer into water soluble materials. Using this technique, it was confirmed that PHB, polypropiolactone (PPL) and PCL degraders are widely distributed in different environments (Suyama et al., 1998). Majority of the strains that are able to degrade PHB belong to different taxa such as Gram-positive and Gram-negative bacteria, *Streptomyces* and fungi (Mergaert and Swings, 1996). It has been reported that 39 bacterial strains of the classes *Firmicutes* and *Proteobacteria* can degrade PHB, PCL, and PBS, but not PLA (Suyama et al., 1998). Only a few PLA degrading microorganisms have been isolated and identified. The population of aliphatic polymer-degrading microorganisms in different ecosystems was found to be in the following order: PHB = PCL > PBS > PLA (Pranamuda et al., 1997).

10.2 Role of Biodegradable Polymers in Environment and Health Benefits

Scientists have discovered ways to use bacterial derived plastics to help benefit our society, both medically and environmentally. Two examples of recent studies find that sutures (made from polymers) and butanediol (BDO); can both be produced from genetically engineered bacteria.

Previously surgeries have resulted in wound infections due to poor sutures. The stitches doctors used were weak and brittle, sometimes failing to last the whole healing process. According to Technology Review, one in five patients' abdominal wounds break open after their surgeries as a result of a prematurely dissolved suture (Bourzac 2007). Scientists have discovered a new stronger suture made from genetically engineered bacteria that absorbs after the wound has been healed. The Food and Drug Administration (FDA) has recently approved these polymer sutures to be used in medical procedures. The sutures are more flexible, durable, and safer than the previous sutures. In addition, these sutures are thermoplastic, allowing them to melt under extreme heat and to solidify when cooled. These polymers are also able to stretch and mold into any position, making it easier for doctors during surgeries (Bourzac, 2007).

1,4-Butanediol (BDO) is an important commodity chemical used to manufacture over 2.5 million tons annually of valuable polymers, and it is currently produced exclusively through feed-stocks derived from oil and natural gas (Yim et al., 2011). The scientists instead of using the more expensive oil or natural gas to make BDO, have developed strains of bacteria that can produce butanediol with as little as water and sugar (Biello, 2008). Previously, oil and natural gas was used to produce BDO that was an energy-intensive process. While combining *Escherichia coli* with water and sugar to make this chemical compound, the use of energy noticeably decreases. Utilizing these three resources allows for a more efficient production process to be conducted under normal temperature and pressure. These genetically engineered bacteria are produced in large fermentation tanks similar to biofuel production from algae (Biello, 2008).

11 ECONOMIC PROSPECTS OF BIODEGRADABLE POLYMERS

The present scenario of plastic industry can be transformed by improving the efficiency of the production of bacterial derived plastics. Currently, the plastics industry relies almost completely on crude oil to produce petrochemicals, but certain firms are starting a push towards the increased dependence on organically generated plastics. There are two challenges involved in making bacterial derived plastics economically competitive with petrochemicals: achieving efficient production of feedstock chemicals, and achieving efficient production of these bioplastics. Both of these challenges revolve around giving enough bacteria the proper conditions to mass-produce PHA's (The Economist, 2004).

For economic purposes, bacteria can be engineered to metabolize glucose to produce plastics. One pivotal step forward for the industry would be to find proper enzymes to break down cellulose, which is abundant in agricultural waste, in order for bacteria to mass-produce a glucose source for plastic production (The Economist, 2004).

One method being used to break down cellulose is DNA microarrays. These are special chips that show what genes are active at a given moment, to work out what other enzymes the (microbes) are making when they digest cellulose (*The Economist*, 2004). If this group of enzymes can be identified, it would narrow down the search for the most efficient, active enzymes that can produce a cost-effective feedstock for plastic production.

12 CONCLUSION

Bioplastics from microbes are an important and exciting new field in microbial biotechnology which promises to help save the environment as well as slow the depletion of non-renewable resources. Although huge work is going on bioplastics production from plant sources like carbohydrates but utilizing microorganisms for producing bioplastic, is still a technology in its infancy stage and many tests and experiments are needed before bioplastics are credited as a feasible approach and replacement for the incredibly versatile polypropylene. The genetic manipulation and engineering of microorganisms opens up an enormous potential for the biotechnological production of biopolymers with tailored properties suitable for high-value medical application such as tissue engineering and drug delivery. Many new innovative schemes are being tried. For example, anaerobically-digested dairy waste is used to grow the algae used for biodiesel production. Bioplastic production can then be integrated into this process by using the carbon- and nutrient-rich dairy waste to harvest PHA-producing organisms that occur naturally in the dairy waste. Similarly microbes flourishing in activated sludge, antibiotic producers, and photosynthetic bacteria, plant growth promoting rhizobia, halophiles and hydrocarbon degraders can also be used as PHA producers besides carrying out their normal physiological roles.

Acknowledgements

The authors are highly grateful to Dr. Ashok K. Chauhan, Founder President and Mr. Atul Chauhan, Chancellor, Amity University UP, Noida, India for their continuous motivation, encouragement, guidance, and financial support.

References

Abdelwahab MA, Flynn A, Chiou BS, Imam S and Orts W (2012). Thermal, mechanical and morphological characterization of plasticized PLA-PHB blends. *Polymer Degradation and Stability* 97(9): 1822-1828.

Akar A, Akaya EU, Yesiladali SK, Celikyilmaz G, Cokgor EU, Tamerler C, Orhon D and Cakar ZP (2006). Accumulation of polyhydroxyalkanoates by *Microlunatus phosphovorus* under various growth conditions. *Journal of Industrial Microbiology and Biotechnology* 33: 215-220.

Akaraonye E, Keshavarz T and Roy I (2010). Production of poly-hydroxy-alkanoates: the future green materials of choice. *Journal of Chemical Technology and Biotechnology* 85: 732–743.

Aldor I and Keasling J (2003). Process design for microbial plastic factories: metabolic engineering of polyhydroxyalkanoates. *Current Opinion on Biotechnology* 14: 475- 483.

Alias Z and Tan IKP (2005). Isolation of palm oil-utilising, polyhydroxyalkanoate (PHA)-producing bacteria by an enrichment technique. *Bioresource Technology* 96:1229-1234

Balamurugan M (2012). Chitosan: A perfect polymer used in fabricating gene delivery and novel drug delivery systems. *International Journal of Pharmacy and Pharmaceutical Sciences* 4(3): 54-56.

Bates C (2011). Life on Mars closer than ever for man claims scientist creating artificial life that feeds off carbon dioxide. *Daily Mail Online* 24. 8. 2011, n. page. Web. 9 May. 2012.

Beccari M, Dionisi D, Giuliani A, Majone M and Ramadori R (2002). Effect of different carbon sources on aerobic storage by activated sludge. *Water Science and Technology* 45(6): 157-168.

Biello D (2008). Turning Bacteria into Plastic Factories. Scientific American.

Bourzac K (2007). Plastic made by Bacteria Commercialized. Technology Review, MIT.

Budde CF, Riedel SL, Hübner F, Risch S, Popović MK, Rha CK and Sinskey AJ (2011). Growth and polyhydroxybutyrate production by *Ralstonia eutropha* in emulsified plant oil medium. *Applied Microbiology and Biotechnology* 89:1611-1619.

Carothers WH and Hill JW (1932). Studies of polymerization and ring formation. XV. Artificial fibers from synthetic linear condensation super polymers. *Journal of the American Chemical Society* 54: 1579-1596.

Ceyhan N and Ozdemir G (2011). Poly-β-hydroxybutyrate (PHB) production from domestic wastewater using *Enterobacter aerogenes*12Bi strain. *African Journal of Microbiology Research* 5(6): 690-702.

Chandra R and Rustgi R (1998). Biodegradable polymers. *Progress in Polymer Science* 23: 1273-1335.

Clementi F (1997). Alginate production by *Azotobacter vinelandii*. *Critical Reviews in Biotechnology* 17: 327-361.

Dalal J, Sarma PM, Lavania M, Mandal Ak and Lal B (2010). Evaluation of bacterial strains isolated from oil-contaminated soil for production of polyhydroxyalkanoic acids (PHA). *Pedobiologia* 54(1): 25-30.

Dircks K, Henze M, Van Loosdrecht MCM, Mosbaek H and Aspegren H (2001). Storage and degradation of poly-β-hydroxybutyrate in activated sludge under aerobic conditions. *Water Research* 35(9): 2277-2285.

Epure V, Griffon M, Pollet E and Averous L (2011). Structure and properties of glycerol-plasticized chitosan obtained by mechanical kneading. *Carbohydrate Polymers* 83: 947-952.

Fernández D, Rodríguez E, Bassas M, Viñas M, Solanas AM, Llorens J, Marqués AM and Manresa A (2005). Agro-industrial oily wastes as substrates for PHA production by the new strain *Pseudomonas aeruginosa* NCIB 40045: Effect of culture conditions. *Biochemical Engineering Journal* 26: 159-167.

Galindo E, Peña C, Núñez C, Segura D and Espin G (2007). Molecular and bioengineering strategies to improve alginate and polyhydroxyalkanoate production by *Azotobacter vinelandii. Microbial Cell Factories* 6:1-16.

Gasser I, Muller H and Berg G (2009). Ecology and characterization of polyhydroxyalkanoate-producing microorganisms on and in plants. *FEMS Microbiology Ecology* 70(1): 142-150.

Gilmore D (2005). PHAs: Chemistry, History, and Biology. Dr. David F. Gilmore: Environmental Microbiology Research Lab. Arkansas State University. Web. 7 May 2012.

Griffen J (2009). An article-Prof. Plastic alter ego: *Arabidopsis thaliana. Pacific Standard: The Science of Society.*

Guo W, Song C, Kong M, Geng W, Wang Y and Wang S (2011). Simultaneous production and characterization of medium chain-length polyhydroxyalkanoates and alginate oligosaccharides by *Pseudomonas mendocina* NK-01. *Applied* Microbiology and *Biotechnology* 92(4): 791-801.

Hernández R, Orive G, Murua A and Pedraz J (2010). Microcapsules and micro-carriers for in situ cell delivery. *Advanced Drug Delivery Reviews* 62 (7-8): 711-730.

Kahar P, Tsuge T, Taguchi K and Doi Y (2004). High yield production of polyhydroxyalkanoates from soybean oil by *Ralstonia eutropha* and its recombinant strain. *Polymer Degradation and Stability* 83: 79-86.

Kannan LV and Rehacek Z (1970). Formation of poly-beta-hydroxybutyrate by Actinomycetes. *Indian Journal of Biochemistry* 7(2): 126-129.

Keshavarz T and Roy I (2010). Polyhydroxyalkanoates: Bioplastics with a green agenda. *Current Opinion in Microbiology* 13: 321–326.

Khanna S and Srivastava AK (2005). Recent advances in microbial polyhydroxyalkanoates. *Process Biochemistry* 40: 607–619.

Khazir S and Shetty S (2014). Bio-based polymers in the world. *International Journal of Life Sciences Biotechnology and Pharma Research* 3(2): 35-43.

Khor E and Lee YL (2003). Implantable applications of chitin and chitosan. *Biomaterials* 24: 2339-2349.

Kumar BS and Prabhakaran G (2005). Production of PHB (bioplastics) using bio-effluents as substrates by *Alcaligenes eutrophus. Indian Journal of Biotechnology* 5: 76-79.

Kumbhakar S, Singh PK and Vidyarthi AS (2012). Screening of root nodule bacteria for the production of polyhydroxyalkanoate (PHA) and the study of parameters influencing the PHA accumulation. *African Journal of Biotechnology* 2(31): 7934-7946.

Lopez J, Javier N, Juan H, Maria C, Carlos C and Marcelo V (2012). Biosynthesis of PHB from a New Isolated *Bacillus megaterium* Strain: Outlook on Future Developments with Endospore Forming Bacteria. *Biotechnology and Bioprocess Engineering* 17: 250-258.

Lugg H, Sammons RL, Marquis PM, Hewitt CJ, Yong P, Paterson-Beedle M, Redwood MD, Stamboulis A, Kashani M, Jenkins M and Macaskie LE (2008). Polyhydroxybutyrate accumulation by a *Serratia sp. Biotechnology Letters* 30(3): 481-91.

Marsudi S, Unno H and Hori K (2008). Palm oil utilization for the simultaneous production of polyhydroxyalkanoates and rhamnolipids by *Pseudomonas aeruginosa. Applied Microbiology and Biotechnology* 78(6): 955-961.

Maskow T and Babel W (2000). Calorimetrically recognized maximum yield of poly-3-hydroxybutyrate (PHB) continuously synthesized from toxic substrates. *Journal of Biotechnology* 77(2-3): 247-253.

Mergaert J and Swings J (1996). Biodiversity of microorganisms that degrade bacterial and synthetic polyesters. *Journal* of *Industrial Microbiology* 17: 463-469.

Moe ST, Draget KI, Skjak-Braek G and Smidsrod O (1995). Alginates. In: Dekker M (ed) Food polysaccharide and application. Dekker, New York, pp 245–286.

Mumtaz T, Yahaya NA, Abd-Aziz S, Rahman NA, Yee PL, Shirai Y and Hassan MA (2010). Turning waste to wealth-biodegradable plastics polyhydroxyalkanoates from palm oil mill effluent- a Malaysian perspective. *Journal of Cleaner Production* 18:1393-1402.

Nair LS and Laurencin CT (2007). Biodegradable polymers as biomaterials. *Progress in Polymer Science* 32: 762-798.

Ohkita T and Lee SH (2006). Thermal degradation and biodegradability of poly (lactic acid)/corn starch biocomposites. *Journal* of *Applied Polymer Science* 100: 3009-3017.

Ojumu TV, Yu J and Solomon BO (2004). Production of polyhydroxy alkanoates, a bacterial biodegradable polymer. *African Journal of Biotechnology* 3 (1): 18-24.

Panda B, Jain P, Sharma L and Mallick N (2005). Poly-β-hydroxybutyrate accumulation in *Nostoc muscorum* and *Spirulina platensis* under phosphate limitation. *Journal of Plant Physiology* 162(12): 1376-1379.

Pranamuda H, Tokiwa Y and Tanaka H (1997). Polylactide degradation by an *Amycolatopsis* sp. *Applied and Environmental Microbiology* 63: 1637-1640.

Reddy CS, Ghai R and Rashmi Kalia VC (2003). Poly-hydroxy-alkanoates: An overview. *Bioresource Technology* 87: 137–146.

Reemer J (2009). Advances in the synthesis and extraction of biodegradable polyhydroxyalkanoates in plant systems – A review. *MG Basic Biotechnology* 5.1: http://ejournal.vudat.msu.edu/index.php/mmg445/article/viewArticle/383/349

Safak S, Mercan N, Aslım B and Beyatli Y (2002). A study on the production of poly-beta-hydroxy butyrate by some eukaryotic microorganisms. *Turkish Electronic Journal* of *Biotechnology* Special issue 11-17.

Saharan BS, Grewal A and Kumar P (2014). Biotechnological production of poly-hydroxy-alkanoates: A review on trends and latest developments. *Chinese Journal of Biology* volume 2014, Hindawi Publishing Corporation, Article ID 802984, 18 pages.

Sanchez-Garcia MD, Gimenez E and Lagaron JM (2008). Morphology and barrier properties of solvent cast composites of thermoplastic biopolymers and purified cellulose fibres. *Carbohydrate polymers* 71(2): 235-244.

Suriyamongkol P, Weselake R and Narine S (2007). Biotechnological approaches for the production of polyhydroxyalkanoates in microorganisms and plants-A review. *Biotechnology Advances* 25: 148-175.

Suyama T, Tokiwa Y, Oichanpagdee P, Kanagawa T and Kamagata Y (1998). Phylogenetic affiliation of soil bacteria that degrade aliphatic polyesters available commercially as biodegradable plastics. *Applied and Environmental Microbiology* 64: 5008-5011.

The Economist (2004). Sea of Dreams: Genetically modified microbes will lead to a revolution in Industrial Biotechnology, Orlando.

Teng WL, Khor E, Tan TK, Lim LY and Tan SC (2001). Concurrent production of chitin from shrimp shells and fungi. *Carbohydrate Research* 332: 305-316.

Tokiwa Y, Calabia BP, Ugwu CU and Aiba S (2009). Biodegradability of Plastics. *International Journal of Molecular Sciences* 10: 3722-3742.

Trainer MA and Charles TC (2006). The role of PHB metabolism in the symbiosis of rhizobia with legumes. *Applied Microbiology and Biotechnology* 71(4): 377-386.

Verlinden RAJ, Hill DJ, Kenward MA, Williams CD and Radecka I (2007). Bacterial synthesis of biodegradable polyhydroxyalkanoates. *Journal* of *Applied Microbiology* 102: 1437-1449.

Volova TG and Kalacheva GS (2005). The synthesis of hydroxybutyrate and hydroxyvalerate copolymers by the bacterium *Ralstonia eutropha. Microbiology* 74: 54-59.Volova T (2004), Polyhydroxalkanoates: Plastics of the 21st Century. 1st ed. Hauppauge, NY: Nova Science Publishers. 1-25. Web.

Weiner RM (1997). Biopolymers from marine prokaryotes. *Trends in Biotechnology* 15: 390-394.

Williams S and Martin D (2005). Applications of Polyhydroxyalkanoates (PHA) in Medicine and Pharmacy. *Biopolymers Online.* DOI: 10.1002/3527600035.bpol4004.

Yan Q, Sun Y, Ruan LF, Chen J and Yu PHF (2005). Biosynthesis of short-chain-length-polyhydroxyalkanoates during the dual-nutrient-limited zone by *Ralstonia eutropha*. *World Journal of Microbiology and Biotechnology* 21:17-21.

Yilmaz M and Beyatli Y (2005). Poly-b-hydroxybutyrate (PHB) production by a *Bacillus cereus* M5 strain in sugar beet molasses. *Zuckerindustrie* 130: 109-112.

Yim H, Haselbeck R, Niu W, Pujol-Baxley C, Burgard A, Boldt J, Khandurina J, Trawick JD, Osterhout RE, Stephen R, Estadilla J, Teisan S, Schreyer HB, Andrae S, Yang TH, Lee SY, Burk MJ and Van Dien S (2011). Metabolic engineering of *Escherichia coli* for direct production of 1,4-butanediol. *Nature Chemical Biology* 7(7): 445-452.

Index

A

Aerobic metabolic pathway 81
Aeroponics 119
Agriculture and microbes 61
Algae for nutrients harvesting 276
Algal biofertilizers 73
Algal remediation 280
Algal wastewater treatment 270
Alginates 312
AM fungi
  classification of 216, 221
  community structure of 224
  diversity of 224
  from India 216
  functions of 224
AM-mediated bio-protection 238
Anaerobic metabolic pathway 82
Antagonistic biocontrol agents 179
Antibiotics 259
  challenges and impact 265
  classification of 261
  early history 260
  in environment 262
  modern history 260
  production and uses of 262
  resistance 263
  working of 261
Anti-microbial plant oils 52
  apricot oil 52
  calendula oil 53
  castor oil 53
  neem oil 52
  sesame oil 54
  sunflower oil 54
  walnut oil 53
  wheatgerm oil 54
Anti-microbials
  from plants 50
Arbuscular mycorrhizae (AM) 108
Axenic culture 122

B

Bacillus thuringiensis 74
Bacteria
  as biopesticides 181
  as insecticides 74
  in leaching 76
Bacterial isolates, screening and characterization of 100
Bacterial polyester fermentation 314
Bacterial polyesters 303
Beauveria bassiana 75
Bio-control agents, production of 183
Biodegradable plastics
  properties of 305
Biodegradable polymers 304
  economic prospects of 318
  role of 317
Biodiesel, production of 273
Bioelectricity 79
Bioethanol 78
  production of 275
Biofuels 77
Biogas 78
  production of 275
Bio-hydrogen 78
Biomining 75
Biopesticide
  application of 186
  microbes used as 180
Bio-plastic producing bacteria 302
Bioplastics 301
  advantages of 315

- limitations of 316
- types of 306

Biopolymers, classification of 302
Bioprotection efficacy 236
Bioremediation 82
Bioremediation practices 83
Blue-green algae 32
- as biofertilizers 27
- general features of 23
- in aquaculture ponds 25
- toxins of 30

Blue-green algal blooms 31, 36
- control and treatment 36, 38
- health risk of 32, 35

Brundtland Commission 60
Bt gene for insect control 182

C

Catecholate siderophores 68
Chemical pesticides 185
Chitin 313
Cinnamon, effects of 56
Co-metabolic pathway 82
CpGV
- advantage of 184
- disadvantages of 184

Cyanobacteria
- and soil fertility 28
- nitrogen contribution by 26
- occurrence and abundance of 26

Cytoplasmic polyhedrosis viruses 75

E

Earthworm
- beneficial role of 12

Endotoxins 30

F

Fluoride pollution 89
Fluoride remediation
- conventional methods of 92
- need for 92

Fluoride toxicity 90
- bioremediation of 92, 93
- molecular mechanism of 91

Fluoride, effects of 90

G

Glomeromycota division 213
Granulosis viruses 75

H

H. Pylori 55
Heavy metal remediation 278
Heliothis armigera 74
Hepatotoxins 30
Host plants, global diversity of 223
Hydroponic technique 117
Hydroxamate siderophores 68

I

Immobilized algae
- wastewater remediation and 279

Inoculum, production of 111, 112
Integrated pest management 176
Iron chelators 67
Iron oxidizers 66
Iron oxidizing bacteria 67

K

Keoladeo National Park,
- physicochemical and bacteriological studies of 136

Lactic acid fermentation 314

M

Medicinal plants
- bhringraj (*Eclipta alba*) 48
- daruharidra (*Berberis aristata*) 49
- for microbial control 45
- garlic (*Allium sativum*) 47
- gokharu (*Tribulus terrestris*) 49
- kutki (*Picrorrhiza kurroa*) 50
- neem (*Azadiracta indica*) 47
- punarnava (*boerhavia diffusa*) 50
- turmeric (*Curcuma longa*) 48

Microbial enhanced oil recovery 77
Microbial fuel cell 79
Microbial life
- biological evolution of 2
- chemical evolution of 2

emergence of 2
Microbiology
future perspectives of 7
history of 3
Microorganisms
as inoculants 72
as renewable energy sources 75
for pollution control 80
in insect control 73
under abiotic stress 70
under biotic stresses 70
Most probable number (MPN) technique 291
Mycorrhizae, types of 212
Mycorrhization, degree of 237
Mycorrhizosphere 241

N

N: P ratio 23
Neurotoxins 30
Non-chemical disease
and pest management 175
Non-symbiotic nitrogen fixers 73
Nutrient film technique 117

O

Organic chemicals, recovery of 276

P

Pesticide
bioremediation of 279
hazards of 172
history of 168
production and usage of 169
types of 169
use of 185
working of 170
PGPR isolates, morphological characteristics of 100
Phoslock 39
application of 40
benefits of 40
working of 40
Phosphate solubilizers 64, 73
Phosphate solubilizing microbes 65
Phyto-antimicrobial compounds 52
Phytoplankton
nutrient requirements of 23
Placket-Burman design 155
Plant Growth Promoting Rhizobacteria (PGPR)
and crop productivity 62
isolation of 99
types and their genera 63
Plant growth regulators 69
Pollutants, biodegradation of 80, 81
Poly(lactic acid) 312
Polyhydroxy Butyrate (PHB) 311
Poly-Hydroxy-Alkanoate (PHA) 301, 306
production of 308, 309
Poly-Lactide (PLA) 312
Polymer-degrading microorganisms 316
Pot cultures 111
Potassium solubilizers 65
Pseudomonaeae 181
Response surface methodology 155, 159
Sand–based media 115
Siderophore producing microorganisms 68
Siderophores 67
Signalling molecules 221
Soil microbial communities 226
functional resilience of 226
Soilless media 116
Solution based - media 116
SSF technique 154
Sulphur oxidizers 66
Sulphur oxidizing bacteria 66
Sustainable agriculture 62
Symbiotic nitrogen fixers 72

T

Thiosulphate utilizers 291
Toxic cyanobacteria
occurrence of 29
Toxin bioremediation 277

V

Vascular plants, mycorrhizal associations of 223

Vermicompost
  application of 19
  benefits of 12
  concept of 12
Vermicomposting methods 13, 15
  bag method 18
  bamboo pit method 17
  composite pit method 17
  merits and demerits of 18
  open bed method 16
  pit/tank method 17
  platform method 16
Vermitechnology 12
Viruses as insecticides 74

W

Wastewater
  composition of 270
  microbiological composition 271
  treatment of 271